FUNDAMENTAL CONCEPTS OF MATHEMATICS

MERRILL MATHEMATICS SERIES

Erwin Kleinfeld, Editor

FUNDAMENTAL CONCEPTS OF MATHEMATICS

George J. Langbehn

Thomas G. Lathrop

Carl J. Martini

*State College at Salem,
Salem, Massachusetts*

Charles E. Merrill Publishing Company

A Bell & Howell Company

Columbus, Ohio

Copyright © 1972, by Charles E. Merrill Publishing Co., Columbus, Ohio. All rights reserved. No part of this book may be reproduced in any form, electronic or mechanical, including photocopy, recording, or any information storage and retrieval system without permission in writing from the publisher.

International Standard Book Number: 0-675-09203-5

Library of Congress Catalog Card Number: 76-150994

1 2 3 4 5 6 7 8 9 10—79 78 77 76 75 74 73 72

Printed in the United States of America

PREFACE

This book had its beginning in efforts to supplement various texts which failed to suit our classroom needs completely. Frequently revised and extended as a result of classroom experience, this material has become the sole text used in many of our freshman classes. During this entire growth process we have tried to maintain certain definite points of view toward the material.

For example, each of the major topics considered is important in contemporary mathematics. While it is undeniably true that the meaning of the term "modern" varies greatly depending upon context and upon who is using it, it is also true that much is made these days of the logical structure of mathematical systems. Hence a large part of the material is placed within the framework of an axiomatic system, and we hope that this along with the selection of topics will give any reader an improved feeling for some of the more important meanings of the term "modern mathematics."

We also agree with the many critics of the type of course which gives a little attention to almost every concept worthy of name-dropping but provides no extensive coverage of any topic. Accordingly the number of major topics in this book is limited and each is treated in enough depth that the student can get into meaningful exercises of more than an introductory nature. Our students seem to come away from the course with more than just a little vocabulary.

Further, this book reflects our conviction that logic receives far too little attention in most elementary texts. Underscoring the prospective teacher's need for work in this area is the fact that classroom materials are now available which introduce topics from logic starting at about the fifth grade. We also feel that command of the basic principles of deductive inference can increase understanding of almost any contemporary treatment of mathematical topics. Defining a few terms and displaying a few truth tables merely provides the student with an introduction to the *language* of deductive reasoning. Only after some substantial practice with the major rules of inference, does the student reach the point where he can understand the *use* of logic in deductive proof. Chapter 2 provides the requisite treatment of elementary logic.

The fact that concepts are stressed over techniques is another major feature of the book. This approach seems highly appropriate for prospective teachers and general liberal arts students as well. It also means that topics are treated differently enough from what most students have experienced that they have something to learn even when a topic itself is essentially repetitive. Students who have had discouraging experiences in trying to learn the techniques presented in high school mathematics have frequently done rewardingly good work when asked to concentrate on the concepts involved.

An added advantage of this feature is that it minimizes the amount of mathematics preparation needed to do the work successfully. The basic prerequisite for most of the book is about a year of high school algebra and some informal work with geometric concepts. Willing students with this minimum background can proceed successfully. Some formal work with principles of geometry is an added prerequisite for Chapter 8. Only Sections 6c, 6f, 6i require background equivalent to three years of standard high school mathematics. In order that students with less than this training can also do some work in Chapter 6, results obtained in these three sections are summarized in §6g and §6j so that all three can be omitted without loss of continuity.

With these background prerequisites in mind, various one-semester courses can be constructed by choosing those chapters and/or sections most suitable for a particular purpose. For example, we cover a portion of Chapter 8 with our own classes because they include Elementary Education majors who come with the requisite background in geometry. In any selection of material, the following relationships among the chapters should be considered.

Chapter 2 is central to our development. Skills gained there are reinforced in the following chapters by making explicit use of inference rules in short or mathematically simple arguments. Principles and terminology introduced in this chapter are used in all the succeeding work.

As for Chapter 1, the language of §1e is used regularly in much of the following material, and occasional reference is made to the outline of number properties given in §1f. As much or as little use may be made of the rest of it as may suit the purpose at hand. Consideration of this chapter may be delayed until §2k is taken up, if desired.

Chapter 3 or the equivalent is prerequisite to the succeeding chapters in the sense that the language of sets is used throughout. Any student with a previous introduction to such material should at least read through the first five sections to be certain that notation, terminology, and definitions used are familiar.

Sections 4a, 4b are prerequisite to Chapter 5. Except for these sections, Chapters 4 through 8 are reasonably independent and may be taken in any order without disturbing continuity.

Exercises are graded so that practice problems appear at the beginning of each set, exercises in establishing and extending the theory at the end. Answers to about half of the exercises, usually the odd ones, are found in the *Appendix* along with hints for working some of the problems and doing some of the proofs. A few exercises marked with an asterisk are either somewhat more difficult than the majority or need special explanation for most effective use. Hints for these will be found in the *Instructor's Manual* along with answers to the remaining exercises and outlines of proofs.

Elements important to the development are numbered for easy location and cross reference. Definition (4-11) is the eleventh definition stated in Chapter 4. Figure 7.3 is the third figure mentioned in Chapter 7. Theorem (3e-5) is the fifth theorem given in Section 3e. Exercises and margin references are identified by Arabic numbers in parentheses. Items from a list of things within a chapter are labeled with small Roman numerals. We hope that this scheme of providing each reference entity with its particular type of identification will make for quick location of desired items.

No work of this kind would be published without the aid of many persons besides the authors. Accordingly our thanks go to all the typists, reviewers, editors, and others who have helped produce this book. We owe a particular debt of gratitude to our colleagues and students who have used previous versions. They have supplied many ideas for improvement. Very special thanks go to Professor Harold Harutunian for his numerous contributions and to Professor Peter C. Wong for checking some of the material in Chapter 7.

CONTENTS

1	**THE NATURE OF MATHEMATICS**	**1**
1a	Historical Retrospect	1
1b	Forms of Reasoning	4
1c	The Axiomatic Method	7
1d	Mathematics and Language	9
1e	The Pattern of Formal Axiomatics	16
1f	Structures and Models	18
2	**ELEMENTARY LOGIC**	**25**
2a	Deductive Argument	25
2b	Axioms for Logic	28
2c	Compound Propositions	33
2d	Simple Arguments	46
2e	Properties of Compounds	52
2f	Valid Argument Forms	62
2g	General Arguments	72
2h	Quantified Propositions	89
2i	Applications	96
2j	The Algebra of Propositions	103
2k	Summary of Chapters 1 and 2	105
3	**SETS**	**113**
3a	The Set Concept	113
3b	Relations Between Sets	119
3c	Properties of Equality and Inclusion	125
3d	Complementation	135
3e	Operations on Sets	139
3f	The Algebra of Sets	156
3g	Quantified Arguments	160

ix

		3h	Miscellaneous Exercises	170
4			**RELATIONS AND FUNCTIONS**	**175**
		4a	Relations	175
		4b	Cartesian Products	179
		4c	Equivalence Relations	182
		4d	Functions	186
		4e	Reverse Relations	193
		4f	Graphing	196
		4g	Curve Sketching	202
5			**AN INTRODUCTION TO PROBABILITY**	**209**
		5a	What is Probability?	209
		5b	Sample Spaces and Events	210
		5c	Probability	213
		5d	More Properties of Probability	216
		5e	The Formula for $P(A \cap B)$	220
		5f	Summary	222
6			**THE NUMBER SYSTEM**	**225**
		6a	Counting	225
		6b	The Natural Numbers – Axioms	230
		6c	Order Properties	243
		6d	Arithmetic of the Natural Numbers	247
		6e	The Integers	252
		6f	Major Theorems for $\mathscr{I}$	263
		6g	The Number Line	269
		6h	The Rational Number System	273
		6i	Division with Rationals	283
		6j	Decimals	291
		6k	The Real Number System	300
7			**NUMERATION**	**305**
		7a	Early Numeration Systems	305
		7b	The Indo-Arabic System	316
		7c	Numeration Bases Other than Ten	317

	7d	Common Algorithms for Arithmetic	325
	7e	Alternative Algorithms	329

8 GEOMETRY 335

	8a	What and Why?	335
	8b	Modernizing Euclidean Geometry	339
	8c	Straight Lines and Their Subsets	343
	8d	Planes and Space	350
	8e	Partitions	356
	8f	Congruence and Measure	364
	8g	Triangles	376
	8h	Quadrilaterals	384
	8i	Plane Curves	388
	8j	Space Loci	390
	8k	Constructions	394
	8l	Miscellaneous Exercises	399

SELECTED ANSWERS AND HINTS 403

INDEX 449

1
THE NATURE OF MATHEMATICS

§1a   HISTORICAL RETROSPECT

There are many mathematical developments which can rightfully be called modern. Most obvious to the layman, perhaps, is some of the new material which has been developed since about 1800. Included here are such topics as the algebra of sets, Boolean algebras in general, algebraic structures, the algebra of vectors, and linear programming. Both the new content and the continual striving for increased precision have resulted in many changes in terminology and notation.

The vast changes which the subject matter of mathematics has undergone from early times to the present are only one of the important transformations in the field. Many persons fail to realize that the approach toward development and study of the subject has changed just as dramatically. Today each branch of the subject is developed and studied using the systematic procedures of logical reasoning. It was not always so.

It appears that counting was the first mathematical activity of consequence which any human being mastered. Some primitive peoples certainly learned to do one or another form of counting far back in preliterate ages. It was not until permanent settlements began to appear that additional mathematics would have been of much use.

Very little is known of Indic and Chinese mathematics in ancient times, but we do know that the Egyptians who lived in the Nile Valley and the Sumerians and later Chaldeans, who inhabited Mesopotamia, developed many concepts and techniques beyond counting. These peoples collected a surprisingly large amount of arithmetic facts for use in keeping records, in calculating taxes, and in commerce. They also learned to use geometric principles in surveying, architecture, and astronomy. The need to resurvey fields after yearly floodings of the Nile is a well-known example of a practical problem which was successfully handled by early development of mathematical techniques.

As early as 2780 B.C. a large store of mathematical knowledge had been accumulated. The work of the great Imhotep who lived about that time and designed and built the "step pyramid" provides a fine example. Certainly considerable use was made of mathematics in overcoming the sheer mechanical problems of quarrying and cutting the stone used and then actually erecting the huge edifice and the shrines, secondary tombs, stonehouses, and other structures associated with it. Ability to use linear measurement was displayed in cutting the stones to the right size to make a square base. The stone slabs were cut with parallel sides and ends. Also, the maze of storage rooms and burial chambers under the pyramid were laid out in a well-ordered design which formed a symmetric and harmonious whole. The entire complex gives evidence of the application of many geometric principles.

The accomplishments of the Egyptians and Mesopotamians provided later peoples with a large collection of results which could be used as the basis for further development. However, it seems that from the dim beginnings of human culture until about 600 B.C. almost no attempt was made by these or other peoples to develop mathematics by any method other than trial-and-error or observation of chance results.

The Greeks added new ingredients to the study of mathematics. They not only developed many new results to add to the knowledge they had acquired from more ancient cultures, but also completely changed the character of the study of the subject by making the first known attempt to develop mathematical arguments by the systematic application of logical reasoning. Just *why* they made this departure is something on which historians are unable to agree. As was the case in almost all ancient cultures, the wealthy citizens of the landowner-slave society in ancient Greece had time to devote to various topics of cultural interest. Also, the trial-and-error mathematics of the Mesopotamians and Egyptians was imprecise. The Greeks may have tried especially hard to avoid the pitfalls of this approach.

Whatever may have been the reason, the Greeks soon became skilled in logic and the art of critical thinking and applied these skills to the study of mathematics. As a result, they concentrated more on building upon a solid logical foundation than on applying their results as they made their contributions to the sciences they had inherited.

This concentration and the fact that the Greek mathematicians put their primary efforts into the study of geometry had an important effect on the history of mathematics. For more than 2,000 years since the time of the Greeks, geometry was the only branch of mathematics regularly studied using the principles of logical reasoning. Various peoples continued to develop the other branches, but most often their methods were the same inexact trial-and-error ones used in earlier times by the Sumerians, Egyptians, and others. In developing and learning mathematics other than geometry, the main method, especially in the beginning stages of such effort, was to treat a discipline as a collection of rules and tricks for solving particular problems. Until very recently this has been especially true of arithmetic and the algebra of numbers.

The Greek mathematicians also began to stress the use of abstraction. An abstract concept is one from which all superfluous ideas have been removed until only the essential associations remain. For example, the earlier peoples had regarded a point as a physical dot, and a line as the actual edge of an object. In Greek hands a point became a dimensionless indicator of position, and the concept of line became something like a ray of light which had neither breadth nor thickness but continued indefinitely in two directions.

This increased use of abstraction, the insistence on logical reasoning, and the development of many new mathematical results all combine to make the Greek contribution especially important. For long after their era, progress was slow. Not until about 1600 did the body of mathematics again begin to grow rapidly.

In the nineteenth century many revolutionary developments in mathematics, such as the development of non-Euclidean geometries and the successful conclusion of the search for an acceptable foundation for calculus caused a major change in approach to the whole subject. One part of this change has been the spread of the use of the principles of logical deductive reasoning into *all* branches. Some mathematicians have even come to regard the use of this twenty-six-hundred-year-old deductive approach as the most important single feature of "modern" mathematics.

From the standpoint of schools, modern developments in mathematics take somewhat different forms. For example, the shifts in grade level at which topics are first studied have often been little short of dramatic. Many topics from the "new math" now appear in elementary grades or even kindergarten although only a few years ago they were first met in high school or college. The set concept is a case in point.

More important perhaps is a new approach to the subject itself and a new attitude toward the teaching of mathematics. There is now more emphasis on precision and the structural aspects of mathematics and correspondingly less on the mechanical, manipulative aspects. We explore the *why* as well as the *how* of the way we do things. The discovery method and a variety of other learning and teaching techniques are being utilized to promote and maintain interest. The resulting increase in efficiency of teaching and learning means that more mathematics can be taught at any grade level.

In spite of all this, the *basic* content of the mathematics taught in the elementary school has undergone little change, even though enthusiasm for new approaches sometimes clouds the fact that ancient facts such as those in the multiplication table must still be learned. Some of the best new techniques are actually ancient, only their use in the present setting being new.

The "discovery" method, for example, was used by Socrates, and writing numbers using numerals to bases other than ten was already ancient when *he* was born.

EXERCISE SET

1. What abstract property is shared by an auto wheel, a dime, and a basketball?
2. What abstract property is shared by snow, cirrus clouds, and good bond typing paper?
3. Explain how the mathematics of the Greeks differed in spirit from that of the Sumerians and Egyptians.
*4. Does a reading of this section and the first part of the *Preface* seem to indicate that up-grading is the goal of "modern" mathematics? Explain. (An example of up-grading is starting algebra in the seventh grade instead of waiting until the ninth.)

§1b   FORMS OF REASONING

We have stated that the Greek mathematicians began to make extensive application of logical reasoning in their work; that is, they felt that many mathematical results ought to be proved rather than merely accepted on the basis of trial-and-error or observation. Thales of Miletus (640?–546? B.C.) seems to have been the first person of note to have approached mathematics in this essentially different way. Hippocrates (about 440 B.C.) was able to organize at least a part of geometry into a fairly extensive chain of theorems. Euclid (about 300 B.C.) completed this work and added many improvements. Pythagoras (582?–500? B.C.), Eudoxus (?), and Archimedes (287?–212 B.C.) were among others who made important contributions both to the content of mathematics and to the use of logical reasoning in obtaining mathematical results. Aristotle (384–322 B.C.) developed the rules used in reasoning into a complete system of logic.

Today we follow in the footsteps of these men by organizing each branch of mathematics into a chain of proved theorems. Thus, it is important to be certain what we mean by proof. Almost anyone would probably agree that *proof is a method used to show that a statement is true beyond question*, but often it is not easy to tell what we mean by saying that a particular statement is true. Thus we are actually asking, "How can we tell when a particular statement is true?"

Some persons might answer that a statement is true if it agrees with common sense. Consider the statement: "Given two objects of the same size and shape, the heavier one necessarily will fall faster than the lighter one if they are dropped from the same point." A few centuries ago almost anyone would have agreed that this is a true statement; it was just common sense. Finally Galileo (1564–1642) showed that both objects would actually fall at the same rate. In this and many other cases, common sense has not shown itself to be a trustworthy guide to what is true and what false.

Other persons might feel that observation of results could be used to show that a particular principle is true. Let us examine a few cases of the use of observation as a standard for the acceptability of certain principles. Through ages on end people observed that the land on which they lived seemed to be flat. Many persons accepted this vast accumulation of observation as a proof that the earth actually was flat. Many of them were not convinced that this conjecture was faulty even when Magellan's ship had completed its "impossible" voyage.

Consider the numeral .1414.... If you are told that this is the decimal representation of a particular number, what are the next two digits in the sequence? Would they be a 1 and a 4 in that order? Observation of the numeral indicates that this might be quite reasonable. What if you are now told that this symbol is actually another name for $\sqrt{2}/10$? Some calculation or the use of a table of square roots will show you that the fifth digit is a 2—not a 1 at all.

Now in this case you only had two opportunities to observe that the ...14... might continue to repeat indefinitely. Perhaps two observations are too few. Would you be willing to accept some result as proven if you could find 999 cases in which the result is correct? Examine the statement "All natural numbers are less than 1000." You can certainly observe 999 cases for which the statement is true! However, it is obviously false.

These and a host of other possible examples show that we must be very careful how we treat the results of observation. For one thing, what seems to be a pattern may be merely a coincidence. For another, it is difficult to be sure we have considered *all* the factors which govern a particular phenomenon. Thus the results of observation can only indicate what may be *probable*.

When we assume on the basis of observation that a given set of circumstances will produce a particular result, we say that we are using inductive **reasoning.**

As the preceding discussion has shown, common sense and inductive reasoning are not acceptable ways to prove statements. This is not to say that inductive methods are not important to us. Every science makes at least some use of inductive reasoning. The entire subject matter of each of the natural sciences is based essentially on observation, but many theories have had to be discarded because someone found an exception to what was once felt to be a general rule. In this regard Albert Einstein once said: "No amount of experimentation can prove me right, but a single experiment might prove me wrong."

In mathematics we use inductive reasoning to help us determine principles which seem at least plausible in relation to the patterns we observe. Once a seemingly plausible relationship has been found, we state it carefully. Such a statement is called a *conjecture*. We then try to prove or disprove the conjecture. If we can do the former, we classify it as a *theorem*. If the latter is possible, we label it false. The kind of reasoning used when we prove or disprove conjectures is called deductive reasoning.

In deductive reasoning we establish the truth of a given conjecture by an acceptable argument which is supported solely by statements previously established as true.

EXERCISE SET

1. In the Middle Ages many persons believed that the earth was flat. Was the conclusion "The earth is flat" arrived at by deductive or inductive reasoning? Explain.

2. Do you think that people who lived near a port in the Middle Ages were as likely to have been so thoroughly convinced that the earth was flat as others were?

3. If you were to run a mouse through a maze 1,000,000 times and discover that each of these times he took nearly two minutes to reach the cheese, what conclusion can you reach inductively?

4. People long ago believed that the sun actually "rose" and "set." That is, they believed that it moved through the sky around the fixed earth. State some of the reasons why this idea first was accepted inductively and then was given up.

*5. Look up *Fermat's last theorem* in a book on the history of mathematics or the foundations of mathematics. Is this principle classified correctly? Explain.

§1c THE AXIOMATIC METHOD

We have previously stated the basic outline of the deductive method. Known true statements are used to prove others true. However, the thoughtful person might well ask, "How do we know when a particular statement is true and thus available for use in proving other statements true?"

Answering this question is not as easy as it might seem at first glance. After a little reflection a person would realize that whenever he tries to determine the truth or falsity of a certain statement in a careful way, he makes a judgment by comparing the statement with some standard.

For example, a policeman uses the standard provided by the law in order to judge whether the statement

(1) This motorist is driving too fast

is true or false. The sentence

(2) $6 + 7 = 1$

can be judged true or false according to rules of arithmetic which we learn at an early age. In this case the particular set of rules is the needed standard. Such a standard for judging whether statements are true or false can be called a *truth standard*.

Returning to (1), we note that each community tries to establish laws which suit its particular situation best. That is, the law is a truth standard which varies from community to community so that (1) might be labeled "true" in one community but "false" in another where speed laws are different. In some incautious community which had no speed laws whatever, neither label would be meaningful. Similarly, a Christian, using the Bible as a standard, may regard a particular statement as true. Yet a Moslem, using the Koran as his standard, may regard the same statement as false.

These examples show that a truth standard is *arbitrary*. As a mathematical example let us consider (2) again and note that it is a perfectly true everyday statement. You don't think so? Certainly it is 1 o'clock seven hours after it is 6 o'clock!

Thus (2) *is* true according to the truth standard provided by clock arithmetic. Of course, you would have been right to label it false according to the rules of ordinary grocery store arithmetic. This illustrates the fact that we must not only have an arbitrary truth standard available in any logical discussion, but also must make clear exactly which standard is being used.

A truth standard must, of course, be given in terms of statements, and the fact that the standard is arbitrary means that we are arbitrarily labeling each such statement true without proof. Now, it might seem that it would be ideal to prove every reasonable statement. However, we must remember that deductive reasoning requires that the steps in an argument be supported by statements previously established as true. Imagine yourself in the situation of trying to prove the very first statement in any discussion. How would you make a start? At that point you would have no known true statement to use to support a logical argument. Thus it should be clear that we must have on hand a few statements arbitrarily labeled "true" before we can make that first proof. These are the statements of the truth standard. They are called *primary statements*.

Where do we obtain primary statements for use as an initial truth standard? The laws of a legal code are passed by citizens of a community or by special governing bodies. The laws of a religion are often those proclaimed by some great teacher of the past. The natural scientist uses inductive reasoning based on whatever observations he can make to help him set up his system of primary statements which he often calls the "basic laws" of his branch of science. Then he constantly strives to make better observations, some of which may show him that he must discard one or more of his particular system of primary principles and replace it by others.

In mathematics we ask only that the primary statements be accepted for use by all those concerned. We do not ask them to represent anything in nature or be "obviously" true. Of course, our reason for studying a particular group of primary statements and their results usually has some practical purpose, but it may be that they merely offer at least some promise of being especially challenging, artistic, or just entertaining rather than useful. Some mathematicians have worked out particular systems of primary statements and studied their results just for the fun of doing it.

This arbitrary nature of the first true statements in a branch of mathematics is the fundamental way in which the subject differs from the natural sciences. Mathematicians may decide that a particular system of primary principles is uninteresting or unfruitful for any of several reasons, but they are never compelled to check against any outside reference as is the natural scientist.

Once a truth standard is accepted, its primary statements can be used to prove other statements deductively. All the statements accepted as true or proved true at a given time then form a new truth standard against which the truth or falsity of new conjectures is judged. The resulting sequence of statements is called a *deductive discourse*, a *logical discourse*, a *logical system*, or a *theory*.

Each primary statement of a discourse is called an *axiom* or *postulate*. Aristotle tried to make a difference between the two, but most logicians and mathematicians agree that according to present day standards there is no meaningful distinction. The word "assumption" is also used, but it is semantically unfortunate. The primary statements are not in any sense "assumed" to be true in the usual meaning of the word. Each is simply labeled true arbitrarily, and flatly asserts that the discourse is to have that particular property. Here the term *axiom* will most often be used for this purpose.

In summary, each branch of mathematics is a logical discourse based upon axioms which are arbitrarily labeled true. The nature of that branch is simply what those axioms assert that it is—nothing more, nothing less. This approach to building up the subject matter of mathematics is often called the *axiomatic method*.

EXERCISE SET

1. Examine the following sentences to see whether you can label each *true* or *false*. Explain your answers.

    a) $7 + 8 = 15$ b) $5 + 9 = 2$ c) $2 \cdot 3 = 11$

2. Any measurement scale is a kind of truth standard. Give an example of a statement which is true relative to one such standard but false relative to another.

3. Let the following two statements be axioms for some discourse.
    (i) All x's are y's.
    (ii) All y's are z's.
    What would seem like a reasonable conjecture to try to prove in this discourse?

4. In what way do a mathematical system and a religion resemble each other?

5. State the fundamental logical difference between mathematics and a natural science.

*6. Is it possible for you to prove your own existence?

§1d   MATHEMATICS AND LANGUAGE

Regardless of the subject we are discussing, whether the discussion is written or oral, or whether we are using inductive or deductive logic, we are always faced with the limitation of having to use the correct semantics and grammar of a particular vernacular to state our principles and express our thoughts effectively.

Just as they do in everyday speech, some words which are used in an axiomatic discourse stand for more than one thing. Consider the sentence

(1)   He is tall.

He refers to any one of many persons. Such a word which stands for any of several recognizably different things is called a *variable*. Further examples are the *S* in

S is a set

and the term *a table* in

A table has four legs

In contrast, a sentence such as

5 is the successor of 4

contains no variable. We feel that 5 and 4 are each symbols for a single specific thing. A symbol which stands for one particular thing is called a *constant*. The term *Sheila* in

Sheila is a ship

is a more difficult matter. We are tempted to ask "Sheila who?" but, although one must admit that there are many different Sheilas and that one of them might be the name of a ship, for the sake of having examples to work with we shall always assume that a proper name refers to one particular person or thing and is, therefore, a constant.

The problem of using variables in a discourse will be discussed in a later section. In these initial discussions of the axiomatic method and its results we shall consider only sentences stated in terms of constants.

Let us now look at some of the specific problems which must be faced in order to introduce useful terms into a discourse. First of all we must distinguish between two categories of terms. Some we shall use with their everyday English meanings. Other terms will have meanings peculiar to the discourse being developed at the moment. These we shall call *specialized terms*.

Sometimes a well-known word is used with a restricted meaning as a specialized term in some discourse. For example, the word "tree" has a more restricted meaning in botany than in everyday usage. Other meanings for tree such as "a device for helping shoes hold their form" or "a stand with pegs on which to hang hats or other garments" or "the set of relationships between a person and his ancestors" obviously are not, in spite of ordinary usage, part of the botanical concept. And unless the botanical definition is learned carefully it might not be obvious that the use of the word *tree* in the term *banana tree* is also incorrect since this plant is not classified botanically as a tree.

Thus we make an effort to delineate carefully meanings of specialized terms. In this process it might seem desirable to give an explicit definition of each specialized term we use, but this is impossible just as it is impossible to prove all statements in a discourse. Since the specialized terms are the only ones which have a meaning particular to a given discourse, only a sentence which contains another such term can give us any information about a particular specialized term. Now imagine yourself trying to define the *first* specialized term of the discourse. Since there is no specialized term available for use, the task is impossible. Some specialized terms must be left undefined.

Anyone who has used a dictionary has run into the "circular" pattern which results whenever one tries to define terms without using at least a few initial terms which are undefined in the explicit sense. Looking up one word, one finds a term which explains it, and looking this up in turn eventually leads back to the original word, thus completing a circle. This is certainly of no use if the terms involved are specialized in meaning.

In order to avoid circularity, then, we agree that a few specialized terms must be left undefined in any logical discourse. These undefined terms are often called the **primitive terms** *of the discourse.*

Moritz Pasch (1843–1930) seems to have been the first person to realize that undefined terms are logically necessary in using the axiomatic method. Some of the undefined terms met in studying mathematics are *point*, *set*, and *successor*.

In spite of this limitation on the use of definition, it should be obvious that we must understand clearly how each specialized term we use in a discourse is related to each other such term. In any discourse we learn the properties of the primitive terms from the context in which they first occur. This situation has something in common with the way you probably learned the everyday meaning of a word such as "tree." From the time you were little you saw trees, felt them, smelled them, and watched them produce flowers or cones, grow, or bend in a strong wind. You may even have run into, or fallen out of, one or two. In the end you had no need to look up a definition in order to have a fairly clear idea of the usual meaning of the term. In other words, you learned the meaning of *tree* from association with the physical object which it symbolizes, i.e., from context.

The method of determining the meaning of a term from its context is called **implicit definition.**

What then is the context in which we find the undefined primitive terms of an axiomatic system? Since the axioms are the first true statements in such a logical development, and since the primitive terms are the first specialized terms in that development, it would seem reasonable for the latter to appear in the axioms. To avoid circularity, in fact, the primitives are the only specialized terms which should appear in them. As one axiom after another is stated, we learn more and more about the primitive terms. We know that we have stated enough axioms for a branch of mathematics when the properties of the primitives have been specified sufficiently so that these terms are usable without misunderstanding. Once a set of primitive terms has been listed, and a set of axioms given which tell how these terms are related, the discourse can be expanded by adding definitions and making proofs.

In contrast to the process of implicit definition described above, all other specialized terms to be introduced into a discourse are to be defined *explicitly*. That is, the properties of nonprimitive specialized terms are specified in a separate statement made for just that purpose. In order to avoid the circularity mentioned above, the words which may appear in an explicit definition are primitives, *previously* defined terms, or nonspecialized terms of the vernacular in use.

> A definition *is an arbitrary agreement on a name for a particular concept introduced into a discourse. Hence it is not subject to proof.*

A good explicit definition must exhibit certain essential characteristics in order to be usable and logically acceptable. Some of these, including the main ones already mentioned, are:

(i) A definition must be a complete sentence.
(ii) One part of the sentence must name the terms being defined and place it in a known category.
(iii) The other part(s) of the sentence must state the property which distinguishes the term from the remaining things in that category.
(iv) A definition should not be redundant; i.e., it should not place more requirements than necessary on the term being defined.
(v) Apart from the term being defined, the only words which can be used in a definition are previously defined specialized terms, primitive terms of the discourse, or nonspecialized terms of the vernacular in use.
(vi) A definition must be consistent with the axioms.
(vii) A definition must be reversible.

The first four characteristics are intended to make a definition usable. The remaining three are requirements of logic. Note that (v) prevents circularity as mentioned previously. The concept of consistency (vi) will be discussed in the following section.

EXAMPLE 1: Discuss the following definition relative to characteristics (i) through (vii).

(2)   A *triangle* is a polygon with only three sides.

Solution: The specialized terms in the sentence are *triangle*, *polygon*, *three*, *sides*. To be a good definition, then, *polygon*, *three*, and *sides* must have been previously defined or labeled as primitives. The term being defined is *triangle*, the known category is *polygon*, and the distinguishing characteristic is *only three sides*; so Characteristics (i), (ii), (iii), and (v) have been satisfied. Reversing (2) yields the statement, "A polygon with three sides is a triangle." This is as true as the original sentence. Thus Characteristic (vii) is satisfied.

Example 2 illustrates what is meant by Characteristic (iv).

EXAMPLE 2: Does the following definition satisfy Characteristic (iv)? "A *rectangle* is a parallelogram all of whose angles are right angles."

Solution: In most courses in plane geometry it is proved that

(3)   the consecutive angles of a parallelogram are supplementary.

It would be enough to say, then, that a rectangle is a parallelogram with one right angle since it can be deduced from (3) that all its angles must be right angles. Thus, the proposed definition does not satisfy Characteristic (iv).

The concept of reversibility will be discussed more carefully in §2a, but the following example should help clarify the idea.

EXAMPLE 3: Consider the sentences, "Two intersecting straight lines are perpendicular" and "Two intersecting straight lines determine right angles." Connect the two sentences with the conjunction *if* in two ways and decide whether the resulting compounds are true or false according to your knowledge of geometry.

Solution:

(4)   If two intersecting straight lines are perpendicular, then they determine right angles.
(5)   If two intersecting straight lines determine right angles, then the lines are perpendicular.

Does (4) seem true? It probably does. What about (5)? Both statements are considered true in geometry.

Whenever a compound statement remains true when the word if *is placed before either clause, it is said to be* reversible.

Thus either (4) or (5) could be a definition as far as Characteristic (vii) is concerned.

To see that not all statements are reversible, consider the following case.

EXAMPLE 4: Is the following reversible?

(6)   If two angles are right angles, then they are congruent.

Solution: Using your previous knowledge as a truth standard, this should seem true. In fact, it is one of the axioms Euclid stated for his geometry. Reversing the clauses we obtain:

(7)   If two angles are congruent, then they are right angles.

Does this seem true? Examine Figure 1.1. You may remember that α and

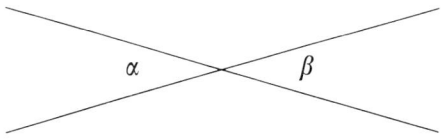

Figure 1.1

β are called vertical angles. They are congruent. Certainly we would not want to claim that they are necessarily right angles. This axiom, then, is simply not reversible. In general, *neither axioms nor theorems are reversible*. In an axiomatic discourse only definitions must have this property.

The axiomatic method especially demands clear and correct usages. Recalling from §1a that increased stress on deductive reasoning as embodied in the axiomatic method is a prime characteristic of "modern" mathematics, it is not surprising that another important characteristic of modern mathematics is the attempt to improve the precision of statements, especially definitions. Yet, every human vernacular is noted for imprecise use of words and for use of idiomatic expressions. In addition, it is almost always true that at least a part of our meaning in conversations is conveyed by gesture, facial expression, voice inflection, or innuendo. Observing Characteristics (i) through (vii) for definitions and corresponding principles for other statements provides maximum help in understanding concepts in spite of the linguistic difficulties.

Since no precise communication is possible unless the terms of a discourse are understood precisely by everyone, it should be evident that the students'

§1d   *Mathematics and Language*   15

part is to learn each definition *quickly* and *carefully*. Failure to do this can only result in a hazy grasp of concepts.

EXERCISE SET

1. Tell which terms in the following sentences are the variables or state that no variable is used.
   a) Five is an even number.
   b) She is Mary's best friend.
   c) $x + 5 = y + 3$

2. Suppose that in some development in plane geometry the terms *line segment, end point, equidistant, polygon, angle,* and *congruence of angles* are primitives or have been defined (assuming their usual properties) and are the only specialized terms available for use at this point. Make logically acceptable definitions of the following entities in the order listed so that they have their usual properties. The counting numbers from 1 to 10 may be used.
   a) trapezoid   b) regular polygon
   c) parallelogram   d) rectangle

3. Discuss the differences in meaning of the words *term* and *statement* as they are used in this section.

4. Which of these statements might be a definition as far as reversibility is concerned? Use your experience as a rough truth standard for determining whether each of the clauses in the given sentence is true or false.
   a) If a triangle is isosceles, then it has two congruent sides.
   b) If a straight line intersects a curve at one point, then it is tangent to the curve.
   c) If a vehicle is an automobile, then it has four wheels and a motor.

5. State which terms would have to be primitive or previously defined in order for the given statement to be a good definition from the standpoint of noncircularity. (The italicized term is the one being defined, and the subject of its discourse is given in parentheses.)
   a) An integer is *even* if and only if it has a factor of 2. (Mathematics)
   b) An account is *payable* if and only if its total is entered in the third column. (Accounting)

*6. Acceptably define *horse* in the sense we usually use the term.

*7. Is the statement "A living thing is any thing capable of both movement and respiration" an acceptable definition of *living thing* in the sense we usually mean the term? Bending, internal flow, etc., are to be considered as types of motion.

§1e THE PATTERN OF FORMAL AXIOMATICS

We have stated that a deductive logical discourse is expanded from the initial axioms by making definitions and proving conjectures. Constructing such proofs is, however, a complicated activity. For this reason the Greeks recognized the need for a systematic procedure.

A system of rules for correctly constructing and analyzing arguments is called a **logic.**

Aristotle was the first person to work out such a logic in detail. The rules he stated were undoubtedly not all invented by him. Most of them had probably been in use for as long as man had existed, and experience had undoubtedly shown them to yield fruitful results when correctly applied to arguments in many fields. Aristotle detailed a system for their correct use and pointed out some errors which resulted from using them incorrectly. In the study of philosophy other logics are taken up, but Aristotelian logic is the only one in common use. It is the one used almost exclusively in mathematics.

With the choice of a suitable logic to provide rules for proof and disproof of conjectures, theorems can be added to the discourse. This addition completes the deductive structure. The parts of such a structure are thus primitive terms, definitions, axioms, a logic, and theorems. Other names for an axiomatic discourse are *logical system* and *The Pattern of Formal Axiomatics*. The following outline reviews the structure.

The Pattern of Formal Axiomatics
  (i) The discourse contains specialized terms called *primitives* which are deliberately left undefined.
  (ii) All other specialized terms are *defined* by using the primitive terms.
  (iii) The discourse contains statements about the primitives called *axioms* which are deliberately left unproved but accepted as true.
  (iv) All other statements in the discourse are *deduced* from the axioms using a set of rules called a *logic*.
  (v) These new statements are called *theorems*, and each theorem is accompanied by a statement which asserts that it was logically deduced from the axioms, e.g. Q.E.D. ("Quod erat demonstrandum" which means "as was to be shown.")

It is often said that a mathematical theory is a body of results obtained from the interaction of a set of axioms and a logic. The Pattern provides us with three types of true statements: axioms, acceptable definitions, and theorems. There are three types of terms used in these statements: everyday terms, primitive terms, and defined terms. The theorems should form the largest part of any such structure, but since they are proved using the chosen logic, their nature is determined in part by it as well as by the axioms of the discourse.

Aristotelian logic used in a discourse requires that the statements of the structure be consistent.

A discourse is consistent **when no statement appears which is simultaneoulsly true and false.**

Suppose, for example, that we were discussing the natural numbers and that we could prove both that $2 + 3 = 5$ and that $2 + 3 = 1$. The properties of equality would then require that $1 = 5$ be true. If we could also prove by a different method that $1 = 5$ is false, we would have the same statement being simultaneously true and false. The system is inconsistent. When such a situation develops we say that a contradiction exists. Using Aristotle's logic, then, *a set of axioms must be internally consistent*. This, then, is the single logical restriction on arbitrary choice of axioms.

Recall also that Characteristic (vi) of a good definition given in the preceding section provided that each definition introduced into a discourse must be consistent with the axioms. If any proposed definition causes a contradiction to appear in the discourse, the offending definition must be discarded in order to keep the system consistent. One rule which helps assure consistency is that a definition must never bring any new entity into existence in a discourse. *Only an axiom or a theorem may state existence*. Thus we have an important logical restriction on choice of defining statements.

In this book we restrict ourselves to consideration of Aristotelian logic. When we use the terms *logic* or *principles of logic* we shall mean Aristotelian logic.

EXERCISE SET

1. Outline the way in which the various parts of the Pattern of Formal Axiomatics are related to each other.
2. Under what conditions, if any, would the following sentence be an ac-

18   *The Nature of Mathematics*

ceptable definition: "A *gaba* exists if and only if it is baked from two blabas"?

§1f   STRUCTURES AND MODELS

Mention has been made of the Greek use of abstraction to arrive at the essential, idealized nature of concepts. In a geometric example, we use dots and strokes on paper to represent the essential nature of physical relationships. Then we carry the abstraction process a step farther and idealize the dots and strokes to the abstract mathematical concepts "point" and "line" respectively. Note that both a dot and a stroke have measurable width. Each also has measurable thickness even though it would be microscopic. However, we associate neither width nor thickness with the mathematical idealizations point and line. Thus, in our abstract concepts, a point is visualized as a dimensionless marker of position; a line, something like a continuous extension of such dimensionless points whose direction of extension is determined by a pair of given points.

This process of divesting a concept of all unessential characteristics has been such a successful approach to the study of mathematics that it has become another of the important modern features of the subject. That is, there is now a tendency to concentrate on the *structure* of each mathematical system as displayed in the basic principles which govern the system. For example, each of the sentences

(1) $$6 + 7 = 1$$

and

(2) $$6 + 7 = 13$$

is true according to some truth standard. From the structural point of view it is not the facts displayed in (1) and (2) which are important. Rather, one asks what fundamental differences exist between the two arithmetics.

One such difference can be immediately noted. Clock arithmetic requires only twelve different symbols to write down any specific results whereas ordinary arithmetic requires infinitely many.

Along with the differences we would also find that some of the properties of the two arithmetics are the same. For example, we have $7 + 6 = 1$ in clock arithmetic so that

(3) $$6 + 7 = 7 + 6$$

holds just as it does in the case of grocery store arithmetic. We say that the sentence

(4) $$a + b = b + a$$

in which the variables a, b stand for numbers, represents a basic principle of each system.

When sufficient basic abstract properties such as (4) of each arithmetic have been stated so as to characterize it completely, we then call it an abstract structure. As an example of such a structure we state the basic properties of the system known as the *real numbers*. We choose this particular system because all of you have had considerable previous experience with use of its basic principles even though in the past you may not have regarded properties of numbers as a collection of abstract basic principles.

To state the properties of the real numbers, we must first state the basic properties of equality as abstract statements. In all these statements the letters $a, b, c, \ldots$ are variables which stand for real numbers. The term given in parentheses is a standard name for that property.

E-1: (reflexive) For each real number a, $a = a$.

E-2: (symmetric) For each pair a, b of real numbers, if $a = b$, then $b = a$.

E-3: (transitive) For each triple a, b, c of real numbers, if $a = b$ and if $b = c$, then $a = c$.

The following properties apply specifically to real numbers, showing how they act under the operations of addition and multiplication.

R-1: (closure for addition) For each pair a, b of real numbers, there is a real number x such that $a + b = x$.

R-2: (uniqueness for addition) For each triple a, b, c of real numbers, if $a = b$ then $a + c = b + c$.

R-3: (commutativity for addition) For each pair a, b of real numbers, $a + b = b + a$.

R-4: (associativity for addition) For each triple a, b, c of real numbers, $(a + b) + c = a + (b + c)$.

R-5: (solvability for addition) For each pair a, b of real numbers in that order, there is a real number x such that $a = b + x$.

R-6: (existence) At least two distinct real numbers exist.

R-7: (closure for multiplication) For each pair a, b of real numbers, there is a real number x such that $a \cdot b = x$.

R-8: (uniqueness for multiplication) For each triple a, b, c of real numbers, if $a = b$ then $a \cdot c = b \cdot c$.

R-9: (commutativity for multiplication) For each pair a, b of real numbers, $a \cdot b = b \cdot a$.

R-10: (associativity for multiplication) For each triple a, b, c of real numbers, $(a \cdot b) \cdot c = a \cdot (b \cdot c)$.

R-11: (solvability for multiplication) For each pair a, b of real numbers in that order and $b \neq 0$, there is a real number x such that $a = b \cdot x$.

R-12: (distributivity, multiplication over addition) For each triple a, b, c of real numbers, $a \cdot (b + c) = (a \cdot b) + (a \cdot c)$.

R-13: (trichotomy) For each pair a, b of real numbers, one and only one of the following holds:
  (i) $a = b$
  (ii) $a < b$
  (iii) $a > b$
  (Note: $a > b$ is simply another way of writing $b < a$.)

R-14: (transitivity of order) For each triple a, b, c of real numbers, if $a < b$ and if $b < c$, then $a < c$.

R-15: For each triple a, b, c of real numbers, if $a < b$, then $a + c < b + c$.

R-16: For each triple a, b, c of real numbers such that $c > 0$, if $a < b$, then $a \cdot c < b \cdot c$.

R-17: (completeness) If any series of real numbers has an upper bound, then it has a least upper bound.

The exact meaning of each of these seventeen statements will not be clear to the beginner, but (3) is an example of R-3. Much more extensive and detailed discussions of the real numbers will be given in Chapter 6. Here we are only making the point that the real numbers are a structure having properties R-1 through R-17. Similar structures, such as other kinds of numbers, might lack one or more of these properties or have properties not possessed by the real numbers.

Like many other mathematical structures, the structure delineated by R-1 through R-17 has a special name. It is called a *complete ordered field*. The student of mathematics will go on to spend some time in studying several such structures.

Since some reference will be made to these real number properties in the material preceding Chapter 6, two more examples of their application to particular real constants are given.

EXAMPLE 1: Consider $7(3 + 2)$. To evaluate this expression we write

$$7(3 + 2) = 7 \cdot 5$$
$$= 35$$

By the distributive property R-12 we could have written

$$7(3 + 2) = 7 \cdot 3 + 7 \cdot 2$$
$$= 21 + 14$$
$$= 35$$

The Nature of Mathematics

ιe increased use of abstraction inherent in our present day concentra-
on the structure of a system over and above mere cataloguing of detailed
erties has enabled mathematicians of this century and the last to
gnize relationships among many entities once thought to be quite dis-
lar. A similar development has taken place in problem solving.
nce people learned to solve problems by practicing the techniques needed
lve that type of problem. This approach demanded knowledge of many
gories of problems and the requisite techniques for dealing with
ι. Now we make a greater effort to stress the *pattern* displayed by a
lem. That is, we try to solve a problem by determining its essential
ιre and then using our knowledge of the basic principles (structure)
he particular system to arrive at a solution. This concentration on pat-
ι as opposed to technique is remarkably efficient—especially in this day of
:hanical aids for solving problems.

EXERCISE SET

Consider E-1 through E-3 and R-1 through R-17 as axioms for the real numbers; what are the primitive terms of the discourse?

The real number 0 such that $a = a + 0$ for each real a is called a **neutral element** (or **identity element**) **for addition** of real numbers. The real number 1 such that $a = a \cdot 1$ for each real a is called a **neutral element** (or **identity element**) **for multiplication** of real numbers. How do we know from the axioms that these particular real numbers exist?

Consider the following axiomatic discourse. Primitive terms: x, y.

A-1: At least two distinct x's exist.
A-2: For any two x's, there is one and only one y associated with both of them.
A-3: For any y, there is at least one x not associated with it.
Theorem 1: At least three x's exist.
Theorem 2: Any two distinct y's have at most one x associated with them.
Theorem 3: At least three y's exist.

Illustrate the relationships in the axioms. Then use an illustration to show that each theorem in turn is consistent with the axioms and the theorems which precede it.

Make an application of the axiomatic discourse of Exercise 3 by letting $x \rightarrow$ committee member and $y \rightarrow$ committee. What do the theorems tell us about this system of committees and committee members?

§1f Structures and Models

EXAMPLE 2: Is there a real number x such that 5
solvability property R-5 states that this number ⏃
perience we know that $x = {}^-2$.

E-1 through E-3 can be considered as axioms for ec
R-17 can be considered as axioms for real numbers.
course can be built on them. Since we have not stuc
at this point, no attempt will be made to demonstrate
three theorems are given only as an illustration
number statement which would be found in an axic
the real numbers. Existence of particular real number
established previously.

Theorem (1f-1): For each real number a there is
that $a + {}^-a = 0$.

Theorem (1f-2): The real numbers 0, 1 have the

Theorem (1f-3): For each triple a, b, c of real nu
$c < 0$, then $a \cdot c > b \cdot c$.

Recall now that the axioms for any mathematical s\
labeled true. Thus the structure inherent in each systen
a part of any possible application. This is not mea
applications should be of no importance. Most of the s
important because they do have a variety of fruitful ap|

What is often misunderstood is the relationship betw
system and the applications to which it may lend itself
plication of a mathematical system to a physical problen
the relationships among the basic concepts of the pro
If the structure of the physical problem agrees with t
mathematical system to within the accuracy of measuren
that the real problem is a concrete model of the abstract a
The properties of the abstract structure then tell us proper
model.

For example, when physical concepts such as dot, ed
on are made to correspond to point, line and plane of th
matical system called Euclidean geometry, many useful a
made. Also the "real" model can be used inductively to h
relationships among the entities of the abstract structure. T
we use when we draw figures in solving a geometric probl

An interesting fact is that particular "real" relationsh
a model of more than one abstract system. For example
of the last century developed several *different* geometries, a
can have the "real" world as a model as far as we can pres

5. Make an application of the axiomatic discourse of Exercise 3 by letting $x \to$ point and $y \to$ line segment. What do the theorems tell us about this system of points and line segments? Draw a representation of this system.

6. Make an application of the axiomatic discourse of Exercise 3 by letting $x \to$ primitive term and $y \to$ axiom. What do the theorems tell us about the system of axioms and primitive terms?

7. Make an application of the axiomatic discourse of Exercise 3 by letting $x \to$ ticklish tiger and $y \to$ tank. What do the theorems tell us about the system of ticklish tigers and tanks? How do the primitive terms here differ from those of Exercise 2 and 3?

8. A number $\bar{a}$ such that for each a, $a + \bar{a} = 0$, is called an **additive inverse**. A number $\hat{a}$ such that for each a, $a \cdot \hat{a} = 1$ is called a **multiplicative inverse**.
   a) How do we know that additive inverses exist among the real numbers according to R-1 through R-17?
   b) How do we know that multiplicative inverses exist among the real numbers according to R-1 through R-17?

2
ELEMENTARY LOGIC

§2a DEDUCTIVE ARGUMENT

We previously stated that an *argument* in a deductive discourse is a sequence of known true statements which allows us to establish that a given conjecture is true relative to these known statements. In using the axiomatic method, the statements to be labeled true at any point are the axioms of the discourse, the defining statements, and the previously proved theorems. Knowing which statements may be labeled true in a discourse is highly necessary, but it is not the only factor of importance in constructing acceptable deductive arguments.

To make a beginning at discovering other factors which play a role in deductive argument, let us take an atlas as a truth standard and look up the town of Portsmouth. It would, of course, be listed among the towns of New Hampshire. New Hampshire would be listed as one of the United States. Also, the United States would be listed as a part of North America. However, we probably would not find anywhere the direct statement that Portsmouth is in North America. In spite of this, is it acceptable to conclude that Portsmouth is in North America?

The given atlas facts and the conjecture above may be written in the following traditional argument layout.

EXAMPLE 1: (i)   Portsmouth is a town in New Hampshire.
            (ii) New Hampshire is in the U. S. A.
            (iii) The U. S. A. is in North America.
            _____
            (iv) Portsmouth is in North America.

The conjecture (iv) is called the *conclusion* of the argument. The statements used deductively to establish a label of true or false for the conclusion in the sequence of an argument are called *premises*. Thus (i), (ii), and (iii) are premises in Example 1.

Returning to the argument itself, we once again ask whether (iv) seems like a reasonable conclusion based on the given premises. It probably does since it is a well-known geographical fact. However, in an axiomatic discourse one often does not know beforehand how to label the conclusion, and in discussing principles of logic themselves, we shall want to consider all possible cases simultaneously. Hence the important question is, "Are we willing to accept the conclusion as true if we accept the premises as true?" If, for example, no atlas were handy and we did not know the geographical facts, would the argument in Example 1 still seem acceptable?

This point may be easier to consider if we examine an argument which is not built upon well-known facts.

EXAMPLE 2: Premises: (i) All zoys are glachos.
(ii) All glachos are ootsters.
(iii) No shfnexi is an ootster.
Conclusions: (i) All zoys are ootsters.
(ii) No zoy is a shfnexi.

Certainly we know nothing about zoys, glachos, ootsters, and shfnexis and cannot, therefore, judge whether any of these five statements is true or false. Nevertheless, each argument should seem reasonable. That is, *each conclusion should seem true if we accept the premises as true*. In some way the construction of the argument, i.e., how premises and conclusions are related, causes us to accept the conclusion as true once we are given that the premises are to be regarded as true.

What if one or more of the premises of an argument were known to be false? Consider the following case.

EXAMPLE 3: Premises: (i) A giraffe is a flowering plant.
(ii) A flowering plant is a living thing.
Conclusion: A giraffe is a living thing.

Using experience as a rough truth standard, most persons would probably agree that Premise (i) is false, Premise (ii) is true, and that the conclusion is true. Suppose, however, that we knew absolutely nothing about a giraffe. Then we would not know how to label Premise (i) and would not know whether the conclusion were true or false. Would the argument seem acceptable in that case? It probably would. Again we see that it is the way in which premises and conclusions are related which determines acceptability of an argument.

It may be useful to consider one more example.

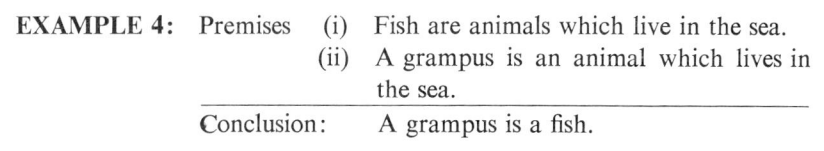

EXAMPLE 4: Premises (i) Fish are animals which live in the sea.
(ii) A grampus is an animal which lives in the sea.
Conclusion: A grampus is a fish.

As before, in order to analyze the argument, assume that the premises are true. Is the conclusion then true without possible exception? If you answered "yes" without pausing, perhaps you should read through the argument again. Have we been told that every animal which lives in the sea is a fish? No. Hence it is possible that a grampus could be a sea animal and still not be a fish. It is not absolutely necessary that the conclusion be true even though the premises are to be regarded as true. The premises and conclusion here are in some way differently related from those in preceding cases.

The way in which premises and conclusion of an argument are related is called the *structure* of the argument. In contrast, the term *content* refers to the meaning of the statement, i.e., whether it is labeled true or false. It is important not to confuse content and structure.

If the structure of an argument is acceptable, then we say that we are using *valid reasoning*. This is not meant to be a strict explicit definition of the term "valid." In §2d explicit criteria for validity are stated. For now, it is important only to see that Example 4 is an example of invalid reasoning whereas Examples, 1, 2, and 3 illustrate valid reasoning.

EXERCISE SET

1. Are validity and truth the same? Explain carefully.
2. If the Portsmouth mentioned in Example 1 above is actually the one in England (i.e., if "Portsmouth, England" is substituted for "Portsmouth" wherever it occurs in Example 1), does the reasoning still seem to be valid?

Determine whether or not the arguments in Exercises 3 through 8 seem logically valid. (Answer as best you can at this point.) Then use your experience as a rough truth standard to determine whether each premise and conclusion is true or false.

3. Premises: (i) San Francisco is in Massachusetts.
              (ii) Massachusetts is in the United States.
   Conclusion: San Francisco is in the United States.

4. Premises: (i) If a zosterops is a bird, it is a two-legged animal.
              (ii) A zosterops does not have two legs.
   Conclusion: A zosterops is not a bird.

5. Premises: (i) All business men are high jumpers.
              (ii) All high jumpers are boy scouts.
   Conclusion: All business men are boy scouts.

6. Premises: (i) A zosterops is not a bird or it has two legs.
              (ii) A zosterops is not a bird.
   Conclusion: A zosterops is two-legged.

7. Premises: (i) All tired tigers are ticklish.
   (ii) All men are ticklish.
   Conclusion: All men are tired tigers.

8. Consider the argument of Exercise 7 with the conclusion replaced by "All tired tigers are men."

9. Form one or more conclusions for the following premises. Does each argument thus produced seem valid or invalid?

   Premises: (i) All pink panthers pass prelims.
   (ii) This pink panther has taken a prelim.

10. Using the axioms of Exercise 3, §1c, as premises, does the conclusion "all x's are z's" seem to be the result of a valid or invalid argument?

§2b AXIOMS FOR LOGIC

The structure of arguments can differ widely. Since complex combinations of premises can occur, it becomes very necessary to have some systematic procedure for constructing valid arguments and for analyzing given arguments to determine their validity. As we have indicated, the requisite procedure is embodied in rules, called inference rules, which make up what we call a logic. In turn, logic can be discussed somewhat axiomatically.

Before it is possible to develop inference rules, it is necessary to take a closer look at the words "true" and "false." These words are specialized terms in logic.

The word "sentence" in its usual meaning is a grammatical concept, but not all sentences can meaningfully be labeled true or false. Consider: "Read this book" and "How old are you?" These are both sentences, but labeling either one true or false would be meaningless. It seems as if this would be the case for all sentences which are commands, wishes, or questions. How about a declarative sentence, usually called a "statement"?

Consider the following statements.

(i) Five is a special number.
(ii) Salem is north of here.
(iii) It will rain in Brunswick, Maine, on June 29, 2001.
(iv) This sentence is false.

Assume a truth standard from mathematics for (i). Some kind of mathematical meaning would still have to be given to the word "special" before any label of "true" or "false" would be applicable.

In the case of (ii), the label to be applied would certainly depend on your geographical location relative to the town and on which particular Salem was meant. Also, imagine your predicament if you were within the boundaries of the indicated Salem at the time the labeling was to take place!

It may be that (iii) is true; it may be that it is false. In either case we are not going to be able to label it at present. Under the assumption that, at some time in the future, we shall be able to label future time statements, we sometimes use them as examples or exercises for the sake of variety.

Try to label (iv). The resulting difficulties should be illuminating.

What these examples show is that time, considerations from within a discourse, and/or logic itself affect the labeling of statements. It is simply not possible to label all statements true or false. The property of being true or false, or having logical usability, is not a grammatical concept.

We shall use the special term proposition *to denote the kind of sentence which can be used in a logical discourse.*

Having determined three initial specialized terms for logic, we start the discourse by taking these terms as undefined and stating axioms which delineate their properties. Thus *true*, *false*, and *proposition* are our primitive terms. The following axioms will serve to determine their properties implicitly. They are among the basic principles which Aristotle used and discussed and are sufficient to serve our purpose here. A fifth axiom will be stated in §2h.

To simplify discussion, we shall call either label true or false a *truth value*. Truth values in the plural refers to all the possible applications of these labels to a proposition.

> **L-1:** Throughout any discourse a proposition retains its truth value.
> **L-2:** The only truth values assignable to a proposition are *true* and *false*.
> **L-3:** Each proposition has exactly one truth value.
> **L-4:** Whenever the possible truth values of one proposition are the same as those of a second in a discourse, either may replace the other at any point of that discourse.

L-1 may look innocuous, but we would not want a proposition to be a kind of statement which could change its truth values in the middle of a discourse. More important, we know that a proposition may be labeled true in relation to one truth standard and false in relation to another. Hence this axiom requires that any sentence which we are going to call a proposition in a discourse must be used only in relation to a single truth standard.

At first glance it may seem that we do not need both L-2 and L-3, but L-2 is to be taken as meaning that in Aristotelian logic we have only two labels available instead of possibly a third, a fourth, or others. L-2 does not state that we might not use both available labels at once. That is, such labels as "true-false" or "true-false-true" or other combinations are possible as far as L-1 and L-2 alone are concerned.

L-3 excludes the possibility of using any such multiple labels. According to this axiom the labels must be used singly. This is, in effect, the requirement of consistency discussed in §1e. Whenever this logic is used, all principles introduced into a discourse (including axioms and definitions) must be stated in such a way that the labels "true" and "false" can be applied and also applied singly.

To recapitulate, L-2 and L-3 together assure us that there are two and only two labels for use and that these are to be used singly. Aristotelian logic cannot be used as the logic for any discourse in which one might desire to label some statements "partly true" or apply some other such label. For this reason, Aristotelian logic is often called "two-valued" in contrast to some others.

To consider Axiom 4, the *Axiom of Substitution*, it is convenient to have a representation for individual propositions.

> **We shall use a single letter such as $p, q, r, \ldots$ to represent a proposition whose truth value is not known.**

L-4 states that a proposition p may be substituted for a proposition q or vice versa if both have the same truth values. The following definition is pertinent.

Definition (2–1): Any propositions p, q which have the same truth values in each possible case are said to be **logically equivalent**. We write $p \subseteq q$.

In terms of this definition, L-4 states that either of two logically equivalent propositions may be substituted for the other in a discourse.

A device called a truth table invented by Charles S. Peirce (1839–1914) is very useful for displaying the possible truth values of any proposition which does not have a single known truth value. The table for a proposition p is shown in Figure 2.1. Here and in all that follows we use capital T, F, to stand for the respective labels. Recall that by Axiom L-3 the cases shown by the rows are not simultaneously possible. Hence, each row is to be considered a case by itself, which mutually excludes each other case.

	p
Row 1	T
Row 2	F

Figure 2.1

§2b   *Axioms for Logic*

Since a proposition is a sentence, modifiers can be used to change meaning, provided that we do not contradict our axioms. The most useful modifier for our purpose is negation.

It is logically acceptable to consider a truth table as a definition. Then the verbal statement it accompanies only describes the relationships shown in the table. Nevertheless, each such statement should be learned carefully in order to make reference to tables unnecessary.

Definition (2–2): For any proposition p, the proposition $\sim p$, called the **negation of** p, is false if p is true and is true if p is false.

p	$\sim p$
T	F
F	T

Figure 2.2

The proposition $\sim p$ is often read "tilde p" and frequently translated as *not-p* or *it is not the case that p*. Is $\sim p$ actually a proposition for any proposition p? To determine this point examine the truth table. Figure 2.2 shows the truth values of $\sim p$. The double vertical bars separate the derived form(s) from the given proposition(s). According to L-2 and L-3, the rows are considered to represent mutually exclusive possibilities, thus $\sim p$ is a proposition.

The form "It is not the case that p" could be called the *basic form for negation*. In general, various derived forms will be called *simplified forms for negation*. Any such derived form must agree in meaning with the basic form to be correct.

> **For example, a simplified form of the negation of a simple proposition (not a compound), in which no word like "all," "each," "some," "at least one," or similar terms of quantity appear, is obtained by grammatically negating the verb.**

EXAMPLE 1: Let p stand for "John is lacing his shoes." Write a correct negation of p.

Solution: The translation of the basic form of $\sim p$ is

(1)   It is not the case that John is lacing his shoes.

Using the preceding rule, we obtain an equivalent simplified form,

John is not lacing his shoes.

Note that the form "John is unlacing his shoes" is *not* correct, for it does not have the same meaning as the basic form (1).

EXAMPLE 2: Negate proposition x, for which the truth table is in Figure 2.3.

x
F
T

Figure 2.3

x	$\sim x$
F	T
T	F

Figure 2.4

Solution: According to Definition (2-2), the truth value of $\sim x$ in Case (Row) I must be T and in Case II must be F. Figure 2.4 shows the truth values of $\sim x$.

EXAMPLE 3: Let q stand for "Every tiger is a bird." Write a correct negation of q in simplified form.

Solution: The term of quantity "every" makes this a more complex case than the one in Example 1. The basic form of the negation would be "It is not the case that every tiger is a bird." The required simplified form is "Some tigers are not birds." Sentences which contain terms of quantity will be considered in §2h.

In this section we have taken the first steps in reducing arguments to symbolic logical form. Analysis of arguments is easier in logical symbols than in verbal form. The Greeks used the longer verbal forms, and even today logic courses as such often hold to this tradition. Gottfried Leibnitz (1646-1716) made a beginning at developing special symbolic notation, George Boole (1815-1864) developed an extensive system, and Giuseppi Peano (1858-1932) and Bertrand Russell (1872-1970) have made important additions.

EXERCISE SET

1. Make a truth table for $\sim(\sim p)$. It should have three columns, one each for p and the derived forms $\sim p$ and $\sim(\sim p)$. Can you reach a conclusion about $\sim(\sim p)$ relative to Axiom L-4 and Definition (2-1)?
2. Can a "term" be reasonably labeled true or false?
3. Is Axiom L-4 a valid proposition?
4. Is Axiom L-1 a true proposition or a false one? Explain.
5. Which of the following are propositions? Use experience as a rough truth standard and interpret symbols in the most obvious way. If a phrase is not a proposition, state why.

    a) Bertha is a bird.

b) Look before you leap.
c) This lion is lazy.
d) Every number is odd.
e) Open the book.
f) $7 = 4$
g) The children running across the field.
h) How old is this car?
i) $3 < 8$
j) $x \neq 3$

6. Let p stand for the proposition "There is no joy in Mudville" and q be the proposition "In the municipal mortuary Mudville's mayor mambos merrily." What can be said about these propositions if they are both considered true?

7. Write a logical negation of each of a) through j) in Exercise 5 in simplified form or explain why it is impossible.

§2c   COMPOUND PROPOSITIONS

The axioms of the preceding section require only that a proposition have the properties stated in the axioms. Therefore, a proposition may be simple or compound, and, like simple sentences, a compound is a proposition if it is true or false only (L-2)—not both (L-3), and if both components are labeled in relation to the same truth standard (L-1). Any word used to combine two propositions into a compound sentence is called a *connective*.

Truth tables for the compounds are useful in several ways. However, we need to establish a conventional pattern for using them so that each person's work can easily be read and understood by others.

Consider two propositions p and q having unknown truth values. Proposition p could be either true or false. If p is true, q may be either true or false, and the same possibilities for q exist if p is false. These possibilities are shown in the tree diagram of Figure 2.5 in which T is listed before F in each case. The truth table of Figure 2.6 gives us a standard way of using each row to represent one of the four possible cases shown in Figure 2.5. Recall that by L-3 *each row of the truth table represents a mutually exclusive case.*

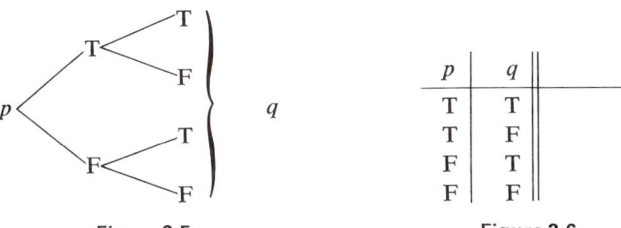

p	q
T	T
T	F
F	T
F	F

Figure 2.5   Figure 2.6

34 *Elementary Logic*

The reader will recall that the primitive terms of this discourse called logic are "true," "false," and "proposition." A truth table employs only these three concepts. Thus the truth table for each of the next three definitions can be considered as the actual defining device. The rest of each definition is a verbal statement of what is found in the table.

Definition (2–3): For any two propositions p, q, the proposition **p $\vee$ q**, called the *inclusive disjunction of p and q*, is false whenever both components are false; otherwise it is true.

p	q	$p \vee q$
T	T	T
T	F	T
F	T	T
F	F	F

Figure 2.7

The usual translation of $p \vee q$ is "p or q." Note that if either one of the component propositions is true or if both are true, then this compound is true. Also, we can see that, by L-2 and L-3, the inclusive disjunction is a proposition.

EXAMPLE 1: Let p stand for "A dog is a canine," q stand for "A cat is a feline." What is the truth value of the compound $p \vee (\sim q)$?

Solution: Using the facts of zoology as a truth standard, p is true and q is true. Therefore, by Definition (2–2) $\sim q$ is false. Thus the disjunction $p \vee (\sim q)$ stands for "A dog is a canine or a cat is not a feline" and is true by Definition (2–3). See Row 2 of Figure 2.7.

Definition (2–4): For any two propositions p, q, the proposition **p $\wedge$ q**, called the *conjunction of p and q*, is true whenever both components are true simultaneously; otherwise it is false.

p	q	$p \wedge q$
T	T	T
T	F	F
F	T	F
F	F	F

Figure 2.8

The compound $p \wedge q$ is usually translated *p and q*. However, *but* and *such that* are also used in certain cases. Examination of Figure 2.8 shows that the conjunction $p \wedge q$ is a proposition. Note also that the term "conjunction" has a meaning different from its usual meaning in grammar.

EXAMPLE 2: Using the same p and q as in Example 1, what is the truth value of $(\sim p) \wedge q$?

Solution: We are given the conjunction "A dog is not a canine and a cat is a feline." By Row 3 of Figure 2.8 this conjunction is false.

The properties specified for the next compound often seem strange to beginners. Remember, however, that the important factor in making a definition is that once it is correctly made (see §1d), we accept it as a true proposition for use in further development of the discourse. There is absolutely no requirement that a definition should not seem strange to some persons. The best possible advice is to treat this definition like any other and simply learn it quickly and well.

Definition (2-5): For any two propositions p, q (in that order), the proposition $\mathbf{p} \Rightarrow \mathbf{q}$, called the *implication*, is false when p is true and q is false; otherwise it is true.

p	q	$p \Rightarrow q$
T	T	T
T	F	F
F	T	T
F	F	T

Figure 2.9

The compound $p \Rightarrow q$ is often read p *arrow* q. Translations are given in Figure 2.10. We note that $p \Rightarrow q$ is indeed a proposition. This compound is often called the *conditional* form. The phrase "in that order" is an important part of the definition. That is, $p \Rightarrow q$ is not the same as $q \Rightarrow p$.

The component propositions p, q of the compound $p \Rightarrow q$ are called the hypothesis and conclusion respectively.

Using these terms, it is useful to remember the provisions of Definition (2-5) as, "an implication is false *only* when its hypothesis is true and its conclusion false; otherwise, it is true."

The table of Figure 2.10 lists a few of the more common translations of "p arrow q." The first nine should be learned carefully, for we shall meet them often in all that follows. The last two are translations which, unfortunately, have not yet gone out of use.

$$p \Rightarrow q$$

if p, then q
p implies (that) q
p only if q
q if p
q because p
p; therefore q
q whenever p
from p it follows that q
q follows from p
p is a sufficient condition for q
q is a necessary condition for p

Figure 2.10

EXAMPLE 3: Let p stand for "$\mathscr{L}_1$ is parallel to $\mathscr{L}_2$" and q stand for "$\triangle ABC$ is isosceles." Translate "$\mathscr{L}_1$ is parallel to $\mathscr{L}_2$ only if $\triangle ABC$ is not isosceles" into logical symbols.

Solution: $p \Rightarrow \sim q$.

EXAMPLE 4: Using the same p and q as in Example 3, translate "$\mathscr{L}_1$ is not parallel to $\mathscr{L}_2$ if $\triangle ABC$ is isosceles."

Solution: $q \Rightarrow \sim p$.

EXAMPLE 5: Using the same p and q as in Example 1, determine the truth value of $\sim p \Rightarrow q$.

Solution: We have $\sim p \Rightarrow q$: If a dog is not a canine, then a cat is a feline. By Definition (2–2) the hypothesis is false. Therefore, by Row 3 of Figure 2.9, the given proposition is true.

EXAMPLE 6: Is the proposition "$2 = 3$ if $5 = 5$" true or false?

Solution: Using experience in ordinary arithmetic as a truth standard, $2 = 3$ is false and $5 = 5$ is true. The proposition can be written in the form "If $5 = 5$, then $2 = 3$"; therefore, Row 2 of Figure 2.9 applies, and the proposition is false.

One additional connective is of considerable importance.

Definition (2–6): For any two propositions p, q, the proposition $\mathbf{p \Longleftrightarrow q}$, called the *bicondition*, is logically equivalent to $(p \Rightarrow q) \land (q \Rightarrow p)$, i.e., $[p \Longleftrightarrow q] \triangleq [(p \Rightarrow q) \land (q \Rightarrow p)]$.

The compound $p \Longleftrightarrow q$ is often read p *double arrow* q. Common translations are given in Figure 2.11. The notation $\Longleftrightarrow$ should seem reasonable from the definition. If we recall that $q \Rightarrow p$ can be written "p if q" and that $p \Rightarrow q$ can be written "p only if q," we see that "p double arrow q" simply expresses the given defining conjunction. The abbreviation "iff" is used to stand for "if and only if."

$$p \Longleftrightarrow q$$

p if and only if q
(if p, then q) and (if q, then p)
p if q and p only if q
p iff q
p is a necessary and sufficient condition for q

Figure 2.11

The truth values of $p \Longleftrightarrow q$ can be obtained from the definition. First the values of $p \Rightarrow q$ are listed as in Figure 2.9. Then recalling that q is the hypothesis and p the conclusion of $q \Rightarrow p$, the truth values of this implication are obtained from Definition (2–5). Finally, the values of the conjunction are determined using Definition (2–4). These steps produce the table in Figure 2.12.

p	q	$p \Rightarrow q$	$q \Rightarrow p$	$p \Longleftrightarrow q$ $(p \Rightarrow q) \wedge (q \Rightarrow p)$
T	T	T	T	T
T	F	F	T	F
F	T	T	F	F
F	F	T	T	T

Figure 2.12

We have set up our truth tables being certain that they are structured according to the requirements of our axioms, and that the information given agrees with our definitions. Hence we can accept facts shown in a table as part of this discourse. On this basis the following theorem has been established.

Theorem (2c-1): The bicondition $p \Longleftrightarrow q$ is *true whenever p, q*, both have the same truth value; otherwise it is false.

EXAMPLE 7: Write statements (3) and (4) of §1d in biconditional form if possible.

Solution: Letting p stand for "Two intersecting lines are perpendicular," and q stand for "Right angles are determined," we see that (3) becomes $p \Rightarrow q$ and that (4) becomes $q \Rightarrow p$. Thus $(p \Rightarrow q) \wedge (q \Rightarrow p)$ stands for "Two intersecting lines are perpendicular iff they determine right angles."

Comparison of the result of this example and the discussion of reversibility in §1d indicates that the following principle must hold.

(1) We call a true bicondition a *reversible proposition*.

Thus

$$2 = 2 \leftrightarrow 3^2 < 0$$

is not reversible because it is a false bicondition. However,

$$2 = 2 \leftrightarrow 3^2 > 0$$

is reversible.

One last compound is listed for the sake of completeness.

Definition (2–7): For any propositions p, q, the proposition $\mathbf{p \underline{\vee} q}$ called *exclusive disjunction*, is false when both components have the same truth value; otherwise it is true.

(The truth table for $\underline{\vee}$ is left to the exercises.) This logical connective corresponds to the usual meaning of English *or*. However, when the word *or* is used in this text, it will be understood to stand for $\vee$ and not for $\underline{\vee}$ unless otherwise specified.

The table in Figure 2.13 provides a summary of the properties of the connectives and modifiers introduced up to this point.

p	q	$\sim p$	$\sim q$	$p \vee q$	$p \wedge q$	$p \Rightarrow q$	$p \leftrightarrow q$	$p \underline{\vee} q$
T	T	F	F	T	T	T	T	F
T	F	F	T	T	F	F	F	T
F	T	T	F	T	F	T	F	T
F	F	T	T	F	F	T	T	F

Figure 2.13

Now that we have introduced the most important connectives, it may be well to note that $\underline{e}$ is not a connective in the sense of $\Rightarrow, \vee, \wedge$, or others. The sentence $p \underline{e} q$ makes a statement about particular propositions, viz., that they have the same truth values whereas the connectives allow us to derive useful new compounds from given propositions.

As an example of a particular logical equivalence, consider the following.

EXAMPLE 8: Find the truth values of the compound $\sim p \Rightarrow q$ and compare with Figure 2.13 to see whether $\sim p \Rightarrow q$ is logically equivalent to one of the propositions listed there.

p	q	$\sim p$	$\sim p \Rightarrow q$
T	T	F	T
T	F	F	T
F	T	T	T
F	F	T	F

Figure 2.14

Solution: The truth table of Figure 2.14 shows the truth values of $\sim p \Rightarrow q$. The truth values of $\sim p$ were obtained using the definition of negation. Those of $\sim p \Rightarrow q$ were then derived according to the definition of implication with $\sim p$ as hypothesis and q as conclusion. Comparison with Figure 2.13 shows that $\sim p \Rightarrow q$ has the same truth values as $p \vee q$. That is,

$$\sim p \Rightarrow q \mathrel{\underline{=}} p \vee q$$

The truth values of the compound $\sim p \Rightarrow q$ just discussed were determined by first working out those of $\sim p$. How did we know that this was the correct order in which to work? Is it not possible that we could first have set down the truth values of $p \Rightarrow q$ and then applied the definition of negation? This procedure would have yielded F-T-F-F in usual vertical order.

The answer to the question just posed is that interpretation of logical compounds is often a matter of establishing conventions and reading the compounds according to those conventions. First, we agree to use parentheses in the same way as in algebra.

That is, in finding a truth value of a compound the truth values of an expression inside the parentheses are worked out first.

Also, we agree that when negation appears with other symbols in a compound, $\sim$ will be regarded as the weakest symbol.

EXAMPLE 9: Find the truth values of the compound $\sim (p \vee \sim q)$.

Solution: Our basic propositions are p and q. The truth values of $p \vee \sim q$ inside the parentheses must be derived first, i.e., the negation of the disjunction is done last. Inside, the symbols $\vee$, $\sim$ both appear. Since no further punctuation appears, $p \vee \sim q$ is to be considered as principally a disjunction. That is, p and $\sim q$ are the individual components of a disjunction. Thus, the truth values of $\sim q$ are obtained first, then those of $p \vee \sim q$, lastly those of $\sim (p \vee \sim q)$. Accordingly, the work is laid out from left to

40 *Elementary Logic*

p	q	$\sim q$	$p \lor \sim q$	$\sim(p \lor \sim q)$
T	T	F	T	F
T	F	T	T	F
F	T	F	F	T
F	F	T	T	F

Figure 2.15

right as in Figure 2.15 where Column 3 was obtained using Column 2 and the definition of negation, Column 4 using Columns 1 and 3 and the definition of disjunction, finally Column 5 using Column 4 and the definition of negation.

We have stated that $\sim$ is considered the weakest symbol in a mixed compound. Our convention for relative strength of symbols in punctuation is stated in the following list where the symbols are listed from strongest to weakest.

$$\Leftrightarrow$$
$$\Rightarrow$$
$$\lor, \land$$
$$\sim$$

Accordingly, the compound

$$\sim p \Rightarrow p \lor q$$

is read as an implication having hypothesis $\sim p$ and conclusion $p \lor q$. The compound

$$p \lor q \land r$$

is ambiguous because the symbols $\lor, \land$ have the same strength.

EXAMPLE 10: Find the truth values of

(2) $$\sim p \land q \Rightarrow \sim q$$

first, as principally a negation; second, as principally a conjunction; third, as principally an implication.

Solution: In order to read the compound $\sim p \land q \Rightarrow \sim q$ as principally a negation we must punctuate it as $\sim(p \land q \Rightarrow \sim q)$. To read (2) as principally a conjunction, we punctuate it as $\sim p \land (q \Rightarrow \sim q)$. This is a conjunction having components $\sim p, q \Rightarrow \sim q$. No further punctuation is needed in order to read (2) as principally an implication because $\Rightarrow$ is stronger than $\land$ and $\sim$. That is, as it stands (2) would be read as an implication having hypothesis $\sim p \land q$ and conclusion $\sim q$. Truth values of these compounds are shown in the table of Figure 2.16. Obviously, the punctuation makes a great deal of difference.

§2c Compound Propositions

p	q	$\sim p$	$\sim q$	$p \vee q$	$\sim p \vee q \Rightarrow \sim q$	$\sim (p \vee q) \Rightarrow \sim q$	$q \Rightarrow \sim q$	$\sim p \vee (q \Rightarrow \sim q)$	$\sim p \vee q$	$\sim p \vee q \Rightarrow \sim q$
T	T	F	F	T	F	T	F	F	F	T
T	F	F	T	T	T	F	T	F	F	T
F	T	T	F	T	T	F	F	T	T	F
F	F	T	T	F	T	F	T	T	F	T

Figure 2.16

Using negation and the compounds of this section, it is possible to translate English sentences with reasonable logical accuracy. However, English communication makes use of facial expressions, gestures, voice inflections, and a variety of special words and phrases to convey shades of meaning which cannot be translated logically. For example, a speaker may state,

(3)   It is raining.

To translate this sentence into logical symbols in some context, one might let p stand for it. Now how would one translate

(4)   It is raining hard

in this context? If "hard" is merely used here to provide emphasis in contrast to (3), p is the likely translation again. But if the discussion is meant to make some essential distinction between the ideas of "raining" and "raining hard," then we would probably use two logical symbols and let one stand for each of (3), (4).

EXAMPLE 11: Translate into logical symbols: (a) It did not rain yesterday; (b) It is raining today.

Solution: Let p stand for "It is raining today" and q stand for "It rained yesterday." If only the fact that "it is raining" is of importance in the problem, then we can translate (b) by p and (a) by $\sim p$. However, there is usually an essential contrast being made between the times when it rained. If so, (a) will be translated as $\sim q$ and (b) as p.

EXAMPLE 12: Using the same (a), (b), and logical symbols as in Example 11, translate: (c) It rained yesterday only if it did not rain today; (d) It rained yesterday if it is raining today.

Solution: Referring to Figure 2.10, we see that (c) translates as

$$q \Rightarrow \sim p$$

Careful reading of the table will also indicate that (d) translates as

$$p \Rightarrow q$$

EXAMPLE 13: Translate: "Judy went to the fair but did not see Michael."

Solution: Let p: Judy went to the fair; let q: Judy did not see Michael. We have not directly defined a connective to be translated as "but." Note, however, that the sense of this proposition is that Judy did both things simultaneously. This suggests that the basic meaning *and* is intended. It is in fact part of English grammar to replace *and* with *but* before a negative clause. Thus the desired translation is

$$p \wedge q$$

EXAMPLE 14: Translate "I shall be in Boston at 10 o'clock or I shall be in Salem then" into logical symbols.

Solution: We have stated that we use the inclusive disjunction symbol $\vee$ for *or* unless otherwise indicated. Here, of course, both clauses cannot be simultaneously true. That is, the exclusive disjunction $p \veebar q$ would be true if we let p: I shall be in Boston; q: I shall be in Salem. Now note that if $p \veebar q$ is true, $p \vee q$ will also be true. Thus, either $p \veebar q, p \vee q$ can be used to translate the given sentence logically. It is standard to use

(5) $$p \vee q$$

The preceding example shows one reason why we do not need to make much use of a separate symbol for the exclusive meaning of *or*. When a situation arises in which an exclusive disjunction is known to be false, or the truth values of the components are not known, then $p \vee q$ cannot be used in place of $p \veebar q$. In such cases we can replace $p \veebar q$ by the equivalent of

p or q but not both

which can be symbolized by

(6) $$(p \vee q) \wedge \sim (p \wedge q)$$

The fact that we can translate the exclusive sense of *or* by (5) or (6) means that we do not need to practice with special logical techniques for using $p \veebar q$ in arguments.

We need to take special care with the grammatical conjunction *that*.

(7) Unless there is punctuation which indicates otherwise, the grammatical conjunction *that* is considered to apply to *all* that follows in the sentence.

EXAMPLE 15: Translate "It is not the case that Olin ate a banana if Elden played his violin" into logical symbols.

Solution: Let p: Olin ate a banana; q: Elden played his violin. According to (7) we must translate the sentence in the form

$$\sim (\quad)$$

The implication $q \Rightarrow p$ will appear inside the parentheses, yielding

$$\sim (q \Rightarrow p)$$

as the correct translation.

EXERCISE SET

1. Make a tree diagram similar to the one in Figure 2.5 but for three propositions p, q, r. Then use this pattern to construct a truth table for the three components p, q, r.

2. List the hypothesis and conclusion of each proposition or tell why it is impossible.
   a) The earth is a planet ⇒ the moon is a satellite.
   b) $3 \cdot 5 = 15$ implies that $9 + 4 = 1$.
   c) Bertha is a bird ⇐ Manfred mangles mastadons.
   d) We are happy and no one is sad.
   e) An angle is an acute angle iff its measure is less than the measure of a right angle.

3. Punctuate each compound so that it is principally a conjunction.
   a) $p \wedge \sim q \vee p$
   b) $p \Rightarrow \sim q \wedge q$
   c) $p \Longleftrightarrow q$
   d) $p \Longleftrightarrow q \vee r$

4. Punctuate each compound so that it is an implication.
   a) $p \Rightarrow q \vee r$
   b) $\sim q \Rightarrow p \wedge q$
   c) $p \Longleftrightarrow q$
   d) $\sim s \vee x \Rightarrow \sim y$

5. Punctuate each compound so that it is basically a conjunction or a disjunction.
   a) $p \Rightarrow q \vee r$
   b) $\sim q \Rightarrow p \wedge q$
   c) $\sim s \vee x \Rightarrow \sim y$
   d) $p \Longleftrightarrow q \vee r$

6. Determine the truth value of each proposition. Assume the most obvious truth standard in each case.
   a) $2 + 3 = 11$ only if $4 + 5 = 9$
   b) It is not the case that Andrew Johnson was president or Samuel Adams was president.
   c) All canaries sing sourly if Mary is merry.
   d) A square is a quadrilateral or a circle is a triangle.
   e) [(q implies p) or (p and q)] if p.

7. Write the specified compounds in verbal form if p stands for "Roy is a boy" and q stands for "Pearl is a girl."
   a) $\sim p \Rightarrow q$
   b) $p \vee \sim q$
   c) $p \Longleftrightarrow \sim q$
   d) $\sim p \wedge \sim q$
   e) $\sim [(p \Rightarrow q) \wedge p]$
   f) $p \vee q \Rightarrow p \wedge q$

8. (a) through (f). If both p, q are to be regarded as true in Exercise 7a through 7f respectively, what is the truth value of the given compound?

9. Determine the possible truth values of the following compound propositions. Consider t as a known true proposition and f as a known false

proposition. In each case compare the possible truth values with those of other propositions (simple or compound) to see whether the given proposition is logically equivalent to some other proposition.

a) $p \vee p$
b) $p \wedge t$
c) $\sim p \vee q$
d) $p \vee f$
e) $p \wedge \sim p$
f) $p \wedge p$
g) $p \Rightarrow p$
h) $p \Rightarrow \sim p$
i) $p \Rightarrow t$
j) $f \Rightarrow p$
k) $\sim p \Rightarrow p$
l) $p \vee \sim q$
m) $p \vee (p \Rightarrow q)$
n) $(p \vee q) \Rightarrow q$
o) $p \vee (\sim p \wedge q) \Rightarrow (q \vee \sim p)$
p) $\sim (p \Leftrightarrow q)$
q) $(p \vee q) \Rightarrow p$
r) $(p \wedge q) \Rightarrow p$
s) $p \Rightarrow (p \vee q)$
t) $p \vee \sim p$
u) $\sim p \Leftrightarrow q$
v) $p \wedge (p \Leftrightarrow q)$
w) $\sim (p \vee q)$
x) $(\sim p \vee q) \wedge (\sim q \vee p)$
y) $(q \wedge \sim p) \vee \sim q$
z) $(p \vee q) \wedge \sim (p \wedge q)$

10. Determine the truth values of the given compound. In each case you will need a truth table constructed according to your result in Exercise 1.

   a) $q \vee r \Rightarrow r \wedge s$
   b) $(\sim p \wedge q) \Rightarrow r$
   c) $[(p \Leftrightarrow q) \wedge (\sim p \vee s)] \Rightarrow \sim q \wedge \sim s$

11. Make truth tables for the following propositions and compare the possible truth values with each other and with those for $p \Rightarrow q$.

   a) $q \Rightarrow p$
   b) $\sim p \Rightarrow \sim q$
   c) $\sim q \Rightarrow \sim p$

12. Translate the following sentences into logical symbols.

   a) A function is continuous if it is differentiable.
   b) John's name will appear on the Dean's list only if he is an honor student.
   c) Four is a number such that it is even or four is a number such that it is odd.
   d) Since John is 35 years old and a citizen, he is eligible to be president.
   e) It is not the case that Francis is a boy and Frances is a girl.
   f) The price of food will rise if and only if wages increase and the cost of packaging increases.

13. Show that $p \wedge q \subseteq q \wedge p$.

§2d SIMPLE ARGUMENTS

Now we are in a position to begin analysis of simple arguments to determine their validity. First it is necessary to formalize explicitly the concept of argument.

Definition (2–8): An **argument** is a compound whose principal connective is $\Rightarrow$.

The hypothesis or any conjoined components of the hypothesis of an argument are the **premises.**

EXAMPLE 1: Is each of the following compounds an argument? Explain.
a) $p \lor (q \Rightarrow r)$ b) $p \lor q \Rightarrow r$ c) $p \land (q \lor r) \Rightarrow \sim q$

Solution: a) The compound $p \lor (q \Rightarrow r)$ is not an argument, for its principal connective is $\lor$, not $\Rightarrow$.
b) The compound $p \lor q \Rightarrow r$ is principally an implication; therefore, it is an argument by the definition. Its single premise is $p \lor q$, and its conclusion is r.
c) The compound $p \land (q \lor r) \Rightarrow \sim q$ is also an argument. It has two premises: $p, q \lor r$. Its conclusion is $\sim q$.

In the arguments considered in §2a, the horizontal bar took the place of the arrow. Thus

(1) $$\frac{p}{q}, \quad p \Rightarrow q$$

are to be regarded as two ways of writing the same thing. To distinguish between them, the one on the left is called *traditional form* or *vertical form* and the one on the right *horizontal form*.

We have already seen that an argument can have several premises. Yet only the hypothesis p is illustrated in each form of (1). Definition (2–8) gives us the clue to what we actually mean by such notation. When more than one premise occurs, the premises are considered to be conjoined. That is,

$$\begin{array}{c} p_1 \\ p_2 \\ \underline{p_3} \\ q \end{array}$$

would be written in horizontal form as

(2) $$p_1 \land p_2 \land p_3 \Rightarrow q$$

Since the definition of conjunction only states what we mean when conjoining two components, we agree that

$$p_1 \wedge p_2 \wedge p_3 \leftrightarrows (p_1 \wedge p_2) \wedge p_3$$

That is, we would form the conjunction of the first two components and then conjoin the result with p_3. With this convention (2) would be written

$$(p_1 \wedge p_2) \wedge p_3 \Rightarrow q$$

Corresponding conventions can be asserted for four or more components.

Now recall that when looking at each argument in §2a you were asked to test it for validity by regarding the premises as true and then seeing whether, as a result, the conclusion seemed true. In applying logic to a branch of mathematics or other axiomatic discourse, we make certain that the premises of an argument can be labeled true and then expect to show by valid argument that a particular conclusion is true. In logic itself we want to consider all possible truth values of premises and conclusions, but with applications in mind we still want this same relationship of premises to conclusion to hold for the valid case.

Definition (2–9): An argument is **valid** iff there is *no* case in which the conclusion is false at the same time that each premise is true.

The reader should note carefully that this definition does *not* state that we require only true premises for a valid argument. Rather, it states that a valid argument must have particular patterns of truth values. For example, if one of the premises of an argument were known to be false, the argument would be valid, for then there would be no case in which the premises were simultaneously true.

As Definition (2–9) suggests, a truth table is a useful device for analyzing arguments having a small number of basic components (regardless of the number of premises). In such cases we proceed as follows.

> **We make a table which shows the truth values of each premise and the conclusion. Then we cross off each row containing an F entry for any premise. After we have done this, the only rows remaining will have an entry of T for each premise. The argument is valid only if the truth value of the conclusion is T in each of these remaining rows. If all the rows are crossed out, this means that the premises are never true simultaneously, and the argument is still valid.**

EXAMPLE 2: Determine the validity or invalidity of the following argument by truth table analysis.

Premise (a):   If it is raining, then the road is wet.
Premise (b):   It is raining.
Conclusion: The road is wet.

48 *Elementary Logic*

Solution: If we let p stand for "It is raining" and let q stand for "The road is wet," then the argument can be symbolized:

(3)
$$\begin{array}{c}\text{(a) } p \Rightarrow q \\ \underline{\text{(b) } p} \\ q\end{array} \quad \text{or} \quad [(p \Rightarrow q) \wedge p] \Rightarrow q$$

We construct a truth table in which each premise and the conclusion has a separate column which is labeled correspondingly. We use the symbol for *therefore*, $\therefore$, to mark the conclusion. We must find the cases in which the premises $p \Rightarrow q, p$ are simultaneously true. Hence the F in Row 2 of Column 3 is not needed, and the whole row is crossed out. The F's under p in Row 3 and 4 are not needed either. These rows are also crossed out. Now the table looks like this:

(b)	$\therefore$	(a)	
p	q	$p \Rightarrow q$	
T	boxed T	T	
~~T~~	~~F~~	~~F~~	because of Premise (a)
~~F~~	~~T~~	~~T~~	because of Premise (b)
~~F~~	~~F~~	~~T~~	because of Premise (b)

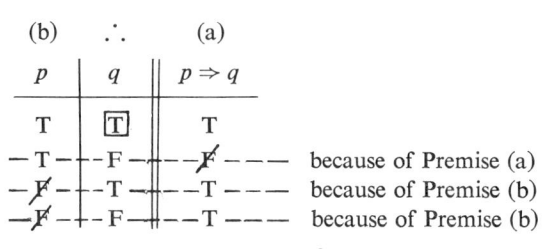

Figure 2.17

All rows except Row 1 have been discarded. This row shows each of the premises true. We see that the conclusion q is true in this case, and that there is no case in which each premise is true and the conclusion false. Thus the reasoning is valid by Definition (2-9). Boxing in the truth value(s) of the conclusion for the cases in which the premises are simultaneously true serves to pinpoint the truth value entries which the definition requires us to examine.

EXAMPLE 3: Do a truth table analysis of the following argument:

Premise (a): If this is a shfnexi, then it is not an ootster.
Premise (b): This is not an ootster.
Conclusion: This is a shfnexi.

Solution: Letting p stand for "This is a shfnexi" and q stand for "This is not an ootster," the symbolic translation is

(4)
$$\begin{array}{c}\text{(a) } p \Rightarrow q \\ \underline{\text{(b) } q} \\ p\end{array}$$

Analyzing by truth table, we have

§2d   Simple Arguments   49

	∴	(b)	(a)
p	q		$p \Rightarrow q$
T	T		T
T	F		F ----- because of Premises (a) and (b)
F	T		T
F	F		T ----- because of Premise (b)

Figure 2.18

Note that the conclusion p could be either true or false when the premises are simultaneously true. That is, there is a case in which each premise is true and the conclusion false. Thus the argument is invalid.

EXAMPLE 4: Examine the following symbolic argument by truth table to determine its validity.

(a) $p \Rightarrow q$
(b) $\sim p \vee \sim q$
$$p$$

Solution: We construct the following table.

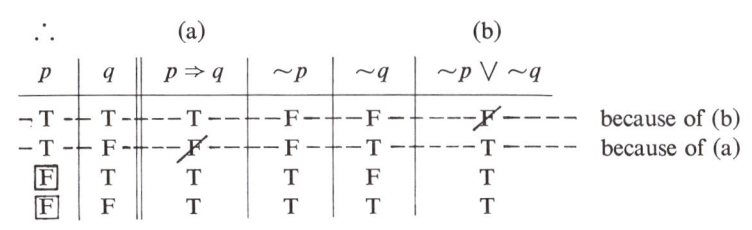

	∴		(a)			(b)	
p	q		$p \Rightarrow q$	$\sim p$	$\sim q$	$\sim p \vee \sim q$	
T	T		T	F	F	F ---- because of (b)
T	F		F	F	T	T ---- because of (a)
F	T		T	T	F	T
F	F		T	T	T	T

Figure 2.19

The argument is invalid, because its conclusion is false in both cases that the premises are simultaneously true.

EXAMPLE 5: Analyze the following argument by truth table.

(a) If it rained yesterday, then it is raining today.
(b) It is not raining today.
(c) It rained yesterday.

It did not rain yesterday only if it is raining today.

Solution: Letting p stand for "It rained yesterday" and q stand for "It is raining today," the symbolic translation is

(a) $p \Rightarrow q$
(b) $\sim q$
(c) p
___
$\sim p \Rightarrow q$

Constructing the requisite truth table, we have:

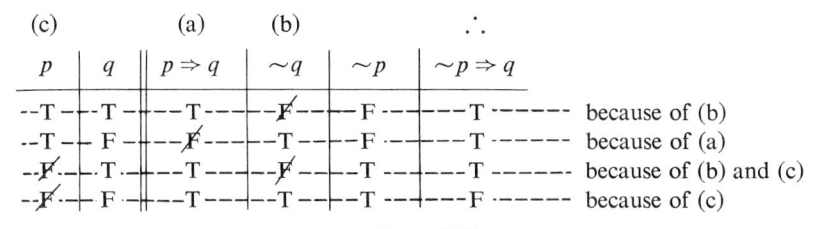

p	q	(a) $p \Rightarrow q$	(b) $\sim q$	$\sim p$	∴ $\sim p \Rightarrow q$	
T	T	T	~~F~~	F	T	because of (b)
T	F	~~F~~	T	F	T	because of (a)
~~F~~	T	T	~~F~~	T	T	because of (b) and (c)
~~F~~	F	T	T	T	F	because of (c)

Figure 2.20

This argument is an example of one whose premises are so related that they can never be simultaneously true. That is, in each case (row) there is at least one premise which has truth value F. Thus, every row is crossed out. However, there is no case in which the conclusion is false and the premises are simultaneously true. The argument is valid.

We call the type of argument in which the premises are never simultaneously true *inconsistent*. Example 5 is an illustration of an inconsistent argument.

EXERCISE SET

1. Show how the specified definitions meet the requirements of §1d for a good definition. Omit discussion of Requirements (iv) and (vi).

    a) (2–8) b) (2–9)

2. Tell whether each compound is an argument. If any compound is an argument, list its premises.

    a) $p \Rightarrow (q \Rightarrow r)$ b) $(p \Rightarrow \sim q) \lor (x \Rightarrow y)$
    c) $p \Leftrightarrow q$ d) $p \land (\sim p \lor q) \Rightarrow p \land \sim q$

3. Use a truth table to show that the argument of
    a) Exercise 4, §2a is valid
    b) Exercise 6, §2a is invalid

Determine whether the following symbolic arguments are valid or invalid.

4. (a) $\sim p$
   (b) $p \Rightarrow q$
   $\overline{p \vee \sim q}$
5. (a) $p \Longleftrightarrow q$
   (b) $p \vee \sim r$
   (c) $\sim q$
   $\overline{\sim p}$

Analyze the arguments in Exercises 6 through 16 to determine whether they are valid or invalid. First write each in logical symbols, then construct a truth table. Tell whether the argument is consistent or inconsistent. Each conclusion is introduced by the term *therefore*.

6. (a) It rained yesterday and it is raining today.
   (b) It rained yesterday iff it is not raining today.
   Therefore, it did not rain yesterday.

7. If this is ACE polish, then it is good polish.
   It is known that this is good polish.
   Therefore, this is ACE polish.

8. $\mathscr{L}_1$ is parallel to $\mathscr{L}_2$ or $\mathscr{L}_3$ is parallel to $\mathscr{L}_2$.
   $\mathscr{L}_1$ is parallel to $\mathscr{L}_2$.
   Therefore, $\mathscr{L}_3$ is not parallel to $\mathscr{L}_2$.

9. $\mathscr{L}_1$ is parallel to $\mathscr{L}_2$ or $\mathscr{L}_3$ is parallel to $\mathscr{L}_2$.
   $\mathscr{L}_3$ is not parallel to $\mathscr{L}_2$.
   Therefore, $\mathscr{L}_1$ is parallel to $\mathscr{L}_2$.

10. Spiros does not eat spinach or Millie drinks milk.
    Millie does not drink milk.
    Therefore, Spiros does not eat spinach.

11. Spiros does not eat spinach and Millie does not drink milk.
    Therefore, it is not the case that Spiros eats spinach or Millie drinks milk.

12. It is not the case that Spiros eats spinach and Millie drinks milk.
    Therefore, Spiros does not eat spinach.

13. If Pat's major is mathematics, she is required to take both physics and logic.
    Pat is not required to take logic.
    Therefore, Pat's major is not mathematics.

14. If soap is scarce, then the linen is dirty.
    Prices remain the same and the linen is not dirty.

Prices do not remain the same or soap is scarce.
Therefore, soap is not scarce.

15. If it rained yesterday, it will rain today.
It will rain today or it will rain tomorrow.
It never rains both yesterday and tomorrow.
If it rains today, it will rain tomorrow.
Therefore, it will not rain today.

16. Use the premises of Exercise 15 leading to the conclusion "It did not rain yesterday."

17. A suitor promises, "I shall kill myself only if you refuse to marry me." The lady in question refuses him. Is suicide his only honorable way out? Explain.

18. The following is a part of a reader's letter which a newspaper received:

"I read a letter in the [name of paper] saying, 'People with too much money talk to themselves.' Well, it isn't so. I have a husband who talks to himself and he hasn't a red cent."

Is this writer's conclusion "It isn't so" a valid deduction if we take her statement "I have . . . ." as a premise?

19. Beginning with the premise "If an object is less dense than water, then it will float on water," write an argument using one of the following statements as a second premise. Use the obvious conclusion in each case. Show whether the argument thus produced is valid or invalid.

   a) Cork is less dense than water.
   b) Aluminum is less dense than water.
   c) A needle can be made to float.
   d) Water is denser than oil.

20. If $p \Rightarrow q$ is a true proposition and if $q \subseteq t$, what can be concluded about the truth value of p?

21. a) If p is a true proposition, which rows can be crossed off in the table for $p \Rightarrow q$? What can we then conclude about the truth value of q?
   b) If p is true and if $p \Rightarrow q$ is a true proposition, which rows can be crossed off in the table for $p \Rightarrow q$? What can be concluded now about the truth value of q?

§2e PROPERTIES OF COMPOUNDS

As we stated in the preceding section, analyzing arguments by truth table works well for simple cases. Techniques for handling more complex cases include some use of the concept of tautology and more use of the principle of logical equivalence of compounds.

In taking up the first of these, let us recall that ordinarily it is impossible to determine the truth value of a compound proposition unless we know the truth values of its components. However, as seen in Exercise 9, §2c, there are some compound propositions which are true and some which are false *regardless of the truth values of their components*.

Definition (2–10): A compound proposition which is always true is called a **tautology**.

Definition (2–11): A compound proposition which is always false is called a **contradiction**.

We say that two propositions are **contradictory** iff their conjunction is a contradiction.

The following three theorems were proved in Exercise 9, §2c.

Theorem (2e-1): $p \Rightarrow p$ is a tautology.

Theorem (2e-2): $p \vee \sim p$ is a tautology.

Theorem (2e-3): $p \wedge \sim p$ is a contradiction.

EXAMPLE 1: Determine the truth value of the proposition "The sun never shines only if the sun never shines."

Solution: The given proposition has the form $p \Rightarrow p$. Hence by Theorem (2e-1) its truth value is T.

EXAMPLE 2: Determine the truth values of $\sim (p \vee q) \wedge p$.

Solution: The derivation columns of the truth table of Figure 2.21 are filled in using the definitions of $\vee$, $\sim$, and $\wedge$ in order. We see that $\sim (p \vee q) \wedge p$ is a contradiction, i.e., the propositions p, $\sim (p \vee q)$ are contradictory.

p	q	$p \vee q$	$\sim(p \vee q)$	$\sim(p \vee q) \wedge p$
T	T	T	F	F
T	F	T	F	F
F	T	T	F	F
F	F	F	T	F

Figure 2.21

Theorem (2e-4): $[(p \Rightarrow q) \wedge p] \Rightarrow q$ is a tautology.

Proof:

p	q	$p \Rightarrow q$	$(p \Rightarrow q) \wedge p$	$[(p \Rightarrow q) \wedge p] \Rightarrow q$
T	T	T	T	T
T	F	F	F	T
F	T	T	F	T
F	F	T	F	T

Thus the proposition is indeed a tautology by Definition (2-10). Q. E. D.

Two important facts about the arrow connective $\Rightarrow$ were established in Exercises 9i and 9j, §2c. They are restated in the following theorem.

Theorem (2e-5): An implication which has a false hypothesis is a tautology and an implication which has a true conclusion is a tautology.

EXAMPLE 3: Using experience as a truth standard, determine the truth value of
(i) $2 = 4$ only if grass is not green;
(ii) $2 = 4$ only if $5 = 5$.

Solution: (i) and (ii) are both true by Theorem (2e-5).

The following theorem shows us that the tautology concept can be useful in analyzing arguments.

Theorem (2e-6): An argument having premises $p_1, p_2, \ldots, p_n$, and leading to a conclusion c is a valid argument iff $p_1 \wedge p_2 \wedge \ldots \wedge p_n \Rightarrow c$ is a tautology.

To simplify consideration of this conjecture, let $p_1 \wedge p_2 \wedge \ldots \wedge p_n$ be represented by P. In this notation the proposition to be proved states "$P \Rightarrow c$ is a valid argument iff $P \Rightarrow c$ is a tautology." Since this conjecture is a bicondition, we can prove that it is true by showing that the two propositions

(1) $\qquad P \Rightarrow c$ is a valid argument
(2) $\qquad P \Rightarrow c$ is a tautology

always have the same truth value. See Theorem (2c-1). The proof is left as an exercise.

Now we can see that any argument having the structure of (2e-4) will be valid. This tautology is important enough to have special names. It is variously called *The Rule of Detachment*, or *Modus* (Ponendo) *Ponens*, or *The*

Fundamental Rule of Inference. The last name is derived from the fact that any tautological implication is called a *rule of inference.*

The use of the principle of logical equivalence to simplify arguments is based on Axiom L-4. We establish and list various pairs of logically equivalent propositions. Then either member of the pair may be substituted for the other to change an argument into more manageable form.

Since an argument is an implication and since premises also are often implications, logically equivalent forms of this compound are especially useful. In this connection it is convenient to have names for certain forms which can be derived from $p \Rightarrow q$.

Definition (2–12): For any implication $p \Rightarrow q$, the implications $q \Rightarrow p$, $\sim p \Rightarrow \sim q$, and $\sim q \Rightarrow \sim p$ are called, respectively, the **converse**, **inverse** and **contrapositive** of $p \Rightarrow q$.

EXAMPLE 4: Write the converse, inverse, and contrapositive of "If it is raining, then the road is not wet."

Solution: The converse is: "If the road is not wet, then it is raining." The inverse is: "If it is not raining, then the road is wet." The contrapositive is: "If the road is wet, then it is not raining."

Especially when using the notion of the converse of an implication it is convenient to have a contrasting name for the implication itself. In this connection we shall call $p \Rightarrow q$ the **positive** form.

Referring to Definition (2-6) of the bicondition, we can restate it in these new terms.

A bicondition is the conjunction of the positive and converse forms of an implication.

We now return to the reversibility principle of §1d and the preceding section for this statement.

An implication is reversible if its positive and converse forms are both true.

Reference to both Definition (2-12) and Exercise 11, §2c shows that

$$(\sim q \Rightarrow \sim p) \equiv (p \Rightarrow q),$$
$$(\sim p \Rightarrow \sim q) \equiv (q \Rightarrow p),$$

and that the converse and inverse forms are *not* logically equivalent to the positive form. These facts are summarized in Figure 2.22. Diagonal pairs *are* logically equivalent. Adjacent pairs are *not* logically equivalent.

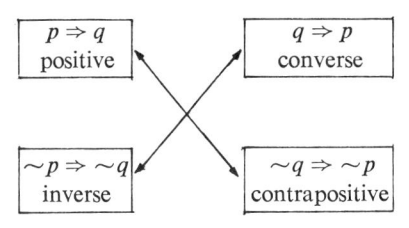

```
        ┌─────────┐              ┌─────────┐
        │ p ⇒ q   │              │ q ⇒ p   │
        │ positive│╲            ╱│ converse│
        └─────────┘ ╲          ╱ └─────────┘
                     ╲        ╱
                      ╲      ╱
                       ╳
                      ╱      ╲
                     ╱        ╲
        ┌─────────┐ ╱          ╲ ┌──────────────┐
        │ ~p ⇒ ~q │╱            ╲│ ~q ⇒ ~p      │
        │ inverse │              │ contrapositive│
        └─────────┘              └──────────────┘
```

Figure 2.22

The facts presented in this illustration were established by truth table. It is fortunate indeed that our basic tautologies, contradictions, and logical equivalences are all provable by this simple means.

EXAMPLE 5: Prove or disprove: $\sim(p \vee q) \stackrel{?}{=} \sim p \vee \sim q$.

| | | | | | **L** | **R** |
p	q	~p	~q	$p \vee q$	$\sim(p \vee q)$	$\sim p \vee \sim q$
T	T	F	F	T	F	F
T	F	F	T	T	F	T
F	T	T	F	T	F	T
F	F	T	T	F	T	T

Solution: Comparing **L** with **R**, we see that the truth values in Row 2 and Row 3 are not the same. Thus the expressions are not logically equivalent and the conjecture has been disproved.

EXAMPLE 6: Prove or disprove: $\sim(p \Leftrightarrow q) \stackrel{?}{=} \sim p \Leftrightarrow q$.

| | | | **L** | | **R** |
p	q	$p \Leftrightarrow q$	$\sim(p \Leftrightarrow q)$	~p	$\sim p \Leftrightarrow q$
T	T	T	F	F	F
T	F	F	T	F	T
F	T	F	T	T	T
F	F	T	F	T	F

Solution: Comparing **L** with **R**, we see that the truth values are the same for each. Thus the expressions are logically equivalent and the conjecture (sometimes referred to as "Griswold's Gremlin") has been proved.

EXERCISE SET I

1. Show that the premises $p \Rightarrow q$, $\sim q, p$ of Example 5, §2d, are contradictory.

§2e   *Properties of Compounds*   57

2. Show that the premises $p \wedge q, p \Longleftrightarrow \sim q$ of Exercise 6, §2d, are contradictory.

3. Show that $[(p \vee \sim q) \wedge \sim p] \Rightarrow \sim q$ is a tautology.

In Exercises 4 through 7 write (a) the converse, (b) the inverse, and (c) the contrapositive of the given implication.

4. A polygon has no diagonals only if it is a triangle.

5. $\mathscr{L}_1 \perp \mathscr{L}_2 \Rightarrow \mathscr{L}_1$ is not $\perp \mathscr{L}_3$.

6. If it is not snowing, then it is raining.

7. Portia is portly if Mickey is sticky.

8. Write the inverse of the contrapositive of "If p then not q."

9. Write the contrapositive of the converse of "If not q then not p."

10. Write the inverse of the converse of "If not p then not q."

11. Prove that the given implications are tautologies.

     a) $p \Rightarrow p \vee q$ b) $p \wedge q \Rightarrow p$

12. Construct truth tables for each of the following compounds and compare them with other tables to see whether the given compound is logically equivalent to some known proposition.

     a) $\sim(p \wedge q)$ b) $\sim p \vee \sim q$
     c) $\sim(p \Rightarrow q)$ d) $\sim(p \underline{\vee} q)$
     e) $q \wedge (q \Rightarrow p) \Longleftrightarrow q$ f) $(p \vee q) \Longleftrightarrow (p \wedge q)$
     g) $(\sim p \Rightarrow q) \vee q$ h) $(p \underline{\vee} q) \wedge (p \Longleftrightarrow q)$
     i) $p \wedge \sim q$ j) $\sim(q \Rightarrow p)$

13. Prove or disprove each conjecture.

     a) $(p \Rightarrow p \wedge q) \underline{e} (p \Rightarrow q)$
     b) $(q \Rightarrow p \wedge q) \underline{e} (p \Rightarrow q)$
     c) $(p \vee q \Rightarrow p) \underline{e} (p \Rightarrow q)$
     d) $(p \vee q \Rightarrow q) \underline{e} (p \Rightarrow q)$
     e) $p \vee \sim q \underline{e} \sim p \vee q$
     f) $\sim(p \wedge q) \underline{e} \sim p \wedge \sim q$
     g) $\sim(p \underline{\vee} q) \underline{e} (p \Longleftrightarrow q)$
     h) $\sim(p \Rightarrow \sim p) \underline{e} (\sim p \Rightarrow p)$
     i) $(p \Rightarrow q) \wedge q \Rightarrow p$ is a tautology

14. Construct a truth table for $p \wedge \sim q \Rightarrow f$, and then examine this proposition for logical equivalence to some well-known compound. Recall that f stands for any known false proposition.

15. Construct the truth table for $p \wedge q \Rightarrow r$ and compare it with the truth table for $p \Rightarrow (q \Rightarrow r)$. What does the comparison show?

16. Repeat Exercise 15 for the compounds $(p \Rightarrow q) \wedge (r \Rightarrow q)$ and $p \vee r \Rightarrow q$.
17. Is the following argument valid? Compare with Exercise 11a.
    (a) Sue is my sister.
    Therefore, Sue is my sister or Brad is my brother.
18. If we know that $(p \wedge q) \Rightarrow r$ is a valid argument, what can we conclude about the argument $p \Rightarrow (q \Rightarrow r)$? See Exercise 15.
19. Suppose we know that the following argument is valid.
    (a) $\triangle ABC \cong \triangle DEF$
    (b) $\mathscr{L}_1$ is not $// \mathscr{L}_2$
    Therefore, a polygon is not a polygon.
    What can we then conclude about the argument
    (a) $\triangle ABC \cong \triangle DEF$
    Therefore, $\mathscr{L}_1 // \mathscr{L}_2$
    See Exercise 14.
*20. Prove Theorem (2e-6).

In this section we list some of the major logical equivalences stated previously and introduce other important ones. In some cases results in previous exercises and examples have indicated that the stated conjectures are indeed theorems. See especially Exercise 9, §2c, and Exercise 12, §2e. Nevertheless, it will be most instructive for the reader to prove them independently at this point. The reader will note that some of the theorems are accompanied by a mysterious word at the right. Each such "word" is actually an abbreviation which is useful in designating a theorem which does not have a special name of its own.

Theorem (2e-7): $(p \Rightarrow q) \underline{e} (\sim q \Rightarrow \sim p)$; that is, an implication and its contrapositive are logically equivalent. CLEP.

The designation CLEP stands for "The *c*ontrapositive form of an implication is *l*ogically *e*quivalent to the *p*ositive form."

EXAMPLE 7: Write another implication logically equivalent to "2 equals 2 only if 7 equals 1."

Solution: Using CLEP we obtain "If 7 is not equal to 1, then 2 is not equal to 2."

We have a symbol for the negation of "equality," namely $\neq$, which is read "does not equal." Using this symbol, the solution of Example 7 could be written "$7 \neq 1 \Rightarrow 2 \neq 2$."

§2e   **Properties of Compounds**   59

We often find it inconvenient to use a disjunction in an argument. It is sometimes possible to produce a more advantageous arrangement by changing the disjunction to an implication according to the pattern of the following theorem. It states the *l*ogical *e*quivalence of a *d*isjunction and an *i*mplication.

Theorem (2e-8): $\sim p \vee q \subseteq p \Rightarrow q.$ LEDI

EXAMPLE 8: Write an implication logically equivalent to the disjunction "$2 \neq 2$ or $7 \neq 1$."

Solution: Using LEDI we obtain "$2 = 2$ only if $7 \neq 1$."

The next four theorems supply the pattern for deriving the simplified form of negation of certain propositions from the basic form.

Theorem (2e-9): $\sim(\sim p) \subseteq p.$ D. N.

This is often called the *Rule of Double Negation* and corresponds to English grammatical usage.

EXAMPLE 9: Write a simplified form of the negation "It is not the case that the given argument is invalid."

Solution: Using D. N., this becomes "The given argument is valid."

EXAMPLE 10: Write a disjunction logically equivalent to the implication "$2 \neq 2 \Rightarrow 7 = 1$."

Solution: Using LEDI, we obtain "$\sim(2 \neq 2)$ or $7 = 1$."
Then using D. N., this becomes "$2 = 2$ or $7 = 1$."

Theorem (2e-10): $\sim(p \wedge q) \subseteq \sim p \vee \sim q.$ DeM ($\wedge$)

This theorem gives us the pattern for the simplified form of the negation of a conjunction. It is called *DeMorgan's Law for Conjunction*.

EXAMPLE 11: Write the correct negation of "2 equals 2 and 7 equals 1."

Solution: The given statement is a conjunction. According to the pattern of Theorem (2e-10), the simplified negation is "2 is *not* equal to 2 *or* 7 is *not* equal to 1."

Looking at the pattern of Theorem (2e-10), one might imagine that the simplified negation of a disjunction would be the conjunction of the negations of its components. The following theorem states that this is indeed the case.

60 *Elementary Logic*

Theorem (2e-11): $\sim(p \lor q) \stackrel{e}{=} \sim p \land \sim q.$ DeM ($\lor$)

This principle is called *DeMorgan's Law for Disjunction*.

EXAMPLE 12: Is "We shall not travel by train or not by bus" the correct negation of "We shall travel by train or bus"?

Solution: The proposed negation is not the one required by the theorem, but this does not directly prove it wrong. However, in Example 5 we did see that $\sim(p \lor q) \stackrel{e}{=} \sim p \lor \sim q$ is false. By this theorem a correct simplified negation is "We shall not travel by train *and* not by bus."

The correct simplified negation form for an implication is by no means intuitively obvious. Certainly neither the inverse nor the contrapositive is the required negation form as an examination of their truth tables would show. The results of Exercises 12c and 12i suggest the following principle which does show the correct simplified negation of implication.

Theorem (2e-12): $\sim(p \Rightarrow q) \stackrel{e}{=} (p \land \sim q).$ NIMP

Proof:

| | | | L | | R |
p	q	$p \Rightarrow q$	$\sim(p \Rightarrow q)$	$\sim q$	$p \land \sim q$
T	T	T	F	F	F
T	F	F	T	T	T
F	T	T	F	F	F
F	F	T	F	T	F

The possible truth values of $\sim(p \Rightarrow q)$ and $p \land \sim q$ are the same. Thus these compounds are logically equivalent.   Q. E. D.

EXAMPLE 13: Write the negation of "If it is snowing then the sky is cloudy" in simplified form.

Solution: By the principle just proved "It is snowing *and* the sky is *not* cloudy" is the correct simplified form of the negation.

The following two facts are called the *commutative properties* of disjunction and conjunction respectively. They say in effect that the order of combining the two components of the compound is immaterial. Proofs are left to the exercises.

Theorem (2e-13): For any propositions p, q,

$$p \lor q \stackrel{e}{=} q \lor p$$

Theorem (2e-14): For any propositions p, q,
$$p \wedge q \equiv q \wedge p$$
Proof of the following principle is also left to the exercises.

Theorem (2e-15): For any propositions p, q,
$$p \Longleftrightarrow q \equiv \sim p \Longleftrightarrow \sim q$$

EXERCISE SET II

21. Write the contrapositive of "If $f(c)$ is an extremum of a function f in some neighborhood of c, then $f'(c) = 0$ or $f'(c)$ does not exist."

22. Negate the following compounds. Use simplified form.
    a) $p \wedge \sim q$
    b) $r \vee \sim s$
    c) $q \Rightarrow \sim s$
    d) $p \vee q \Rightarrow p \wedge q$

23. Rewrite the following as logically equivalent implications.
    a) $q \vee \sim s$
    b) $\sim (q \wedge r)$
    c) $p \vee q$
    d) $\sim [(p \vee s) \wedge q]$

24. Rewrite the following as logically equivalent disjunctions.
    a) $p \Rightarrow \sim q$
    b) $\sim r \Rightarrow \sim s$
    c) $p \vee q \Rightarrow p \wedge q$
    d) $\sim (p \wedge \sim q)$

25. Using simplified form, write correct negations of each of the following propositions.
    a) Spiros eats spinach and Millie drinks milk.
    b) If it rained yesterday, it will rain today.
    c) $\mathscr{L}_1$ is parallel to $\mathscr{L}_2$ or $\mathscr{L}_3$ is parallel to $\mathscr{L}_2$.
    d) This dress will be ruined if it rains.
    e) $2x = 2y$ because $x = y$
    f) Two lines intersect only if they are not parallel.
    g) A natural number which is not even is odd.
    h) The angles are congruent and the lines are parallel.
    i) $3 \leq 8$

26. Prove by a truth table:
    a) Theorem (2e-11)
    b) Theorem (2e-13)
    c) Theorem (2e-14)
    d) Theorem (2e-15)

27. Prove or disprove each proposition:

a) $(p \lor q) \lor r \equiv p \lor (q \lor r)$
b) $(p \land q) \land r \equiv p \land (q \land r)$
c) $p \land (q \lor r) \equiv (p \land q) \lor (p \land r)$
d) $p \lor (q \land r) \equiv (p \lor q) \land (p \lor r)$
e) $p \lor (q \land r) \equiv (p \lor q) \land r$
f) $p \land (q \lor r) \equiv (p \land q) \lor r$

28. If the number of component propositions in a compound is n, how many rows will be needed in a truth table constructed to display the possible truth values of that compound? We have seen, for example, that 2 components require 4 rows.

29. Prove Theorem (2e-7) by using Theorems (2e-8) and (2e-13).

§2f VALID ARGUMENT FORMS

As we have seen, proof of the validity or invalidity of an argument can be accomplished by truth table. This method is quite cumbersome when more than three component propositions are involved, so an alternate method of argument analysis will be developed. This method depends on your ability to recognize simple valid argument forms which we shall prove once and for all by the truth tables.

The first standard argument form to consider is the principle of Theorem (2e-4) which is called the **Rule of Detachment** or *Modus Ponens*. This is symbolized as follows:

(1) $\qquad [(p \Rightarrow q) \land p] \Rightarrow q \quad \text{or} \quad \begin{array}{c} p \Rightarrow q \\ p \\ \hline q \end{array}$

We have seen that this is a tautology and is thus a valid argument by Theorem (2e-6). See also Example 2, §2d.

Many students find that learning a verbal restatement of each valid argument form is helpful both in remembering the pattern of the rule and in seeing how to apply it. A suitable restatement of the Rule of Detachment is "Given an implication and its hypothesis, its conclusion is a valid deduction."

Consideration of the form on the right in (1) should suggest three other possibilities:

(i) (a) $p \Rightarrow q$ (ii) (a) $p \Rightarrow q$ (iii) (a) $p \Rightarrow q$
    (b) q (b) $\sim p$ (b) $\sim q$
    ─────   ─────   ─────
       p $\sim q$ $\sim p$

or in the horizontal form:

§2f Valid Argument Forms

(i) $(p \Rightarrow q) \wedge q \Rightarrow p$
(ii) $(p \Rightarrow q) \wedge \sim p \Rightarrow \sim q$
(iii) $(p \Rightarrow q) \wedge \sim q \Rightarrow \sim q$

If you carefully inspect these arguments using the conjunction symbol as a focus and recall some definitions from §2e, you should realize why suitable names for these are (i) the *converse argument*, (ii) the *inverse argument*, and (iii) the *contrapositive argument*. Also, if you recall the fact that the contrapositive form of an implication is the only one of the three forms which is logically equivalent to the positive form, it will seem reasonable that the assertion that these are valid forms would be correct only for (iii).

Disproof of converse argument (i):

	∴	(b)	(a)
	p	q	$p \Rightarrow q$
	T	T	T
	T	F	F — because of Premises (a) and (b).
	F	T	T
	F	F	T — because of Premise (b).

In Row 3 the conclusion is false and the premises are each true. Thus the converse argument is invalid. Having shown that this form is invalid, any other argument having this form can immediately be classified as invalid *without* use of the truth table. See Example 3, §2d.

Disproof of the inverse argument (ii):

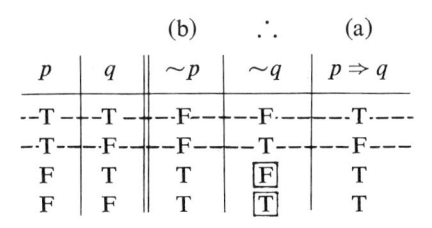

		(b)	∴	(a)
p	q	$\sim p$	$\sim q$	$p \Rightarrow q$
T	T	F	F	T
T	F	F	T	F
F	T	T	F	T
F	F	T	T	T

Q. E. D.

Again the F in Row 3 shows that this argument is invalid. Hence we label any *inverse argument* as *invalid*.

EXAMPLE 1: Determine whether the following argument is valid or invalid.

(a) $2 = 3$ only if $3 = 4$.
(b) $2 \neq 3$.
Therefore, $3 \neq 4$.

Solution: Letting p stand for $2 = 3$ and q for $3 = 4$, the argument translates as:

(a) $p \Rightarrow q$
(b) $\sim p$
Therefore, $\sim q$

This is exactly the inverse argument form (ii), and we have seen that any argument having this form is invalid.

Proof of the contrapositive argument (iii):

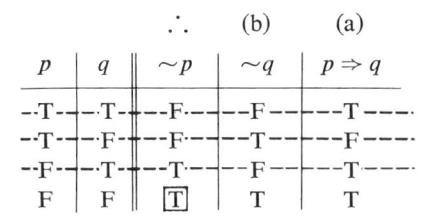

p	q	$\sim p$	$\sim q$	$p \Rightarrow q$
		∴	(b)	(a)
T	T	F	F	T
T	F	F	T	F
F	T	T	F	T
F	F	$\boxed{T}$	T	T

Q. E. D.

We have proved the following theorem.

Theorem (2f-1): $[(p \Rightarrow q) \wedge \sim q] \Rightarrow \sim p$ is valid.

This contrapositive argument form is known variously as *Modus* (tollendo) *tollens* or the **Rule of Contraposition**. This is the only valid form of (2). It can be verbalized as "Given an implication and the negation of its conclusion, the negation of its hypothesis follows."

EXAMPLE 2: Determine whether the following argument is valid or invalid.

(a) $2 = 2 \Rightarrow 3 = 3$
(b) $3 \neq 3$.
Therefore, $2 \neq 2$

Solution: Letting p stand for $2 = 2$ and q for $3 = 3$, the symbolic form becomes:

(a) $p \Rightarrow q$
(b) $\sim q$
Therefore, $\sim p$

Since this is exactly the form (iii) of the contrapositive argument, the argument is valid.

The following table might be of assistance. The resemblance to the table of Figure 2.22 should be noted.

§2f   **Valid Argument Forms**

$$\begin{array}{ll} p \Rightarrow q & \text{Rule of Detachment} \\ \underline{p} & \text{valid} \\ q & \end{array} \qquad \begin{array}{ll} p \Rightarrow q & \text{Converse Argument} \\ \underline{q} & \text{invalid} \\ p & \end{array}$$

$$\begin{array}{ll} p \Rightarrow q & \text{Inverse Argument} \\ \underline{\sim p} & \text{invalid} \\ \sim q & \end{array} \qquad \begin{array}{ll} p \Rightarrow q & \text{Rule of Contraposition} \\ \underline{\sim q} & \text{valid} \\ \sim p & \end{array}$$

Figure 2.23

The converse and inverse arguments are the main invalid forms we learn to recognize at a glance. To handle argument analysis and construction of proofs efficiently we need to learn to recognize several valid forms in addition to the Rule of Detachment and the Rule of Contraposition.

The rule which is perhaps the simplest valid argument form of all is actually a restatement of part of Definition (2-8) of argument. We recall that the premises of an argument are considered to be conjoined. That is, saying p, q are premises of an argument, we have

(3) $$\begin{array}{c} p \\ \underline{q} \\ p \wedge q \end{array}$$

This form is called the **Rule of Conjunctive Addition** or *Rule of Adjunction*.

Another valid form which makes use of conjunction is

$$p \wedge q \Rightarrow p$$

You had an opportunity to show that this implication is a tautology in Exercise 11, §2e. Thus it is a valid form.

Theorem (2f-2): $p \wedge q \Rightarrow p$ or $\dfrac{p \wedge q}{p}$ or $\dfrac{p \wedge q}{q}$ is valid.

As shown, this argument might have been stated $p \wedge q \Rightarrow q$. The argument is saying that if a conjunction is accepted as true, then as a result each of its components is true. We call it the **Rule of Conjunctive Simplification.**

EXAMPLE 3: Consider the argument:
The diagonals of the parallelogram are congruent and its angles are congruent. Therefore, the angles of the parallelogram are congruent.
    Is this argument valid?

Solution: This argument has the form $p \wedge q \Rightarrow q$. Hence it is valid by Theorem (2f-2).

The valid argument forms considered to this point have involved conjunction and implication. In Exercise 11, §2e, you showed that a form using disjunction was also a tautology. We restate:

$$p \Rightarrow p \lor q$$

is a tautology. Thus, this too is a valid argument form.

Theorem (2f-3): $p \Rightarrow p \lor q$ or $\dfrac{p}{p \lor q}$ or $\dfrac{q}{p \lor q}$ is valid.

In words, "If p is any premise, then its disjunction with any other proposition is a valid deduction." The proof is given by way of stressing this form.

Proof:

p	q	$p \lor q$
T	T	T
T	F	T
F	T	T
F	F	F

Argument $p \Rightarrow p \lor q$ is valid by definition.   Q. E. D.

This principle is called the **Law of Disjunctive Addition**.

EXAMPLE 4: A man noticed a child bundled up in a snow suit. "Are you a boy or a girl?" he asked. "Yes," the child replied, "what else would I be?"

Analyzing this exchange, let p be "I am a boy" and q be "I am a girl." The child is one of p or q; therefore, to the question $(p \lor q)$ the answer is "yes" by disjunctive addition. That is, the argument fits either the pattern $p \Rightarrow p \lor q$ or $q \Rightarrow p \lor q$.

Another valid form involving disjunction is the following.

Theorem (2f-4): $[(p \lor q) \land \sim p] \Rightarrow q$ or $\dfrac{p \lor q}{\sim p}$ or $\dfrac{p \lor q}{\sim q}$ is valid.

This argument simply states that given a disjunction and the negation of one of its components, the other component is a valid deduction.

Proof:

p	q	$\sim p$	$p \lor q$
T	T	F	T
T	F	F	T
F	T	T	T
F	F	T	F

The argument is valid by the definition of valid argument.

Q. E. D.

This form is generally called *Modus Tollendo Ponens* or the **Rule of Disjunctive Simplification.**

EXAMPLE 5: Determine whether the following argument is valid or invalid.

> I am a boy or a girl.
> I am not a girl.
> I am a boy.

Solution: Let p stand for "I am a boy" and q stand for "I am a girl." This argument has the form

$$\begin{array}{c} p \vee q \\ \sim q \\ \hline p \end{array}$$

It is valid by Theorem (2f-4).

The next form we consider has three component propositions. The argument is as follows:

Theorem (2f-5): $[(p \Rightarrow q) \wedge (q \Rightarrow r)] \Rightarrow (p \Rightarrow r)$ or

$$\begin{array}{c} p \Rightarrow q \\ q \Rightarrow r \\ \hline p \Rightarrow r \end{array}$$

is valid.

Proof:

p	q	r	(a) $p \Rightarrow q$	(b) $q \Rightarrow r$	∴ $p \Rightarrow r$
T	T	T	T	T	T
T	T	F	T	F	F
T	F	T	F	T	T
T	F	F	F	T	F
F	T	T	T	T	T
F	T	F	T	F	T
F	F	T	T	T	T
F	F	F	T	T	T

The argument is valid.

Q. E. D.

This form is called the **Chain Rule** or *Law of Hypothetical Syllogism.*

EXAMPLE 6: Determine whether the following argument is valid or invalid.

$\triangle ABC \sim \triangle DEF$ if $\mathscr{L}_1$ is not parallel to $\mathscr{L}_2$.
$\triangle ABC \sim \triangle DEF$ only if $\angle A \cong \angle D$.
If $\mathscr{L}_1$ is not parallel to $\mathscr{L}_2$, $\angle A \cong \angle D$.

Solution: Letting p stand for "$\mathscr{L}_1$ is not parallel to $\mathscr{L}_2$," q stand for $\triangle ABC \sim \triangle DEF$, and r stand for $\angle A \cong \angle D$, the argument translates into the symbolic form

$$p \Rightarrow q$$
$$q \Rightarrow r$$
$$p \Rightarrow r$$

This argument is valid by Theorem (2f-5), the Chain Rule.

In summary, the following are frequently used valid argument forms which are used to simplify more complex arguments. They are often called rules of inference and should be learned well. Rules (viii) and (ix) are listed here for completeness. They will be discussed in §2g.

INFERENCE RULES

(i) $\dfrac{p \Rightarrow q,\ p}{q}$ Rule of Detachment (R. D.)

(ii) $\dfrac{p \Rightarrow q,\ \sim q}{\sim p}$ Rule of Contraposition (R. C.)

(iii) $\dfrac{p,\ q}{p \wedge q}$ Conjunctive Addition (C. A.)

(iv) $\dfrac{p \wedge q}{p},\ \dfrac{p \wedge q}{q}$ Conjunctive Simplification (C. S.)

(v) $\dfrac{p}{p \vee q}$ Disjunctive Addition (D. A.)

(vi) $\dfrac{p \vee q,\ \sim p}{q},\ \dfrac{p \vee q,\ \sim q}{p}$ Disjunctive Simplification (D. S.)

(vii) $\dfrac{p \Rightarrow q,\ q \Rightarrow r}{p \Rightarrow r}$ Chain Rule (C. R.)

(viii) $\dfrac{p \wedge \sim q}{f}$ is logically equivalent to $\dfrac{p}{q}$
     Rule of Indirect Inference (I. I.)

(ix) $\dfrac{p \wedge q}{r}$ is logically equivalent to $\dfrac{p}{q \Rightarrow r}$

Rule of Conditional Inference (C. I.)

Many other inference rules are possible. Some are found in the following exercises.

EXERCISE SET

1. Show that the third form of Conjunctive Simplification $p \wedge q \Rightarrow q$ in Theorem (2f-2) is valid.
2. Show by direct appeal to Theorem (2e-6) that the Rule of Contraposition is a valid form.
3. Show by direct use of Definition (2-9) in a truth table that the Rule of Conjunctive Simplification is a valid form.
4. Show by direct appeal to Definition (2-9) that the third form $[(p \vee q) \wedge \sim q] \Rightarrow p$ in Theorem (2f-4) is a valid argument.
5. Prove the Rule of Disjunctive Simplification by appeal to Theorem (2e-6).
6. Prove the Rule of Disjunctive Addition by appeal to Theorem (2e-6).
7. Show by direct appeal to Definition (2-9) that the third form $q \Rightarrow p \vee q$ in Theorem (2f-3) is a valid argument form.
8. Prove that $(x \wedge \sim x) \wedge y \Rightarrow z$ is a tautology. This is sometimes called the *Rule of Contradiction*.
9. Prove Rule (vii) by appeal to Theorem (2e-6).
10. Prove: a) Rule (viii) b) Rule (ix)

The arguments in Exercises 11 through 16 are valid. Supply the reason why each conclusion is justified. Quote a standard inference rule or rules to justify your opinion.

11. Premise (a): $q \vee \sim s$
    Premise (b): $\underline{\sim q}$
    Conclusion: $\sim s$

12. P(a) $q \Rightarrow \sim v$
    P(b) $\underline{\sim v \Rightarrow q}$
    $u \Rightarrow q$

13. $\underline{s \wedge \sim r}$
    $\sim r$

14. P(a) $m \Rightarrow (r \lor s)$
    P(b) $\underline{m}$
    $r \lor s$

15. P(a) $s \Rightarrow \sim q$
    P(b) $\underline{q}$
    $\sim s$

16. $\underline{p \lor q}$
    $r \lor (p \lor q)$

In Exercises 17 through 37 determine whether the given argument is valid and state your reasons. Use a standard inference rule whenever possible. Resort to truth table only when an argument is not constructed so as to conform to a standard inference pattern.

17. The earth is flat implies that the moon is a sphere.
    The earth is not flat.
    The moon is not a sphere.

18. If prices are rising, steelworkers are living high.
    Steelworkers are living high.
    Hence prices are rising.

19. The days are becoming brighter.
    The nights are becoming darker if the days are becoming brighter.
    The nights are becoming darker.

20. The sides are congruent.
    The triangle is not isosceles.
    The sides are congruent and the triangle is not isosceles.

21. If line $\mathscr{M}$ is parallel to line $\mathscr{N}$, then $\measuredangle A \cong \measuredangle B$.
    But $\measuredangle A$ is not congruent to $\measuredangle B$.
    Line $\mathscr{M}$ is not parallel to line $\mathscr{N}$.

22. $\underline{2 = 5 \text{ and } 4 = \pi}$
    $4 = \pi$

23. If angle A is congruent to angle B, then $\overline{AC}$ is congruent to $\overline{BC}$.
    But we know $\overline{AC}$ is congruent to $\overline{BC}$.
    Therefore, angle A is congruent to angle B.

24. If February is warmer than June, then July is colder than December.
    February is warmer than June.
    Therefore, July is colder than December.

25. Cora cooks candy.
    Therefore, candy is sweet or Cora cooks candy.

26. If X passes the final, then he will pass the course.
    X flunks the course.
    Therefore, X did not pass the final examination.
27. If Maine is larger in area than Texas, then Alaska is larger in area than the United States.
    Maine is larger in area than Texas.
    Therefore, Alaska is larger in area than the United States.
28. $a = b$ only if $a \neq b \lor a < b$
    Therefore, if $a = b$, then $a < b$.
29. The weeks are becoming shorter.
    The months are becoming longer implies that the weeks are becoming shorter.
    Therefore, the months are becoming longer.
30. If U gets D or less in all of his courses, he will be in bad standing. U does not get D or less in all of his courses. Therefore, U is in good standing.
31. If John swims fast, then elephants eat eggs.
    If elephants eat eggs, then John is a jerk.
    Therefore, if John swims fast, then John is a jerk.
32. Harry wins the race or comes in second.
    Harry does not win the race.
    Therefore, Harry comes in second.
33. This man is a metal worker or a wood worker.
    He is not a wood worker.
    Therefore, he is not a metal worker.
34. If line $\mathscr{L}$ is not perpendicular to line $\mathscr{M}$, then line $\mathscr{M}$ is not perpendicular to line $\mathscr{N}$.
    But line $\mathscr{M}$ is perpendicular to line $\mathscr{N}$.
    Therefore, line $\mathscr{L}$ is not perpendicular to line $\mathscr{M}$.
35. If I have done these exercises correctly then I understand valid arguments.
    However, I do not understand valid arguments.
    Therefore, these exercises are not correctly done.
36. The triangle is not isosceles or it is not scalene.
    The triangle is scalene.
    Therefore, the triangle is not isosceles.
37. This argument is valid only if I have not forgotten the rule.
    I have forgotten the rule.
    Therefore, this argument is invalid.

In Exercises 38 through 47, obtain a conclusion, if possible, using the given statements as premises. In each case tell which rule of inference is used. The best conclusion is considered to be the one which uses all the premises.

38. Millie drinks milk only if Spiros eats spinach. Millie drinks milk.
39. If Casey strikes out, then there is no joy in Mudville. Casey strikes out.
40. If $4 = 9$ then $5 = 10$. If $5 = 10$ then $9 = 14$.
41. Mickey is sticky if Martin mixes marmalade. Mickey is sticky.
42. Mudville's mayor mambos merrily and Manfred mangles mastodons.
43. Bertha is a bird or Pearl is a girl. Pearl is not a girl.
44. If $\triangle ABC \sim \triangle ADE$, then $\measuredangle A \cong \measuredangle D$. However, $\measuredangle A$ is not congruent to $\measuredangle D$.
45. Harry is hairy. Mary is merry.
46. If the investigation results in problems, it will be a failure. The investigation is needed only if it is not a failure. The investigation is needed.
47. If a shfnexi is a glacho, then an ootster is a glacho.
    A shfnexi exists or a shfnexi is a glacho. No shfnexi exists.
48. If possible, assign truth values to the components p, q, r in such a way that the given compound is true; otherwise, state that this is impossible.

    a) $(\sim p \vee q) \wedge \sim q$ b) $\sim(\sim p) \wedge \sim q$
    c) $\sim(p \vee \sim p)$ d) $(p \Longleftrightarrow q) \wedge (\sim q \vee r)$
    e) $(\sim p \Rightarrow q) \wedge (\sim q \vee r)$ f) $(\sim p \Rightarrow p) \wedge (\sim p \vee q)$

49. Try to assign truth values to the components of the given arguments in such a way as to show that the implication can be false.

    a) $(q \wedge r \Longleftrightarrow q) \wedge (\sim r \Rightarrow p) \Rightarrow r$
    b) $(u \Rightarrow r) \wedge (\sim p) \wedge (\sim p \vee \sim r) \wedge (q \Longleftrightarrow p) \Rightarrow \sim u$
    c) $(p \Rightarrow q) \wedge (\sim q \vee \sim p) \Rightarrow \sim q$.

§2g   GENERAL ARGUMENTS

For the most part the arguments of the preceding sections were simple enough for you to be able to determine when any one of them was valid or invalid by recognizing that it had the same pattern as a standard inference form.

§2g General Arguments

In this section more complex cases which cannot be analyzed by simple recognition and for which a truth table would require far too much labor will be considered. Analysis of such cases is carried out by applying standard inference rules in a sequence of steps.

Recall that an argument is basically an implication. A proof of an implication such as $p \Rightarrow q$ is simply a verification by valid form that $p \Rightarrow q$ is true in *all* cases.

Definition (2-13): The **proof** that $p \Rightarrow q$ is a valid argument consists of a sequence of propositions each of which must satisfy at least one of the following:

(i)   the proposition is a component of p, i.e., one of the premises;
(ii) the proposition is a premise allowed by a particular inference rule;
(iii) the proposition is a valid deduction from a preceding part of the sequence;
(iv) the proposition is true within the discourse, i.e., axiom, definition, or previously proved theorem;
(v)   the last proposition in the sequence is q.

In logic exercises the previously proved theorems of (iv) will usually be one of the tautologies or logical equivalences proved in §2e.

(1) In order to construct such a proof, first choose two premises from among the given ones with which you can reach a conclusion by valid form. Then pick out another premise which you can use with the first conclusion to reach a second conclusion. This process is continued until the stated conclusion of the argument is reached or until some indication that the argument may be invalid is obtained.

EXAMPLE 1: Consider the following argument:

Premise (a) $q \Rightarrow r$
Premise (b) $\sim r \lor s$
Premise (c) $p \Rightarrow q$
Premise (d) u
Premise (e) $\underline{u \Rightarrow \sim s}$
$\sim p$

Propositions (a) through (e) are the premises, and $\sim p$ is the stated conclusion. Determine whether the argument is valid or invalid.

Solution: As stated in (1), we generally try to use a single component premise first. Using this rule, we take the single component Premise (d) u as a beginning and use it with Premise (e) which contains a u.

Elementary Logic

Statements	Reasons
1. u	1. Premise (d)
2. $u \Rightarrow \sim s$	2. Premise (e)
3. $\sim s$	3. Rule of Detachment (R. D.)

Now look through the premises again for something that may be used with $\sim s$ to form another valid argument.

4. $\sim r \lor s$	4. Premise (b)
5. $\sim r$	5. Disjunctive Simplification (D. S.)

Now look through the premises again for something that may be used with $\sim r$ to form another valid argument.

6. $q \Rightarrow r$	6. Premise (a)
7. $\sim q$	7. Rule of Contraposition (R. C.)

Now look through the premises again for something that may be used with $\sim q$ to form another valid argument.

8. $p \Rightarrow q$	8. Premise (c)
9. $\sim q$	9. Rule of Contraposition (R. C.)

Q. E. D.

EXAMPLE 2: What conclusions can we obtain from the following argument? "If the trial results in a hung jury then it is family justice. The trial is desired only if it is not family justice. The trial is desired."

Procedure: First label each of the component propositions and then rewrite the argument in symbolic form.

p: The trial results in a hung jury.
q: It is family justice.
r: The trial is desired.

With these labels the first sentence translates as $p \Rightarrow q$, the second sentence as $r \Rightarrow \sim q$, and the third as r. In symbolic form we have

(a) $p \Rightarrow q$
(b) $r \Rightarrow \sim q$
(c) r

As before, we use the single component premise first, pairing it with another of the premises which allows us to make use of one of the known standard inference rules if possible.

Statements	Reasons
1. $r \Rightarrow \sim q$	1. Premise (b)
2. r	2. Premise (c)
3. $\sim q$	3. Rule of Detachment (R. D.)

Now look back through the premises for something that may be used with $\sim q$ to form another valid argument.

4. $p \Rightarrow q$ 4. Premise (a)
5. $\sim p$ 5. Rule of Contraposition (R. C.)

Now all the premises have been used and the resulting conclusions are in Step 3 and Step 5, i.e., $\sim q$, $\sim p$. Translating each into English, "It is not family justice," "The trial will not result in a hung jury." The latter is considered the best conclusion because all the premises were used to obtain it.

The preceding example illustrates the fact that we use a single component most often with the Rule of Detachment, the Rule of Contraposition, or the Rule of Disjunctive Simplification.

By Conjunctive Simplification we may break a conjunction into its component parts. Therefore,

(2)   in any case where the argument has no single component premise, one should look for a premise that is a conjunction.

When one component of the conjunction is a single component, it can be used as in the preceding examples.

EXAMPLE 3: Show that the following argument is valid.

(a) $p \wedge q$
(b) $p \Rightarrow r$
$\therefore (r \wedge q) \vee s$

Solution:

Statements	Reasons
1. $p \wedge q$	1. Premise (a)
2. p	2. C. S.
3. $p \Rightarrow r$	3. Premise (b)
4. r	4. R. D.
5. q	5. C. S. (from Step 1)
6. $r \wedge q$	6. C. A.
7. $(r \wedge q) \vee s$	7. D. A.

Q. E. D.

Suppose that none of the premises of an argument consists of a single component or a conjunction. How does one begin the analysis? Then we can

(3)   try to use the fact that the negation of an implication or of a disjunction is actually a conjunction and thus make use of (2).

Recall Theorems (2e-11) and (2e-12), i.e., DeM ($\vee$) and NIMP.

Elementary Logic

Now suppose that none of the premises consists of a single component, a conjunction, or a form which is logically equivalent to a conjunction. Suppose further that one must actually attempt to do such a bothersome type of problem. The best approach is usually to use the Rule of Indirect Inference or the Rule of Conditional Inference. However, the following technique can be effective in simplifying the work.

(4) Implications and disjunctions can be effectively used together by recalling Theorem (2e-8), i.e., LEDI. Also, several implications often can be greatly simplified by using the Chain Rule with or without CLEP.

Consider the following argument.

EXAMPLE 4: Determine whether this argument is valid or invalid.

(a) $s \lor \sim q$
(b) $\sim q \Rightarrow p$
(c) $\underline{p \Rightarrow s}$
$\sim s \Rightarrow s$

Solution: Here there is no single component premise and no conjunction. You can solve the problem by using LEDI and Axiom L-4 with the Chain Rule.

Statements	Reasons
1. $s \lor \sim q$	1. P(a)
2. $s \lor \sim q \underline{\equiv} (\sim s \Rightarrow \sim q)$	2. (2e-8), i.e. LEDI
3. $\sim s \Rightarrow \sim q$	3. Substitution, i.e., Axiom L-4
4. $\sim q \Rightarrow p$	4. P(b)
5. $\sim s \Rightarrow p$	5. C. R.
6. $p \Rightarrow s$	6. P(c)
7. $\sim s \Rightarrow s$	7. C. R.

Q. E. D.

These four examples have illustrated methods for handling valid cases. If we see that the relationships among the premises do not seem to be such that we can show the argument to be valid, it may be that the argument is invalid. In simple cases we have been able to demonstrate invalidity by showing that the argument fitted either the inverse or converse forms or by using a truth table. In order to show invalidity in more complex cases, we use a method based on the idea of *truth value assignment*. See Exercises 48 and 49 of Section 2f.

> **To demonstrate invalidity by truth value assignment, we look for one assignment of truth values for the components such that each of the premises is true**

§2g   General Arguments   77

and the conclusion is false. **If such an assignment is possible, the argument is invalid by the definition of valid argument** (2-9).

EXAMPLE 5: Show by the method of truth value assignment that the argument of Example 3, §2d, is invalid.

Solution: Symbolically we had

(a) $p \Rightarrow q$
(b) q
     p

or

$$(p \Rightarrow q) \wedge q \Rightarrow p$$

To make the conclusion false we choose $p \underline{e} f$. Then we have

$$(f \Rightarrow q) \wedge q \Rightarrow f$$

Since we want an assignment which causes *each* premise to be true, we make $q \underline{e} t$ because q is a single component premise. This additional assignment yields

$$(f \Rightarrow t) \wedge t \Rightarrow f$$

By the definition of implication, this becomes

$$t \wedge t \Rightarrow f$$

We have found one case in which each premise is true and the conclusion false. Thus the argument is *invalid*.

EXAMPLE 6: Prove that the following argument is valid or show that it is invalid by truth value assignment.

(6)
(a) $p \vee q \Rightarrow r$
(b) $\sim r \Rightarrow q$
     p

Solution: Again, neither of the premises is a single component or a conjunction. We can try to use LEDI and the Chain Rule.

1. $\sim r \Rightarrow q$ 1. P(b)
2. $(\sim r \Rightarrow q) \underline{e} (\sim q \Rightarrow \sim \sim r)$ 2. CLEP (2e-7)
3. $\sim q \Rightarrow \sim \sim r$ 3. L-4
4. $\sim \sim r \underline{e} r$ 4. D. N. (2e-9)
5. $\sim q \Rightarrow r$ 5. L-4
6. $p \vee q \Rightarrow r$ 6. P(a)

It does not look as though our inference rules lead to the conclusion p. In horizontal form, the argument of (6) reads

(7) $\quad (p \lor q \Rightarrow r) \land (\sim r \Rightarrow q) \Rightarrow p$

Trying $p \rightleftharpoons f$ to make the conclusion false, we obtain

$$(f \lor q \Rightarrow r) \land (\sim r \Rightarrow q) \Rightarrow f \quad \text{by L-4}$$

Now, to make each component of the conjunction true, assign

$$q \rightleftharpoons f, r \rightleftharpoons t$$

whence

$$\begin{aligned}(f \lor f \Rightarrow t) \land (\sim t \Rightarrow f) \Rightarrow f & \quad \text{by L-4} \\ (f \Rightarrow t) \land (\sim t \Rightarrow f) \Rightarrow f & \quad \text{by (2-3)} \\ (f \Rightarrow t) \land (f \Rightarrow f) \Rightarrow f & \quad \text{by (2-2)} \\ t \land t \Rightarrow f & \quad \text{by (2-5)} \end{aligned}$$

We have found a case in which each premise is true and the conclusion is false. The argument is invalid.

When you are trying to do a proof and come to a point beyond which you do not know what steps to take, it may be helpful to check whether or not all the premises have been used. This is *not* to say that all of them *must* be used, but an easy proof may sometimes become difficult or impossible when too few premises are used.

When asked to test premises for consistency, we do so by finding at least one assignment of truth values for which their conjunction is true.

EXERCISE SET I

1. Test the premises of each argument for consistency.

    a) Exercise 16   b) Exercise 18
    c) Exercise 22   d) Exercise 30

Exercises 2 through 7 outline examples of proofs which have some of the reasons omitted. Supply the missing reasons.

2.
    1. $\sim r$ 1. P
    2. $p \Rightarrow r$ 2. P
    3. $\sim p$ 3.
    4. $p \lor \sim q$ 4. P
    5. $\sim q$ 5.

3.
    1. $p \Rightarrow r$ 1. P
    2. $\sim r$ 2. P
    3. $\sim p$ 3.
    4. $p \lor q$ 4. P

§2g   **General Arguments**   79

      5. $\underline{q}$ 5.
      6. $q \vee s$ 6.

4. 1. $\sim q \vee \sim r$ 1. P
    2. $\underline{\sim q \vee \sim r \underset{=}{} q \Rightarrow \sim r}$ 2.
    3. $q \Rightarrow \sim r$ 3.
    4. $\underline{\sim p \Rightarrow q}$ 4. P
    5. $\sim p \Rightarrow \sim r$ 5.
    6. $\underline{\sim p \Rightarrow \sim r \underset{=}{} r \Rightarrow p}$ 6.
    7. $r \Rightarrow p$ 7.
    8. $\underline{r \Rightarrow p \underset{=}{} \sim r \vee p}$ 8.
    9. $\sim r \vee p$ 9.

5. 1. $\sim (r \Rightarrow p)$ 1. P
    2. $\underline{\sim (r \Rightarrow p) \underset{=}{} r \wedge \sim p}$ 2.
    3. $r \wedge \sim p$ 3.
    4. r 4.
    5. $\sim p$ 5.
    6. $\underline{\sim p \Rightarrow q}$ 6. P
    7. q 7.
    8. $\sim q \vee s$ 8. P
    9. $\underline{s}$ 9.
  10. $s \wedge r$ 10.

6. 1. $\sim (p \vee q)$ 1. P
    2. $\underline{\sim (p \vee q) \underset{=}{} \sim p \wedge \sim q}$ 2.
    3. $\sim p \wedge \sim q$ 3.
    4. $\sim q$ 4.
    5. $\underline{\sim q \Rightarrow s}$ 5. P
    6. $\underline{s}$ 6.
    7. $s \vee \sim r$ 7.

7. 1. $u \Longleftrightarrow w$ 1. P
    2. $\underline{(u \Longleftrightarrow w) \underset{=}{} (u \Rightarrow w) \wedge (w \Rightarrow u)}$ 2.
    3. $(u \Rightarrow w) \wedge (w \Rightarrow u)$ 3.
    4. $w \Rightarrow u$ 4.
    5. $\underline{\sim u}$ 5. P
    6. $\sim w$ 6.

In Exercises 8 through 35 prove the argument valid or make a truth value assignment to show that it is invalid. Use of the Rules of Conditional Inference or of Indirect Inference is not expected but may be propitious.

(Symbolic Arguments)

    8. (a) $s \vee r$ 9. (a) $\sim (p \vee r)$

(b) $p \Rightarrow \sim r$
(c) p
$\therefore, s$

10. (a) $\sim (p \lor q)$
(b) $p \lor r$
(c) $s \Rightarrow r$
$\therefore, s$

12. (a) $p \lor q$
(b) $q \Rightarrow r$
(c) $p \Rightarrow s$
(d) $\sim s$
$\therefore, r \land (p \lor q)$

14. (a) $r \Rightarrow \sim p$
(b) $(p \land s) \lor w$
(c) $w \Rightarrow (q \lor u)$
(d) $\sim q \land \sim u$
$\therefore, \sim r$

16. (a) $p \Rightarrow s$
(b) $\sim q$
(c) $s \Rightarrow q$
(d) $r \lor p$
(e) $q \lor \sim r$
$\therefore, s$

18. (a) $\sim u \Rightarrow r$
(b) p
(c) $\sim q \lor \sim r$
(d) $q \Leftrightarrow p$
$\therefore, \sim u$

*20. (a) $p \Rightarrow q$
(b) $q \Rightarrow r$
(c) $(p \Rightarrow r) \Rightarrow s$
(d) $\sim s \lor u$
$\therefore, u$

(b) $q \Rightarrow r$
(c) $q \lor \sim s$
$\therefore, \sim s$

11. (a) $p \land s \Leftrightarrow r$
(b) $\sim r$
(c) p
$\therefore, \sim s$

13. (a) $\sim s \lor \sim r$
(b) $\sim r \Rightarrow \sim q$
(c) $\sim s \Rightarrow p$
(d) $\sim p$
$\therefore, \sim q \land \sim p$

15. (a) $\sim (p \lor \sim r)$
(b) $q \lor p$
(c) $r \Rightarrow s$
(d) $(q \land s) \Rightarrow (u \land s)$
$\therefore, s \land u$

17. (a) $q \Rightarrow r$
(b) $p \Rightarrow q$
(c) $p \lor u$
(d) $u \Rightarrow s$
(e) $\sim r$
$\therefore, s$

*19 (a) $\sim (p \land q)$
(b) $\sim q \Rightarrow u$
(c) $\sim p \Rightarrow u$
(d) $s \Rightarrow \sim u$
$\therefore, \sim s$

*21. (a) $p \Rightarrow q$
(b) $q \lor r$
(c) $p \lor \sim r$
$\therefore, q \lor \sim r$

(Verbal Arguments)

22. Twenty-five is a perfect square. Blackboards are green or 25 is not a perfect square. Therefore, blackboards are green or π is irrational.

23. If Flipper is a mammal, then he gets oxygen from the air. If he gets oxygen from the air, then he has no need for gills. Flipper is a mammal and his habitat is the ocean. Therefore, he has no need of gills.

24. Either this bird is a sparrow or it is a gull. This bird is small. If this bird is small, then it is not a gull. Therefore, this bird is a sparrow.
25. If Tom is seven, then Tom is the same age as June. If John is not as old as Tom, then John is not as old as June. Tom is seven and John is as old as June. Therefore, John is as old as Tom, and Tom is the same age as June.
26. If △ ABC is isosceles, then ∢A is acute. Six is a triangular number. If ∢A is acute, then ∢C is obtuse. ∢C is not obtuse. Therefore, ∢A is not acute or six is not a triangular number.
27. If John is shorter than Bob, then Mary is taller than Jean. Mary is not taller than Jean. If John and Bill are of the same height, then John is shorter than Bob. Therefore, John and Bill are not of the same height.
28. If Al did not win the race, then either Bob was second or Carl was second. If Bob was second, then Al did win the race. If Don was second, then Carl was not second. Al did not win the race. Therefore, Don was not second.
29. If the clock is slow, then Rogers arrived before 12:00 P.M. and he did see Cassidy's horse leave. If Cassidy is telling the truth, then Rogers did not see Cassidy's horse leave. Either Cassidy is telling the truth or he was in the building at the time of the crime. The clock is slow. Therefore, Cassidy was in the building at the time of the crime.
30. If Paul is strong or James is smart, then Diane has dimples. It is false that Diane has dimples or James is smart. Paul is strong and James is smart. Therefore, Diane has no dimples.
31. If the contract was not approved, then the total output will remain as it is. If the total output remains as it is, then we shall not add new staff to the plant. Either we shall not add new staff to the plant or production of this item will be delayed for a month. However, production of this item will not be delayed for a month. Therefore, the contract was approved.
32. Triangle ABC is isosceles if the base angles are congruent. Also △ABC is isosceles or △ABD is scalene. Triangle ABD is scalene or the base angles are not congruent. However, △ABD is not scalene. Therefore, △ABC is isosceles or the base angles are not congruent.
33. If it is not the case that a watched pot never boils only if it is not worth a bird bush, then you can tickle tigers. But you cannot tickle tigers. If a watched pot never boils, then it is not worth a bird bush.

If it is worth a bird bush, then the watched pot never boils. Therefore, a watched pot never boils if and only if it is not worth a bird bush.

34. Chris is in the kitchen with Jim if she is eating crackers. She is not with Jim or she is not eating crackers. Therefore, Chris is not in the kitchen.

35. This rule will be approved today if and only if it is supported by the boss. Either it is supported by the boss or the boss's wife opposes it. If the boss's wife opposes it, then it will be delayed in family deliberations. Therefore, either this rule will be approved today or it will be delayed in family deliberations.

An algebraic proof of an argument usually falls into one of two general categories. In the type studied to this point, use is made of inference rules which require only the given premises to complete the work. Some arguments, however, would be extremely difficult, if not impossible, to prove by this method. Hence, to make analysis of such arguments easier, various inference rules are used which allow us to augment the given premises by other premises.

To see how this latter type of proof form works, recall that any argument has the compound form

(8) $$p \Rightarrow q$$

The stated conclusion of the argument is q. As provided in Definition (2-8), the hypothesis p of the implication is the conjunction of the premises of the argument.

Now consider an argument having premises

$$p_1, p_2, \ldots, p_n$$

where p_1 is the first and p_i is the last of the given premises. Now

$$p \triangleq p_1 \wedge p_2 \wedge \ldots \wedge p_n$$

is the conjunction of the premises, and the complete argument (8) can now be written

(9) $$p_1 \wedge p_2 \wedge \ldots \wedge p_n \Rightarrow q$$

When the given premises are augmented by an additional premise, say x, the total premises are then

$$p_1 \wedge p_2 \wedge \ldots \wedge p_n \wedge x$$

Of course, an argument with these premises has a form different from that of (9). For one thing, the conclusion is very likely to be different from the stated conclusion q of (9). Let us call this new conclusion y. We are now considering an argument

§2g   General Arguments

(10) $$p_1 \wedge p_2 \wedge \ldots \wedge p_n \wedge x \Rightarrow y$$

The premises of this new argument are the stated premises of the given argument of (9) augmented by a proposition x, and its conclusion is the proposition y. In the form of (8) it is written

(11) $$p \wedge x \Rightarrow y$$

At first glance, augmenting the given premises of an argument to produce a new argument may seem like a complication of dubious value. However, in certain cases in which the new argument of (10) is logically equivalent to the given argument of (9) there is a very real advantage. In such cases the Axiom of Substitution, L-4, allows us to substitute the new argument for the given one. If the new argument is easier to prove than the given one, a real gain has been made.

Use of augmented premises requires that the given argument form be altered into the specific pattern required by the particular inference rule used to obtain the additional premise.

Within the altered form the standard rules of inference are still used to obtain conclusions at various steps.

We shall make a special study of two particular rules of inference which allow use of augmented premises. These are the Rule of Indirect Inference (also called Reductio ad Absurdum) and the Rule of Conditional Inference. See (viii) and (ix) in the table of inference rules given in §2f.

The **Rule of Indirect Inference** (viii) was proved in Exercise 10, §2f. It is restated here for reference.

Theorem (2g-1): $(p \wedge \sim q \Rightarrow f) \subseteq (p \Rightarrow q)$ where f is a known false proposition.   I. I.

This inference rule states that the argument

(12) $$p \wedge \sim q \Rightarrow f$$

is logically equivalent to the argument $p \Rightarrow q$. Whenever (12) is easier to prove valid than (8), we may substitute proof of (12) for proof of (8).

Writing I. I. in traditional layout, we have

(13) $$\frac{p \wedge \sim q}{f} \text{ is logically equivalent to } \frac{p}{q}$$

This form states that we may substitute the argument on the left for the one on the right. To do so we augment the given premises by $\sim q$, the negation of the stated conclusion, and then derive a known false statement by our usual methods of inference. Having completed this procedure, we have done the equivalent of proving $p \Rightarrow q$. In the notation of (9) and (10), we replace

$$p_1 \wedge p_2 \wedge \ldots \wedge p_n \Rightarrow q$$

by

$$p_1 \wedge p_2 \wedge \ldots \wedge p_n \wedge \sim q \Rightarrow f$$

The false statement f required by the Rule of Indirect Inference could be the negation of an axiom, a definition, or a previously proved theorem in an axiomatic discourse. In practice with stated arguments, any contradiction will do. The simplest of these is the conjunction $x \wedge \sim x$ of Theorem (2e-3). Since $x \wedge \sim x \subseteq f$, we can restate the Rule of Indirect Inference as

(14) $$(p \wedge \sim q) \Rightarrow x \wedge \sim x \subseteq (p \Rightarrow q)$$

or

$$\frac{p}{\frac{\sim q}{x \wedge \sim x}} \text{ is logically equivalent to } \frac{p}{q}$$

where q and x are any propositions and p is the conjunction of the given premises.

The Rule of Indirect Inference can be very useful when there is neither a single component proposition nor a conjunction among the premises and the stated conclusion q consists of a single component. In such a case $\sim q$ also consists of a single component, and the argument may be much easier to prove as a result of having the use of such an additional premise.

In general, when we attempt to show an argument valid by using indirect inference we must continue the pattern until we reach a contradiction. Once this has been accomplished we are able to state that the conclusion of the original argument is the result of valid inference by substitution. Of course, failure to derive the required false statement f or $x \wedge \sim x$ indicates that the argument may be invalid.

EXAMPLE 7: Prove that the following argument is valid by I. I.

(a) p
(b) $\sim q \Rightarrow \sim p$
$\overline{ q }$

Solution: (Done by indirect inference using (14)).

Statements	Reasons
1. $\sim q$	1. Indirect Premise (I. P.)
2. $\sim q \Rightarrow \sim p$	2. P(b)
3. $\sim p$	3. R. D.
4. p	4. P(a)

§2g   General Arguments

5. $p \wedge \sim p$ 5. C. A. using (3, 4)
6. q 6. I. I.

Q. E. D.

Since we have completed the left member of (14) in Step 5, we have, in effect, completed the right member. Hence the conclusion in the right member must be correct and is stated in Step 6.

The **Rule of Conditional Inference**, (ix) of §2f, was proved in Exercise 10 of that section. It is the basic form for the majority of mathematical proofs.

Theorem (2g-2): $[(p \wedge q) \Rightarrow r] \leftrightharpoons [p \Rightarrow (q \Rightarrow r)]$ or $\dfrac{p \wedge q}{r}$ is logically equivalent to $\dfrac{p}{q \Rightarrow r}$.   C. I.

Note that the conclusion of the argument

(15) $\qquad\qquad\qquad p \Rightarrow (q \Rightarrow r)$

is an implication, $q \Rightarrow r$. Hence C. I. is especially advantageous when the stated conclusion of an argument is an implication as is often the case in mathematical arguments.

In words, this principle states that when we are trying to prove an argument whose stated conclusion is an implication, we may augment the premises by the *hypothesis* of that implication. Then, arriving validly at the conclusion of the implication is equivalent to proving the implication itself.

Writing the argument of (15) in the form of (11), we have

(16) $\qquad\qquad p_1 \wedge p_2 \wedge \ldots \wedge p_n \Rightarrow (q \Rightarrow r)$

Conditional inference allows us to use q, the hypothesis of the conclusion, as an added premise and prove

(17) $\qquad\qquad p_1 \wedge p_2 \wedge \ldots \wedge p_n \wedge q \Rightarrow r$

instead. Whenever q is a single component proposition and there is no single component premise, this can be a real advantage, and we can substitute proof of (17) for proof of (16). Of course, failure to deduce r by valid inference may indicate that the argument of (17), and thus that of (16), is invalid.

EXAMPLE 8: Prove the following argument valid by C. I.

    (a) $s \Rightarrow u$
    (b) $r \Rightarrow p$
    (c) $u \Rightarrow r$
           $s \Rightarrow p \vee q$

Solution: (Done by Conditional Inference).

Statements	Reasons
1. s	1. Conditional Premise (hypothesis)
2. $s \Rightarrow u$	2. P(a)
3. u	3. R. D.
4. $u \Rightarrow r$	4. P(c)
5. r	5. R. D.
6. $r \Rightarrow p$	6. P(b)
7. p	7. R. D.
8. $p \lor q$	8. D. A.
9. $s \Rightarrow p \lor q$	9. Rule of Conditional Inference (C. I.)

Q. E. D.

EXAMPLE 9: Show whether the following argument is valid or invalid.

$$\begin{array}{ll}(a) & z \\ (b) & \sim x \lor (s \land y) \\ \hline & x \Rightarrow y \land z \end{array}$$

Solution: We note that the two premises are not even related. If this argument is valid and if this fact can be shown without resort to a 16-row truth table, C. I. is the obvious inference pattern to try to use because the conclusion is an implication whose hypothesis is a single component only.

Statements	Reasons
1. x	1. Hypothesis (C. P.)
2. $\sim x \lor (s \land y)$	2. P(b)
3. $s \land y$	3. D. S.
4. y	4. C. S.
5. z	5. P(a)
6. $y \land z$	6. C. A.
7. $x \Rightarrow y \land z$	7. C. I.

Q. E. D.

At this point it might be well to review the approach we use in showing validity of an unfamiliar argument

$$p_1 \land p_2 \land \cdots \land p_n \Rightarrow q$$

If there are at most three components among the premises, a truth table appealing directly to Definition (2-9) of valid argument can be used. If the argument is too complex for this approach, we look for a premise which consists of a single component. It can then be used as the first proposition in an algebraic proof sequence. Lacking this, a conjunction, if it yields single components by conjunctive simplification, can be used. Lacking either of

§2g   **General Arguments**

these, the Chain Rule with LEDI can frequently be used. If the premises are too complex for these procedures, we can augment them by an additional premise as allowed by a particular inference rule. Then we must obtain the conclusion required by that rule. If the stated conclusion is a single component, indirect inference can be advantageous. If a stated conclusion is an implication, conditional inference usually offers the best approach.

EXERCISE SET II

These exercises are included for your enjoyment and enlightenment. Do them either by C. I. or I. I. If an argument is invalid, show this by assignment of constants. Those who feel the need of further practice exercises can look in §2k or in Chapters II and III, *First Course in Mathematical Logic*.[1]

In Exercises 36 and 37 supply the missing reasons.

36.
1. $\sim p$ — 1. I. P.
2. $\sim p \Rightarrow \sim u \land r$ — 2. P
3. $\sim u \land r$ — 3.
4. $\sim u$ — 4.
5. $u \lor \sim s$ — 5. P
6. $\sim s$ — 6.
7. $r \Rightarrow s$ — 7.
8. $\sim r$ — 8.
9. r — 9.
10. $r \land \sim r$ — 10.
11. p — 11.

37.
1. p — 1. C. P.
2. $p \Rightarrow \sim s$ — 2. P
3. $\sim s$ — 3.
4. $s \lor q$ — 4. P
5. q — 5.
6. $q \lor u$ — 6.
7. $p \Rightarrow q \lor u$ — 7.

38. (a) $p \Rightarrow u \lor s$
    (b) $\sim u$
    $p \Rightarrow s$

39. (a) $p \Rightarrow u \land s$
    (b) $p \lor s$
    s

40. (a) $u \Rightarrow p$
    (b) $\sim u \lor \sim p$
    p

41. (a) $\sim p \Rightarrow q \land r$
    (b) $q \Rightarrow s$
    $\sim p \Rightarrow s$

---

[1]Patrick Suppes and Shirley Hill, *First Course in Mathematical Logic* (New York: Blaisdell Publishing Company, 1964).

42. $p \lor q$
    $\sim p \Rightarrow \sim q$

43. $\sim w \lor s \Rightarrow w$
    w

44. (a) $q \land r \Rightarrow s$
    (b) r
    $q \Rightarrow s$

45. (a) $u \Rightarrow q$
    (b) $q \Rightarrow \sim s$
    $u \Rightarrow s$

46. (a) $p \lor q \Rightarrow r$
    (b) $p \Rightarrow \sim s$
    $s \Rightarrow r$

47. (a) $s \lor w \Rightarrow \sim s$
    (b) $\sim u \lor s$
    u

48. (a) $\sim(p \land q)$
    (b) $\sim r \Rightarrow q$
    (c) $\sim p \Rightarrow r$
    r

49. (a) $\sim p \Rightarrow \sim s$
    (b) $\sim p \lor r$
    (c) $r \Rightarrow \sim u$
    $s \Rightarrow \sim u$

50. (a) $p \Rightarrow \sim s$
    (b) $s \lor \sim r$
    (c) $\sim(u \lor \sim r)$
    $\sim p$

51. (a) $p \lor q$
    (b) $u \Rightarrow \sim p$
    (c) $\sim(q \lor r)$
    $\sim u$

52. (a) $r \Rightarrow \sim q$
    (b) $u \lor q$
    (c) $u \Rightarrow s$
    $\sim s \Rightarrow \sim r$

53. (a) $p \Rightarrow q \lor r$
    (b) $q \Rightarrow \sim p$
    (c) $s \Rightarrow \sim r$
    $\sim(p \land s)$

54. (a) $u \land r \Rightarrow \sim s$
    (b) $\sim s \Rightarrow u$
    (c) $\sim r \Rightarrow \sim s$
    r

55. (a) $s \lor r$
    (b) $s \Rightarrow \sim p$
    (c) $r \Rightarrow q$
    $\sim p \lor q$

56. (a) $p \Rightarrow q$
    (b) $p \Rightarrow (q \Rightarrow r)$
    (c) $q \Rightarrow (r \Rightarrow s)$
    $p \Rightarrow s$

57. (a) $s \Rightarrow (q \lor r)$
    (b) $q \Rightarrow (r \Rightarrow p)$
    (c) $p \lor s$
    $\sim s$

58. (a) $\sim r \lor \sim p$
    (b) $u \lor s \Rightarrow r$
    (c) $p \lor \sim s$
    (d) u
    $\sim(u \lor s)$

*59. (a) $p \Rightarrow q$
    (b) $r \Rightarrow s \lor p$
    (c) $u \lor v \Rightarrow r$
    (d) $\sim q$
    $\sim s \Rightarrow \sim u$

*60. (a) $p \lor q \Rightarrow r \land s$
    $\sim p \lor r$

61. Olin has eaten 13 bananas and a cheeseburger or he is hungry. If Olin has eaten a cheeseburger or is in the kitchen, then he is not hungry. Therefore, if Olin has eaten 13 bananas, he is not in the kitchen.

62. Either the day is not warm or there are many children swimming. The crickets' chirp is high-pitched. If there are many children swimming, then the crickets' chirp is high-pitched and the ants are running fast. Therefore, the day is not warm or the ants are running fast.

§2h QUANTIFIED PROPOSITIONS

From the preceding sections we recall that a proposition is a statement (i.e., declarative sentence) to which one of two truth values is assignable singly according to some given truth standard. Now we consider declarative sentences whose truth value cannot be immediately determined.

EXAMPLE 1: Consider p: $2x = 3$. Is this a proposition?

Solution: Assuming the truth standard to be ordinary arithmetic, we see that if $x = 1$ or many other values, p is false. If $x = \frac{3}{2}$, however, p is true. In fact, we see that, for all meaningful values of x, p is either true or false, but we can not tell which until an actual substitution of the name of a particular number is made. Hence $2x = 3$ is not a proposition.

The reason that the sentence $2x = 3$ could not immediately be labeled true, false, or meaningless is because x is a symbol which stands for any of many things. Such symbols are very useful in mathematics. As was pointed out in §1d, they also occur in everyday speech.

EXAMPLE 2: Consider "She is blond." Is this a proposition?

Solution: Grammatically this is a sentence. It is also declarative. Can we assign it a truth value? The "science" of complexion analysis would seem to provide us with an adequate truth standard. Yet there is obviously a difficulty. The sentence would be true for some persons who could be referred to as "she" and be false for others. The fact that the word *she* refers to many persons makes it impossible to label the sentence true or false immediately. Hence it is not a proposition.

Symbols such as the letter x in Example 1 are called variables. Also, pronouns such as *she* function grammatically as variables.

We understand that a variable *is a symbol which may be replaced by meaningful names of specific things under discussion.*

A particular truth standard will provide us with meaningful names. Then we may further restrict the meaning of a variable by stating that it is to stand only for certain meaningful names of those specified. It would not be meaningful to substitute Mary for x in $2x = 3$ if the truth standard is arithmetic. Likewise, it would not be meaningful to substitute the name *John* for the variable *she*, for this variable does not stand for male beings. Another term which means "variable" is *place holder*. The blank in the sentence "_____ was the president of the United States" is such a place holder. The triangle in the sentence

$$7 = \triangle + 1$$

is also a place holder. The triangle can be filled in with a name, and such a substitution will be meaningful if the substitute is actually the name of a number. For example,

$$7 = \boxed{4} + 1$$

is a meaningful sentence in arithmetic. It is false in grocery arithmetic, of course, but it has exactly one truth value. Hence it is a proposition.

Definition (2–14): An **open sentence** $p(x)$ is a statement which contains a variable x and which becomes a proposition when a *meaningful* name is substituted for the variable.

The symbol $p(x)$ stands for an open sentence in which the variable is x. Other terms for "open sentence" are *propositional function, condition on the variable*, and *propositional form*.

The sentences "$2x = 3$" and "she is blond" in the preceding examples are open sentences. "$2(1) = 3$" and "Mary is blond" are propositions.

EXAMPLE 3: Is "$x = 3$ and $2 = 2$" a proposition if the truth standard is ordinary arithmetic?

Solution: The given sentence is an open sentence. We can not label it true or false until a specific numeral (number name) is substituted for x. If 3 is substituted, the sentence is true; if other numerals are substituted, it is false.

Now let us consider an open sentence $p(x)$ and the meaningful names which may be substituted for x. Since we have defined $p(x)$ to be either true or false when a meaningful name is substituted for x, there are three distinct possibilities.

(i) $p(x)$ **may be true for each meaningful** x.
(ii) $p(x)$ **may be true for at least one meaningful** x.
(iii) $p(x)$ **may be true for no meaningful** x, **i.e., false for all** x.

Definition (2-15): If an open sentence $p(x)$ is true for each meaningful substitution for x, we write "$\forall x, p(x)$." If $p(x)$ is true for at least one x, we write "$\exists x, p(x)$." If $p(x)$ is true for none of the meaningful substitutions for x, we write "$\forall x, \sim p(x)$."

The symbol $\forall$ is called the universal quantifier. It is often translated by the English words "all," "any," "each," and "every." The symbol $\exists$ is called the existential quantifier. It is often translated by the English terms "some," "at least one," "there is," and "exists."

The sentences considered in Examples 1, 2, 3 would all be true if existentially quantified and false if universally quantified.

In spite of the fact that English translations of the quantifiers have been given, no English word is exactly equivalent to either one. Hence the provisions of Definition (2-15) must be learned with care. Note also that with a quantifier the symbol $p(x)$ is to be read "$p(x)$ is true."

An important fact about quantified propositions is that any universally quantified proposition must also be true existentially. Thus we have:

Theorem (2h-1): If $\forall x, p(x)$, then $\exists x, p(x)$.

Note from Definitions (2-14) and (2-15) that if there is no meaningful substitution for the variable x in $p(x)$, then $p(x)$ is not an open sentence and neither $\forall (x)p(x)$ nor $\exists (x)p(x)$ is a proposition.

Consider the following two examples of quantified sentences.

EXAMPLE 4: Is the sentence "$\forall x, 5x - 4 = 11$" a proposition?

Solution: Yes. Substituting 0 for x, we have $5 \cdot 0 - 4 = 11$ which is certainly false. Hence by Definition (2-15), "$\forall x, 5x - 4 = 11$" is false. Therefore, the sentence is a proposition.

EXAMPLE 5: Is the sentence "$\exists x, 5x - 4 = 11$" a proposition?

Solution: Yes. Substituting 3 for x yields the true sentence $5 \cdot (3) - 4 = 11$. Then $\exists x, 5x - 4 = 11$ is true by Definition (2-15). Hence it is a proposition.

These examples should also make the following axiom seem like a reasonable principle.

L-5: If each variable of an open sentence is modified by a quantifier, the entire sentence is a proposition.

Since sentences such as $\forall x, p(x)$ or $\exists x, p(x)$ are propositions by this axiom, we need some way to negate them. If John said

All birds are blue

you would probably say "That is false." Now how would you convince John of his error? To show John that "all birds are blue" is false, you need exhibit only one bird of non-blue color, e.g., a yellow bird. Symbolically, John said "$\forall (x) B(x)$" (all birds are blue), and you said "$\sim [\forall (x) B(x)]$ because $\exists (x) \sim B(x)$." That is, it is false that all birds are blue because there is one yellow bird.

Generalizing this discussion, the negation of a universally quantified proposition is logically equivalent to the existentially quantified negation of the proposition.

EXAMPLE 6: Returning to the open sentence $5x - 4 = 11$, let us modify the variable by each quantifier and do the same for its negation $5x - 4 \neq 11$. Then label each resulting proposition with its correct truth value.

Solution: (i) $\forall x, 5x - 4 = 11$ is false.
            (ii) $\exists x, 5x - 4 = 11$ is true.
            (iii) $\forall x, 5x - 4 \neq 11$ is false.
            (iv) $\exists x, 5x - 4 \neq 11$ is true.

Note that in this statement either (ii) or (iv) might be considered as logical negations of (i), but (iii) cannot.

John's next statement is

Some men are intelligent

You believe that this, also, is false. How do you convince John in this case? To show that "Some men are intelligent" is false you must show symbolically that

No men are intelligent

John said "$\exists (x), I(x)$," and you said "$\sim [\exists (x) I(x)]$ since $\forall (x) \sim I(x)$," i.e., it is false that some men are intelligent because no men are intelligent. [Note by Definition (2-15) that for logical purposes we translate "No men are intelligent" as "All men are not intelligent."]

Generalizing this discussion, the negation of an existentially quantified proposition is logically equivalent to the universally quantified negation of the proposition.

The following definition should seem reasonable.

Definition (2-16):

$$\sim [\exists x, p(x)] \subseteq [\forall x, \sim p(x)]$$

also,

$$\sim [\forall x, p(x)] \subseteq [\exists x, \sim p(x)]$$

EXAMPLE 7: Let q stand for "Everything is green"; write $\sim q$ in verbal form.

Solution: q has the form: $\forall x, x$ is green. The correct negation is

$$\sim (\forall x, x \text{ is green}) \subseteq (\exists x, x \text{ is not green})$$

In verbal form we have "For some x, x is not green" or "something is not green" or "at least one thing is not green."

Up to this point, we have been concerned with arguments involving propositions without variables. How does our previous work apply to quantified propositions? It is possible to change universally quantified propositions to logically equivalent implications. Consider the proposition:

(i)   All cats are felines

You should be able to agree that (i) is equivalent to

(ii)   If this is a cat, then this is a feline

Symbolically, $\forall (c) f(c) \subseteq c(x) \Rightarrow f(x)$. This illustrates the following useful result.

Universally quantified propositions can be changed to an implication having the same variable in both hypothesis and conclusion.

EXAMPLE 8: Write correct negations of "Some summers are sultry."

Solution: Using Definition (2-16) directly we have

(1)   All summers are not sultry

Changing this to an implication, we have

(2)   If this is summer, it is not sultry

Note that the English of (1) is not too clear. By Definition (2-15) we can also write:

(3)   No summers are sultry

The English here is unmistakable. Hence, (3) is the translation we must use instead of (1).

EXAMPLE 9: Write a correct negation of "All cats are black or white" in simplified form.

Solution: The proposition has the form "$\forall x, x$ is a black cat or a white cat." By Definition (2-16) the negation is "$\exists x, \sim(x$ is a black cat or a white cat)." Now, by DeMorgan's Law we have "$\exists x, x$ is not a black cat and x is not a white cat" or "Some cats are not black and some cats are not white." There is another form available. Can you find it? Recall NIMP.

EXERCISE SET

1. Write correct negations of each of the propositions and reduce to simplified form.

   a) Every integer is even. (two ways)
   b) No differentiable functions are not continuous.
   c) There is a book on the shelf. (two ways)
   d) A cat exists which is not green.
   e) Every integer is even and odd.
   f) $\exists p, p \wedge \sim p$
   g) $\forall x, x^2 + 1 = 0$
   h) $\forall p, (p \wedge q \Rightarrow q)$
   i) For all $x, x < 0$
   j) $\exists p, (p \Rightarrow \sim p)$
   k) $\exists p, p$ is false.

2. Determine the truth value for each of the propositions in Exercise 1, (a) through (k). Then determine whether the negation has the opposite truth value.

3. In what four ways may a collection of words fail to be a proposition?

4. Negate the statement "All's well that ends well."

5. Prove or disprove the converse of Theorem (2h-1).

6. Which of the following sentences is not a proposition? Explain. Label each proposition true or false.

   a) $\forall x, x + 1 = x$
   b) $\exists x, x + 1 \neq x$
   c) $\exists x, x + 1 = x$
   d) $\forall x, x + 1 \neq x$
   e) $\exists x, x + y = 0$
   f) $\forall x, \forall y, x + y = 0$
   g) $\forall x, \exists y, x + y = 0$
   h) $\exists x, \exists y, x + y = 0$
   i) $\exists x, \forall y, x + y = 0$
   j) $\forall u, \forall v, \exists w, \exists y, u + v + w + x + y = 0$

7. Which of the following is a proposition? an open sentence? neither? Explain.

   a) $x = 3$ or $4 = 7$
   b) $x^2 = 1$ and $2 = 3$

c) x is an even number
d) Some numbers are even.
e) $\forall x$, x is even
f) If x is even, then x is odd
g) $x = x + 1$ implies x is odd
h) $x = 7 \Rightarrow 3x = 21$
i) Everything is true.

In Exercises 8 through 15, determine if possible whether the arguments are valid or invalid; if it is not possible, explain why.

8. (a) All prime integers greater than two are odd.
   (b) x is a prime integer greater than two.
   $\therefore$, x is an odd integer.

9. (a) All rectangles are parallelograms.
   (b) All squares are rectangles.
   (c) All parallelograms are quadrilaterals.
   $\therefore$, all quadrilaterals are squares.

10. (a) $\forall x, \forall y, \forall z$, if x is less than y and y is less than z, then x is less than z.
    (b) If x is less than 3, then x is less than π.
    (c) Sin π is less than 1.
    (d) 1 is less than 3.
    $\therefore$, sin π is less than π.

11. (a) $\forall a, \forall b, a = b \lor a < b \lor a > b$
    (b) $5 \neq 4$
    $\therefore$, $5 > 4$

12. (a) All integers not divisible by two are odd.
    (b) 13 is an odd integer.
    $\therefore$, 13 is not divisible by 2.

13. (a) $\forall u, \forall v, \forall w, (u - v = w \Rightarrow v + w = u)$
    (b) $\forall u, \forall v, \forall w, u = v \land v = w \Rightarrow u = w$
    (c) $8 - 4 = 3 + 1$
    (d) $3 + 1 = 4$
    $\therefore$, $4 + 4 = 8$

14. (a) $\forall x, \forall y, \forall z, x(yz) = (xy)z$
    (b) $8 \cdot 3 = 24$
    (c) $2 \cdot 4 = 8$
    (d) $12 = 4 \cdot 3$
    (e) $\forall a, \forall b, \forall c, a = b \Rightarrow a \cdot c = b \cdot c$
    (f) $\forall x, \forall y, \forall z, x = y \land y = z \Rightarrow x = z$
    $\therefore$, $2 \cdot 12 = 24$

15. (a) $\forall x, \forall y, \forall z, x < y \wedge y < z \Rightarrow x < z$
    (b) $4 < 7$
    (c) $e < \pi$
    (d) $\pi < 4$
    $\therefore, e < 7$

§2i APPLICATIONS

The inference rules studied in this chapter are the ones most often used in mathematical arguments. For the following two reasons, however, it may be difficult to recognize that a particular rule has been or should be applied to produce a given result.

For one thing, in order to become familiar with the "ground rules," the beginner does logic problems which concern themselves primarily with form. The content of propositions is purely incidental at this stage. In an application it may be difficult to separate the two without some added practice. Secondly, in stating mathematical arguments we tend to concentrate on the principles of the discourse being developed and often fail completely to make specific mention of inference rules. Therefore, to facilitate learning, we shall present steps of arguments in detail and then note the possible pitfalls.

We develop any branch of mathematics as an axiomatic discourse either explicitly or implicitly. Thus our premises in applications are either accepted as true (axioms and definitions) or already proven true (theorems). A false proposition may occasionally be used to obtain a particular result but such a proposition will be specifically labeled false. The important point is that in either the true or the false case each proposition used in an argument has a single specific truth value.

In the context of application, then, the proposition

(1) $$p \longleftrightarrow q$$

tells us as much as the logical sentence

$$p \equiv q$$

If (1) is a definition, axiom, or theorem, it tells us that p, q have the same truth value, and we may substitute either proposition for the other in that discourse. Thus, in applications, we make no formal use of logical equivalence.

In working out deductive arguments, occasional use is made of compounds which we have not studied directly but which can be derived from familiar compounds. Consider, for example,

(2) $$p \wedge q \Rightarrow r$$

The converse of this implication is, of course,

§2i Applications

(3) $$r \Rightarrow p \wedge q$$

In certain cases one of the forms

(4) $$p \wedge r \Rightarrow q$$

or

(5) $$r \wedge q \Rightarrow p$$

is useful. Either of these forms is called a *partial converse* of (2) because only one of the conjoined components of the hypothesis is interchanged with the conclusion.

EXAMPLE 1: If $\mathscr{L}_1, \mathscr{L}_2, \mathscr{L}_3$ are straight lines, write two partial converses of "If $\mathscr{L}_1 // \mathscr{L}_2$ and $\mathscr{L}_1 \perp \mathscr{L}_3$, then $\mathscr{L}_2 \perp \mathscr{L}_3$," and determine the truth value of each proposition if the truth standard is Euclidean geometry.

Solution: The partial converse of form (4) is "If $\mathscr{L}_1 // \mathscr{L}_2$ and $\mathscr{L}_2 \perp \mathscr{L}_3$, then $\mathscr{L}_1 \perp \mathscr{L}_3$." The partial converse of form (5) is "If $\mathscr{L}_2 \perp \mathscr{L}_3$ and $\mathscr{L}_1 \perp \mathscr{L}_3$, then $\mathscr{L}_1 // \mathscr{L}_2$." Both the given proposition and its two partial converses are true in the plane but false in space. See Figure 2.24.

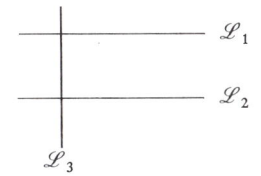

Figure 2.24

Now let us look at some sample proofs in discourses outside logic.

EXAMPLE 2: Prove: If a triangle is equilateral, then it is equiangular.

Solution:

1. $\triangle ABC$ is equilateral.
   1. Hypothesis (C. P.) (restated in terms of some $\triangle ABC$)

2. If $\triangle ABC$ is equilateral, then $\overline{AC} \cong \overline{BC}$
   2. Definition of equilateral

3. $\overline{AC} \cong \overline{BC}$
   3. R. D.

4. If $\overline{AC} \cong \overline{BC}$, then $\triangle ABC$ is isosceles
   4. Definition of isosceles (restated for this $\triangle$)

5. $\triangle ABC$ is isosceles
   5. R. D.

6. If $\triangle ABC$ is isosceles, then $\angle A \cong \angle B$
   6. P. T. (The angles opposite the $\cong$ sides of an isosceles $\triangle$ are also $\cong$.)

7. $\sphericalangle A \cong \sphericalangle B$ 7. R. D.

8. Also $\sphericalangle B \cong \sphericalangle C$ 8. Steps 2-7 repeated using $\overline{AC} \cong \overline{AB}$

9. $\sphericalangle A \cong \sphericalangle B \wedge \sphericalangle B \cong \sphericalangle C$ 9. C. A.

10. If $\sphericalangle A \cong \sphericalangle B \wedge \sphericalangle B \cong \sphericalangle C$, then $\sphericalangle A \cong \sphericalangle B \cong \sphericalangle C$ 10. P. T. (transitive property of $\cong$)

11. $\sphericalangle A \cong \sphericalangle B \cong \sphericalangle C$ 11. R. D.

12. If $\sphericalangle A \cong \sphericalangle B \cong \sphericalangle C$, then $\triangle ABC$ is equiangular   12. Definition of equiangular (restated for this $\triangle$)

13. $\triangle ABC$ is equiangular   13. R. D.

14. If a triangle is equilateral, then it is equiangular   14. C.I.

Q. E. D.

The statements on the left in this example, all part of the argument, have been restated in order to use notation and terminology convenient to the problem at hand. Each "statement" is to be true unless specifically stated to be false. Each "reason" on the right states the principle by which we know its "statement" is true. Thus each "reason" asserts that its "statement" is:

1. a known true proposition
    (a) from the discourse, i.e.,
        (i) an axiom
        (ii) a definition (Steps 2, 4, 12 in Example 2 above)
        (iii) a previously proved theorem (Steps 6, 10),
    (b) from logic, i.e.,
        (i) a tautology;
        (ii) a logical equivalence;
2. a premise allowed by a particular inference rule (Step 1);
3. a conclusion allowed by a particular inference rule (Steps 3, 5, 7, 9, 11, 13, 14).

Inclusion of much painstaking detail can be quite tiresome; therefore, various steps are often omitted from proofs depending on what the writer feels should be obvious to the reader. In most proofs in even fairly elementary texts the inference steps (3, 5, 7, 9, 11, 13, 14) of this example would be omitted. In succeeding sections we shall also begin to follow this practice. For the more advanced, several more steps could be left out, leaving the reader to supply some important links in the argument. However, the beginner usually appreciates seeing considerable detail in a few early examples before pushing on toward maximum abbreviation. Since the use of deduction as an essential method is one of the things which sets mathematics apart from other sciences, it is reasonable to put some effort into understanding the principles which we so often gloss over.

Stress on deduction should not blind us to the fact that inductive reasoning leads the mathematician to many worthwhile conjectures which he then attempts to prove deductively. As an elementary example, let us consider such number sums as $2 + 2, 2 + 4, 4 + 6$. We can see that in each of these cases the sum of two even numbers is even. Is it also possible to deduce this principle for all sums of even numbers? If so, a definition of "even" and "odd" will be needed.

Definition (2–17): An integer* a is **even** iff $a = 2k$ where k is also an integer. An integer b is **odd** iff $b = 2k + 1$ where k is also an integer.

In addition to this defininion, axioms and theorems for numbers would be needed. In §1f a list of axioms specifically for real numbers is given. However, all the axioms that we want to use here hold for integers as well as for real numbers. When one of the axioms is used, its reference number will be given so that you can look up its statement. We now state and attempt to prove our conjecture.

EXAMPLE 3: Prove: If the integers a, b are both even, then the sum $a + b$ is also even.

Proof:

1. a, b are even integers
2. If a, b are even, then $a = 2k_1$, $b = 2k_2$ for some integers k_1, k_2
3. $a + b = 2k_1 + 2k_2$
4. $2k_1 + 2k_2 = 2(k_1 + k_2)$
5. $a + b = 2(k_1 + k_2)$
6. If k_1, k_2 are integers, then $(k_1 + k_2)$ is an integer
7. $k_1 + k_2$ is an integer
8. If $a + b$ is an integer such that $a + b = 2(k_1 + k_2)$ where $k_1 + k_2$ is an integer, then $a + b$ is even

1. C. P.
2. Definition (2-17)
3. Uniqueness of Sum (R-2)
4. Distributive Property (R-12)
5. Transitivity of Equality (E-3)
6. Closure Property (R-1)
7. R. D. on (2, 6)
8. Definition (2-17) restated for this problem

---

*See §3a for a statement describing the integers.

Elementary Logic

9. $a + b$ is even
10. If a, b are even, then $a + b$ is even

9. R. D. on (5, 7, 8)
10. C. I.

Q. E. D.

If we consider such combinations as $2 + 3$ and $3 + 3$, we are led to two more conjectures which are stated in Exercises 9 and 10, respectively. Similarly, consideration of products such as $2 \cdot 2 = 4$, $4 \cdot 6 = 24$, etc., should lead us to the proposition stated in the following example.

EXAMPLE 4: Prove: If a, b are even integers, then the product $a \cdot b$ will be an even integer.

Proof:

1. a, b are even integers
2. If a, b are even integers, then $a = 2k_1, b = 2k_2$ for integers k_1 and k_2
3. $a = 2k_1, b = 2k_2$
4. $a \cdot b = (2k_1)(2k_2)$
5. $(2k_1)(2k_2) = 2[k_1(2k_2)]$
6. $a \cdot b = 2[k_1(2k_2)]$
7. If $2, k_2$ are integers, then $2 \cdot k_2$ is an integer
8. $2k_2$ is an integer
9. If $k_1, (2k_2)$ are integers, then $[k_1(2k_2)]$ is an integer
10. $[k_1(2k_2)]$ is an integer
11. If $a \cdot b = 2[k_1(2k_2)]$ and $k_1(2k_2)$ is an integer, then $a \cdot b$ is even
12. $a \cdot b$ is even
13. If a, b are even integers, then the product $a \cdot b$ is an even integer

1. C. P. (Hypothesis)
2. Definition (2-17)
3. R. D.
4. Uniqueness of Product (R-8)
5. Associative Property (R-10)
6. Transitive Property (E-3)
7. Closure Property (R-7)
8. R. D. on (7, 2)
9. Closure Property (R-7)
10. R. D. on (2, 8, 9)
11. Definition (2-17)
12. R. D. on (6, 10, 11)
13. C. I.

Q. E. D.

Consideration of such products as $2 \cdot 3$ and $3 \cdot 5$ would lead to the conjectures stated in Exercises 11 and 12 respectively.

Again it must be admitted that the proofs shown in Examples 3 and 4 could be shortened considerably and still be completely acceptable to mathematicians. A few more examples will be studied in detail in the following chapters.

Since applications of logical principles make use of the format of an axiomatic discourse, it may be well to mention the matter of consistency once more. From §1e we know that the premises used in any application using Aristotelian logic as part of the discourse *must be consistent*. That is, in spite of that fact that inconsistent arguments are classified as valid from the strictly logical point of view, in applications of logic we use only premises which are consistent.

The reason for the requirement of consistency is easy to show using terms which we have developed since the concept was introduced in Chapter 1. Suppose we were to allow the contradictory premises p, $\sim p$ in some discourse. From Theorem (2e-3) we know that

$$p \wedge \sim p \subseteq f$$

which is any false proposition. Now note that the argument

$$p \wedge \sim p \Rightarrow q$$

is valid and that q is any proposition whatever.

Unless we are very careful to assure consistent premises in an axiomatic discourse, any proposition at all could be proved true.

If we are not careful to assure ourselves that the premises we use in applications are true and that our reasoning is valid, still other difficulties can follow. Only the possibility of reasoning from true premises to a false conclusion has been eliminated by valid reasoning. Recall that an argument having false premises and either a true or a false conclusion is valid as well as an argument having true premises and a true conclusion.

If we assume that it is possible to proceed from any premises to any conclusion by *invalid* reasoning and if we call the conjunction of the premises the hypothesis, then there are eight possible combinations of true (T) or false (F) hypotheses (H) and conclusions (C) with valid (V) or invalid (I) reasoning (R). Figure 2.25 illustrates these cases. Note that this is not a truth

	1.	2.	3.	4.	5.	6.	7.	8.
H	T	T	T	T	F	F	F	F
R	V	V	I	I	V	V	I	I
C	T	F	T	F	T	F	T	F

Figure 2.25

table. As stated, Definition (2-9) eliminates Column 2, but this is the *only* column ruled out.

For example, working from the true premises to a true conclusion does not mean that an argument is validly constructed and, therefore, acceptable. See Column 3.

As another example, people often have a tendency to feel that reaching a true conclusion by valid reasoning somehow guarantees that a premise is true. To show that this is not the case, consider the following argument.

1. $2 = 1$ 1. P
2. $\underline{2 = 1 \Rightarrow 1 = 2}$ 2. E-2 (§1f)
3. $\underline{1 = 2}$ 3. R. D.
4. $3 = 3$ 4. R-2, E-3 (§1f)

In this example a true conclusion was reached by valid steps from a given premise which is known to be false for our number system. Compare Columns 1 and 5 of Figure 2.25. Example 3, §2a, shows another valid argument whose conclusion is true even though the premises are false.

Sometimes we are tempted to feel that arguing from known true premises guarantees a true conclusion. That this is not so can be seen by comparing Columns 1 and 4 of Figure 2.25. We must be careful that our argument is valid. Invalid arguments may lead to amazing conclusions indeed. The following example illustrates what can happen.

    (a) If it is a kitten, it is soft.
    (b) This is soft.
    ∴, this is a kitten.

Premise (a) is true. If whoever is talking is pointing to some feathers, (b) is also true. Nevertheless, the conclusion is false. That the reasoning here is invalid you can verify by recognizing the pattern or constructing its truth table. Compare Column 4 of Figure 2.25.

EXERCISE SET

1. When a pupil is not given full credit on a problem, the teacher often hears, "But teacher, I got the right answer!" Comment on this remark with respect to Columns 3 and 7 of Figure 2.25.

2. Which column in Figure 2.25 is illustrated by the following arguments from §2a?

    a) Example 3   b) Exercise 5

3. a) Rewrite the compound of (2) using $\Rightarrow$ as the only connective.

b) Restate the given proposition of Example 1 using this alternate form.
4. Considering the result obtained in Exercise 3, rewrite "If a is even, then b is even only if $a \cdot b$ is even" in a logically equivalent form.
5. Prove the following valid or invalid.
$$\frac{q \Rightarrow (\sim p \vee r)}{p \Rightarrow (q \Rightarrow r)}$$
6. A theorem of Euclidean geometry states, "The alternate interior angles determined by a transversal on a pair of lines are congruent iff the lines are parallel." Is it possible to have the lines intersect and have the angles congruent? Explain.
7. Write two partial converses of the proposition "If two triangles have congruent bases and congruent altitudes, then they have equal areas." Determine the truth values of each of the three propositions.
8. Write two partial converses of the proposition "If a line is $\perp$ a radius at its point on the circle, it is tangent to the circle." Determine the truth value of each of the three propositions.

In Exercises 9 through 12 prove the stated conjecture, showing use of required inference rules.

9. If a is an even integer and b is an odd integer, then $(a + b)$ is an odd integer.
10. If a, b are odd integers, then the sum $a + b$ is an even integer.
11. If a is even and b is odd, then the product $a \cdot b$ is even.
12. If a, b are both odd, then the product $a \cdot b$ is odd.
13. If a^2 is an even integer, then a is an even integer.

§2j THE ALGEBRA OF PROPOSITIONS

In §2c, §2d, and §2e many facts about conjunction, disjunction, and negation were proved. If we imagine that $\vee$, $\wedge$, and $\sim$ relative to $\underset{\thicksim}{\in}$ correspond to the familiar number operations $+$, $\times$, and $-$ relative to $=$, some of the properties would look familiar when compared with number properties, others would not. For example, Theorem (2e-13) would read $x + y = y + x$ for all numbers x, y. However, there is no number property which would state "For all numbers $x, x \cdot x = x$," which would correspond to the result of Exercise 9f, §2c.

The properties of $\vee$, $\wedge$, and $\sim$ relative to $\underset{\thicksim}{\in}$ form the kind of mathematical structure which we call an "algebra." To aid in comparing this algebra of propositions with the algebra of numbers and other algebras, we list the basic properties of the algebra of propositions here. In all the statements, p, q, r are propositions, t is a true proposition, and f is any false proposition.

The following are the basic properties of this algebra of propositions.

A-1: For all p, q there is a proposition x such that $p \wedge q \equiv x$. §2c

A-2: For all p, q there is a proposition x such that $p \vee q \equiv x$. §2c

A-3: For the operation $\wedge$, $p \wedge t \equiv p$ for all p. Ex. 9, §2c

A-4: For the operation $\vee$, $p \vee f \equiv p$ for all p. Ex. 9, §2c

A-5: For all $p, p \wedge \sim p \equiv f$. Ex. 9, §2c

A-6: For all $p, p \vee \sim p \equiv t$. Ex. 9, §2c

A-7: For all $p, q, p \wedge q \equiv q \wedge p$. (2e-14)

A-8: For all $p, q, p \vee q \equiv q \vee p$. (2e-13)

A-9: For all $p, q, r, (p \wedge q) \wedge r \equiv p \wedge (q \wedge r)$. Ex. 27b, §2e

A-10: For all $p, q, r, (p \vee q) \vee r \equiv p \vee (q \vee r)$. Ex. 27a, §2e

A-11: For all $p, q, r, p \wedge (q \vee r) \equiv (p \wedge q) \vee (p \wedge r)$. Ex. 27c, §2e

A-12: For all $p, q, r, p \vee (q \wedge r) \equiv (p \vee q) \wedge (p \vee r)$. Ex. 27d, §2e

A-13: For each $p, q, r, p \equiv q \Rightarrow p \wedge r \equiv q \wedge r$.

A-14: For each $p, q, r, p \equiv q \Rightarrow p \vee r \equiv q \vee r$.

A-1 and A-2 are called *closure* properties for $\wedge$ and $\vee$. A-3 states that $\wedge$ has the *neutral (identity) element* t, and A-4 that $\vee$ has the neutral element f. A-5 and A-6 state that each proposition p has an *inverse* $\sim p$ for $\wedge$ and for $\vee$.

A-7 and A-8 state that each operation is *commutative*. Note that commutativity involves compounds of only two components with one operation and states that the order of combination does not affect the result.

A-9 and A-10 are called the *associative* properties of conjunction and disjunction respectively. Note that the associative pattern consists of three components compounded by one operation and states that the *grouping* of the component propositions does not affect the result.

A-11 and A-12 are called respectively the *distributive property of conjunction over disjunction* and the *distributive property of disjunction over conjunction*. Note that distributivity makes a statement about three components compounded with *two* operations. Thus it is recognizably different from either commutativity or associativity.

A-13 states that conjoined compounds are *uniquely (well) defined*. A-14 states that disjoined compounds are *uniquely defined*.

Using these properties, many other properties of the algebra of propositions can be proved without truth tables, i.e., algebraically. Some of them follow:

A-15: For all p, q, $\sim(p \wedge q) \subseteq \sim p \vee \sim q$. (2e-10)

A-16: For all p, q, $\sim(p \vee q) \subseteq \sim p \wedge \sim q$. (2e-11)

These are *DeMorgan's* laws.

A-17: For all p, $p \wedge p \subseteq p$. Ex. 9, §2c

A-18: For all p, $p \vee p \subseteq p$. Ex. 9, §2c

These are called *idempotent* properties.

A-19: $\sim(\sim p) \subseteq p$. (2e-9)

This is the Law of *Double Negation*.

Properties resulting from use of $\Rightarrow$, $\Longleftrightarrow$, or $\underline{\vee}$ are not usually considered as part of the algebra of propositions. However, all the properties of $\Rightarrow$, for example, can be worked out algebraically by using the properties above and LEDI.

EXERCISE SET

1. Using the correspondence of operations suggested in the first paragraph of this section, determine which of the properties A-1 through A-19 do and which do not correspond to properties of the algebra of numbers as given by the axioms of §1f and other familiar number principles.
2. Using A-15 and A-19, prove A-16.
3. Using LEDI and A-12, prove that
$$(p \Rightarrow q) \wedge (p \Rightarrow r) \subseteq (p \Rightarrow q \wedge r)$$

§2k SUMMARY OF CHAPTERS 1 AND 2

The very ancient peoples developed mathematical knowledge primarily by inductive methods. Inductive reasoning is certainly useful in helping us determine what is plausible, but it can lead to errors. The Greeks who inherited the work of the more ancient cultures developed much additional content. They also developed the science of deductive reasoning including the principles for constructing valid arguments.

Although the conclusions of deductive reasoning are compelling once the premises are accepted, a special framework is needed to assure that in each discourse the deductive results can with certainty be labeled true. Such a system has the following parts:

(i) Primitive (undefined) terms—These are specialized terms whose properties are given implicitly by the axioms. They remain undefined in the explicit sense in order to avoid circularity.

(ii) Axioms—These are chosen to form the basic truth standard of the discourse. They state the properties of the primitive terms. They are arbitrarily accepted as true because there are no previous true propositions in the system by which to prove them. They must be consistent with each other.

(iii) Definitions—These are propositions which introduce desired new specialized terms into the discourse by stating their properties explicitly. These, too, are accepted as true. They must (a) be consistent with the axioms; (b) avoid circularity; and (c) be reversible. To avoid circularity, the only specialized terms which may be used in a definition are primitive terms and previously defined terms. Special requirements of construction must be observed to make them usable.

(iv) A logic—A system of rules for organizing arguments into valid form for the purpose of extending the theory of a discourse. Aristotelian logic is the one most often used.

(v) Theorems—These are initial conjectures reclassified as true once they are shown to be conclusions of valid deductive arguments. Each step of such an argument is justified by quoting a principle of logic or a principle already known to be true in that discourse. A final statement asserts the fact that the conclusion is implied by the axioms of that discourse (Q. E. D. or other appropriate form).

We no longer regard axioms as obvious or absolute truths nor do we feel that the propositions of a discourse state actual properties of the real world. If the observed properties of the real world agree with properties of some mathematical system to within needed limits of accuracy, we say that the physical system is a model of the mathematical structure. We can then use the latter to help us analyze problems from "reality."

No determination of the truth of a statement is possible unless the arbitrary truth standard for a particular discourse is available. Thus, a statement which is true in one discourse may be false or meaningless in another. For this reason a logic can be applied to various discourses only if it concerns itself with the form or structure of arguments rather than with the content of propositions.

The converse of an axiom or a theorem is true only if stated as a separate axiom or proved separately as a theorem. Since a definition is to be reversible, both its positive and its converse forms are to be regarded as true.

§2k   *Summary of Chapters 1 and 2*   107

The initial truth standard (set of premises) of the discourse is the set of axioms. At any other point in the development the total truth standard consists of the axioms, definitions, and previously proved theorems of that discourse. These propositions are then premises for succeeding deductive arguments.

EXERCISE SET

Miscellaneous Exercises.

1. The names of a few men who are important in the development of mathematics have been mentioned in these two chapters. Write a short paper on one of them discussing his importance to "modern" mathematics in particular.
2. What abstract property is shared by arms, legs, eyes, ears of a normal human being?
3. Are the conclusions of Gallup polls arrived at by inductive or deductive reasoning?
4. Is the following proposition true or false? Primitive terms are called primitive because they are the first terms with which we work.
5. Look up the word "point" in a dictionary. Is the definition circular or noncircular? Explain.
6. State whether each of the following is a good definition according to the principles stated in §2d. Explain how the principles are satisfied.
    a) (2-3)   b) (2-10)
7. If the following lists give the truth value columns for some compounds in usual order, write the compounds in as simple a form as possible.
    a) F T T T   b) T F F T
    c) F T F F   d) F F T F
    e) F F F T   f) T T T T
    g) T T T F   h) T T F F
    i) T F T F
8. Write the converse, inverse, and contrapositive of the given proposition.
    a) It snows only if it is cold.
    b) If two line segments are not similar, then they are congruent.
9. Is the proposition "If each of two numbers is even, then their sum is even" reversible? Explain.

10. Construct truth tables for the following:
    a) $\sim(\sim p \land \sim q) \Leftrightarrow r$
    b) $[(p \Rightarrow q) \land \sim p] \Rightarrow \sim q$
    c) $(p \land q) \land (\sim p \lor \sim q)$
    d) $[(p \Rightarrow q) \land (q \Rightarrow r)] \Rightarrow (p \Rightarrow r)$

11. Prove or disprove each proposition.
    a) $p \lor q \Rightarrow p \land q$ is a tautology
    b) $p \land q \Rightarrow p \lor q$ is a tautology
    c) $p \lor q \Leftrightarrow p \land q$ is a tautology
    d) $[(\sim p \Rightarrow q) \land (\sim r \Rightarrow p)] \Leftrightarrow [\sim(p \lor r) \Rightarrow q]$ is a tautology.
    e) $[(\sim p \lor q) \Rightarrow r] \mathrel{\underline{e}} [p \Rightarrow (\sim q \lor r)]$
    f) $[p \Rightarrow (q \land r)] \mathrel{\underline{e}} (\sim r \Rightarrow \sim p) \land q$
    g) $p \lor q \mathrel{\underline{e}} p \lor (\sim p \land q)$
    h) $p \land (q \lor p) \mathrel{\underline{e}} p$
    i) $p \mathrel{\underline{e}} (p \land q) \lor p$
    j) $\sim(p \Rightarrow q) \mathrel{\underline{e}} p \Rightarrow \sim q$

12. From the Rule of Detachment and LEDI prove the Rule of Disjunctive Simplification.

13. Assuming usual context, determine if possible which of the sentences are true and which are false. If not possible, state why.
    a) If $2 + 3 = 5$, then $4 = 4$.
    b) If $17 - 5 = 3$, then $4 \cdot (7 + 2) = 28$.
    c) If triangles are kites, then whales are fish.
    d) If $x = 3$, then $2x = 6$.
    e) Kittens are crocodiles and grass is green.
    f) Kittens are crocodiles or grass is green.
    g) $x = 3$ or grass is not green.

14. Write correct negations of each of the following propositions in simplified form.
    a) Some dogs are lazy (2 forms).
    b) All cats are animals (2 forms).
    c) Some Berthas are birds.
    d) If some apples have worms then all apple sauce is green.
    e) All burbot fish equal 100,000 and some astonished men equal 1,000,000.
    f) All dogs are dignified if and only if some kittens are cute.
    g) $8 \leq 2$
    h) All prime numbers are odd.
    i) $\exists x, 3 < x < 5$

15. Determine whether or not each of the following collections of words is a proposition. If not, state why. Assume the most obvious truth standard.
    a) Cindy is Sid's sister.
    b) x is Sid's sister.
    c) $\exists\, x$ and $\exists\, y$, $x + y = y + x$
    d) $7 < -2$
    e) Read this book.
    f) I wish she were here.
    g) $2 \not< 3 \wedge 2 \not> 3 \Rightarrow 2 = 3$
    h) Every number is both even and odd.

In Exercises 16 through 52 determine by any logically allowable method whether the given argument is valid or invalid. Assume universal quantification if any is needed.

16. P(a): Mary is timid if she is not a muskrat.
    P(b): Mary is not timid.
    C: Mary is a muskrat.

17. If $x = 2$, then $y \neq 3$. $x = 2$ iff $y = 3$. Therefore, $x \neq 2$.

18. Use the premises of Exercise 17 with "$y = 3$" as conclusion.

19. If Spiros doesn't eat spinach, then Millie drinks milk. Spiros eats spinach or Millie doesn't drink milk. Therefore, Spiros eats spinach.

20. Use the premises of Exercise 19 with "Millie drinks milk" as the conclusion.

21. $\mathscr{L}$ is perpendicular to $\mathscr{M}$ only if $\mathscr{M}$ is not parallel to $\mathscr{N}$. $\mathscr{M}$ is not parallel to $\mathscr{N}$. Therefore, $\mathscr{L}$ is perpendicular to $\mathscr{M}$.

22. If $a = b$ or $b \neq c$, then $a \neq b$. Either $a = b$ or $b = c$. Therefore, $b \neq c$.

23. Use the premises of Exercise 22 with "$a < c$" as conclusion.

24. Use the premises of Exercise 22 with "$a \neq b$" as conclusion.

25. $p \Rightarrow q \wedge r$
    $p \Rightarrow r$

26. $s \vee x$
    $s \vee \sim x$
    s

27. $p \Rightarrow q \wedge s$
    p
    $y \Rightarrow \sim s$
    $\sim y$

28. $x \vee y$
    $y \Rightarrow \sim z$
    z
    $x \vee \sim z$

29. (a) $p \wedge r \Rightarrow p$

30. (a) $p \Leftrightarrow q$

(b) $\dfrac{\sim r \Rightarrow p}{r}$

31. (a) $p \Leftrightarrow q$
    (b) $\dfrac{\sim p \vee q}{p \wedge q}$

33. (a) $p \Rightarrow q$
    (b) $\dfrac{r \Rightarrow s}{(p \vee r) \Rightarrow (q \vee s)}$

35. (a) $p \Rightarrow q$
    (b) $r \vee \sim q$
    (c) $\dfrac{\sim p \vee \sim r}{\sim p}$

37. (a) $p \Rightarrow q$
    (b) $r \Rightarrow s$
    (c) $\dfrac{p \vee \sim s}{q \vee r}$

39. (a) $p \Rightarrow q$
    (b) $\sim q \Rightarrow s \vee u$
    (c) $\dfrac{\sim u}{s \vee \sim p}$

41. (a) $r \Rightarrow \sim q$
    (b) $r \vee q$
    (c) $\dfrac{r \Rightarrow \sim p}{p \Rightarrow \sim r}$

43. (a) $\sim p \Rightarrow s$
    (b) $\sim s \vee u$
    (c) $\sim (q \vee p)$
    (d) $\dfrac{\sim q \Rightarrow \sim u}{p}$

45. (a) $s \vee u$
    (b) $\sim s \vee r$
    (c) $\sim r \vee u$
    (d) $\dfrac{\sim u}{s \Rightarrow p}$

47. (a) $p \Rightarrow q$
    (b) $\sim p \Rightarrow r$
    (c) $\sim q \vee s$
    (d) $\dfrac{\sim r \vee q}{\sim (s \vee r) \Rightarrow p \vee q}$

(b) $\dfrac{p \vee q}{p \wedge q}$

32. (a) $\sim p \vee \sim q$
    (b) $\dfrac{s \Rightarrow \sim q}{p \Rightarrow \sim s}$

34. (a) $u \vee (v \wedge w)$
    (b) $\dfrac{u \Rightarrow w}{w}$

36. (a) $p \Rightarrow q$
    (b) $s \Rightarrow u$
    (c) $\dfrac{p \vee u}{q \vee s}$

38. (a) $w \Rightarrow q$
    (b) $(q \wedge p) \Leftrightarrow w$
    (c) $\dfrac{\sim p \Rightarrow w}{p}$

40. (a) $p \vee q \Rightarrow r$
    (b) $s \Rightarrow \sim q$
    (c) $\dfrac{\sim p \wedge s}{r}$

42. (a) $r \vee q \Rightarrow \sim p$
    (b) $s \Rightarrow q \vee \sim p$
    (c) $\dfrac{\sim r \wedge s}{\sim p}$

44. (a) $r \Rightarrow s$
    (b) $u \Rightarrow p \vee r$
    (c) $q \vee s \Rightarrow u$
    (d) $\dfrac{\sim s}{\sim p \Rightarrow \sim q}$

46. (a) $\sim p$
    (b) $q \vee \sim r$
    (c) $q \Leftrightarrow p$
    (d) $\dfrac{u \Rightarrow r}{u \Rightarrow \sim p \wedge q}$

48. (a) $\dfrac{\sim p}{p \Rightarrow q}$

49. We shall close the student lounge or a riot will break out there. A riot will not break out in the student lounge or the student council will succeed. Hence, if the student council does not succeed, then we shall close the student lounge.

50. A necessary condition that we not close the student lounge is that a riot take place there. We shall close the student lounge. Therefore, no riot will take place there.

51. Jack is healthy or he is sick. For him not to be sick it is sufficient that he take pink pills. If he doesn't take pink pills, then he will be dropped from the team. Therefore, if Jack is healthy, then he will be dropped from the team.

52. If it is warm and soft, then it is alive. If it is alive and has four feet, it is a kitten. If it is an animal, then it is soft and has four feet. It is false that it is not an animal or is warm. Therefore, it is a kitten.

Answer the questions asked in each of Exercises 53 and 54.

53. Consider: If Harry hates hotheads, then Ken kicks kids.

   a) Suppose that the given proposition is true and also that Ken indeed kicks kids. What, then, is known about the statement "Harry hates hotheads"?
   b) Suppose it is known that the given proposition is false. What is then known about Harry and what is known about Ken?
   c) Suppose it is known that the given sentence is true and that "Harry hates hotheads" is also true. What is then known about Ken's foot?

*54. Consider: Jim plays golf on Saturdays if his wife is not weeping and Bob is not fishing. Bob fishes Saturdays if his wife is weeping and Jim is not playing golf. On Saturday either Jim plays golf or Bob does not fish.

   a) On a certain Saturday Bob is fishing and Bob's wife is not weeping, but Jim's wife is weeping. What about Jim?
   b) On another Saturday Jim is not golfing. What about Jim's wife? What about Bob? What about Bob's wife?
   c) If Bob's wife is weeping, what about Jim?
   d) Could there be a Saturday when Jim plays golf and Bob fishes and both wives weep?

In Exercises 55 through 58 draw a valid conclusion from each set of premises or show why none is possible. The optimal conclusion uses all premises.

55. (a) Babies are illogical. (b) No one is despised who can manage a crocodile. (c) Illogical persons are despised. (Lewis Carrol)

56. (a) No ducks waltz. (b) No officers ever declined to waltz. (c) All my poultry are ducks. (Lewis Carroll)

57. (a) All of the old articles in this cupboard are cracked. (b) No jug in this cupboard is new. (c) Nothing in this cupboard that is cracked will hold water. (Lewis Carroll)

58. (a) If there is not a sufficient supply of teachers and if there are not enough books in Salem, the students can't make grades good enough to maintain the population of the college. (b) If there is a shortage of rooms in Salem, then there is both a shortage of desks and a shortage of chalk boards. (c) If there is a shortage of desks, then there is an insufficient supply of teachers. (d) If there is a shortage of chalkboards, then there are not enough books. (e) There is a shortage of rooms at Salem.

59. Consider the following axiomatic discourse in which the primitive terms are "row" and "chair."

    Axiom 1: Any two rows have at least one chair in common.
    Axiom 2: Any two rows have at most one chair in common.
    Axiom 3: Each chair is in at least two rows.
    Axiom 4: Each chair is in at most two rows.
    Axiom 5: The total number of rows is four.

    a) Prove: There are exactly six chairs.
    b) Prove: There are exactly three chairs in each row.
    c) Make an application of this system by translating the primitives according to row → committee; chair → member. What do the theorems tell us in these terms?
    d) Make an application of this system by translating the primitives according to row → axiom; chair → primitive terms. What do the theorems tell us in these terms?
    e) Make another application as in c) and d).

3
SETS

§3a THE SET CONCEPT

The term "set" is used in everyday language in such phrases as "a set of books" or "a set of dishes." However, the frequency of occurrence of the concept of a set is somewhat hidden by the fact that words such as "class," "family," "bunch," "collection," "aggregate" are used with other kinds of things to express the same notion. For example, we say "a *bunch* of grapes" instead of "a *set* of grapes."

(1) Intuitively, we think of a set as *a collection of things which are separate and distinct from other things*. Thus, given some thing, we know whether it belongs to that set or does not.

Mathematically, the set concept is also most useful. However, the mathematical usage varies in some important ways from its everyday counterpart. Hence, we do not define the mathematical notion of set explicitly. Instead, we define it implicitly by stating axioms to specify its properties.

Thus our basic primitive terms are *set* (which we usually symbolize by a capital letter, block or script) and *belongs to* (which we indicate by the pitchfork symbol $\in$). We usually use x as the variable which stands for a thing that belongs to set. Thus the sentence

$$x \in S$$

is read "x belongs to the set S" or just "x belongs to S." Sometimes we read it "x is a *member* of S" or "x is an *element* of S." The sentence

$$x \notin S$$

is read "x does *not* belong to S" or "x is not a member of S" or "x is not an element of S." The sentence

$$S \ni y$$

is read "S *contains* y" or "S has element y."

We state the membership idea of (1) for a particular set either by listing the members of the set or by stating an open sentence

$$p(x)$$

When the latter method is used, $p(x)$ then becomes the rule by which we tell which things are members of the set and which are not according to whether $p(x)$ is true or false when particular names are substituted for x. The truth standard for the particular discourse allows certain substitutions for the variable x and excludes others. Allowable substitutions are further restricted whenever we agree to use the variable to stand for certain things and not others.

However, the method of specifying elements of a set by using substitutions into some $p(x)$ can not be used unrestrictedly. Without going into all the difficulties which can arise, let us simply say that we do not want all the collections of things which could possibly be determined in this way to be called sets.

For a true axiomatic discourse the axioms should make a complete specification of all set properties. The axioms stated below do not do this, but they do allow us to begin a development of many of the more important set properties which find frequent use in almost all of mathematics. At the same time we shall have opportunity to illustrate specific application of some of the principles of the logic from the preceding chapter.

Since we use open sentences to determine members of sets, let us recall that we have three truth possibilities resulting from substituting allowable names for x, viz.,

(i) $p(x)$ *may be true for each substitution for* x;
(ii) $p(x)$ *may be true for at least one substitution for* x:
(iii) $p(x)$ *may be true for no substitution for* x.

Corresponding to each of these three possibilities, we state an axiom for sets. We are talking only about those x's admitted for use by a particular truth standard.

S-1: There is a set I such that for each x, $x \in I$.
S-2: For each $p(x)$ there is a set A such that $x \in A \leftrightarrow p(x)$ is true.
S-3: There is a set $\varnothing$ such that for each x, $x \notin \varnothing$.

The set I of S-1 is called the *Universe of Discourse* and provides a set whose members are determined according to Case (i) above. Basically, the members of I for a particular discourse are determined by the axioms for that discourse. Since all possible members in a discussion belong to this set, we have

(2) $$x \in I \subseteq t$$

where t is a true proposition.

Corresponding to Case (iii) above, Axiom S-3 provides for the existence of a particular set to which absolutely nothing belongs. The symbol $\emptyset$ is a letter borrowed from the Danish alphabet. We call it simply the **empty set** or **null set**. Since $x \notin \emptyset$ is true for all $x \in I$, we also have

(3) $\qquad\qquad\qquad x \in \emptyset \longleftrightarrow p(x)$ is false for each x

Each set A of S-2 corresponds to Case (ii) for particular $p(x)$. Thus the following propositions about sets have the stated truth values for $\emptyset$ and any set A in general relative to a particular universe of discourse I.

$\qquad\qquad\qquad x \in I$ is always (i.e., universally) true.
$\qquad\qquad\qquad x \in A$ is sometimes true, sometimes false.
$\qquad\qquad\qquad x \in \emptyset$ is always (i.e., universally) false.

The following definition provides a special name for the set determined by particular $p(x)$ in accordance with S-2.

Definition (3-1): The set A determined by an open sentence $p(x)$ is called the **truth set** or **solution set** of $p(x)$.

The solution set of any $p(x)$ which is always true is the Universe I, and the solution set of $p(x)$ which is always false is the null set $\emptyset$.

EXAMPLE 1: The truth set or solution set of "x was the first constitutional president of the U.S.A." is a set whose only member is "George Washington."

EXAMPLE 2: What is the solution set of the open sentence $x^2 < 0$ where x is a real number?

Solution: For real numbers, $x^2 < 0$ is universally false. Thus the solution set is $\emptyset$.

There are three common ways to designate a set. In accordance with S-2 we may simply provide a symbol for the set and make a statement that its elements satisfy a particular open sentence.

EXAMPLE 3: Let P be the set of presidents of the U.S.A. The open sentence which determines P is

$\qquad p(x)$: x has been president of the U.S.A.

We know, for example, that when "Thomas Jefferson" is substituted for x, $p(x)$ is true. Thus
$$\text{Thomas Jefferson} \in P$$

A second common way of specifying a set is to list the members of the set within braces as, for example, $\{a, b, c\}$. This is called **roster** notation. With this notation we often do not state the open sentence which determines the set. However, we should remember that there is some rule which has determined the elements of the set.

EXAMPLE 4: Make a roster of the solution set determined by $p(x)$: x is one of the first five letters of the English alphabet.

Solution: $\{d, a, e, c, b\}$.

The reader should note that the obvious roster symbol for the empty set is the pair of empty braces

(4) $\qquad\qquad\{\ \}$

The third common designation for a set is the so-called **set builder** notation

(5) $\qquad\qquad \{x \mid p(x)\}$

The notation of the set builder is read "The set of all things x such that $p(x)$ is true." Thus (5) is a symbol for any set determined according to S-2. We now have three different designations for any particular set: capital letter, roster, and set builder.

Note that (5) makes use of a special symbol for "such that." There are two current symbols for this phrase:

(i) $\quad \cdot \ni \cdot \quad$ A dot, a left pitchfork, and a dot;
(ii) $\quad | \quad$ A vertical bar, used *only* within set braces.

Sometimes a problem will specify that only a portion of a particular set is to be used to obtain substitutions for the variable x of an open sentence $p(x)$. In such a case the following terminology is pertinent.

Definition (3-2): The set of things whose names may be substituted for the variable x of a $p(x)$ in determining a solution set is called the **replacement set** or **scope** of x.

The scope or replacement set of the variable x is sometimes indicated in conjuction with $p(x)$. For a scope S, (5) would then be written

(6) $\qquad\qquad \{x \mid p(x) \land x \in S\}$

In many cases the entire universal Set I will be the scope of a variable. When a replacement set contains a single element, or when we are speaking about a particular element of a specific set, we call that element a *constant*.

EXAMPLE 5: Make a roster of "x is a block capital letter made with curved lines."

Solution: The scope is the set of 26 letters of the alphabet. The roster is $\{B, C, D, G, J, O, P, Q, R, S, U\}$.

Since the concepts of solution set and scope find considerable use with sets of numbers, informal descriptions of some of the most important number sets follow.

§3a   *The Set Concept*   117

(i)   Natural numbers: $\mathcal{N}: \{1, 2, 3, 4, 5, \ldots\}$.
(ii) Whole numbers: $\mathcal{W}: \{0, 1, 2, 3, 4, \ldots\}$.
(iii) Integers: $\mathcal{L}: \{\ldots, -4, -3, -2, -1, 0, 1, 2, \ldots\}$.
(iv) Rationals: $\mathcal{Q}: \left\{\dfrac{a}{b} \,\middle|\, b \neq 0 \wedge a, b \in \mathcal{L}\right\}$, i.e., a rational number is one which can be written as the ratio of two integers.

Some numbers such as π, $\sqrt{2}$, sin 51°, and others cannot be written as the ratio of two integers. These are called irrational numbers and can be symbolized as $\mathcal{Q}'$.

(v)   Real numbers: $\mathcal{R}: \{x \mid x \text{ is rational} \vee x \text{ is irrational}\}$. Some sample real numbers are $\{0, -\sqrt{2}, \frac{37}{4}, \frac{44}{71}, \log 3\}$.

From this point on the script letters $\mathcal{N}$, $\mathcal{L}$, $\mathcal{Q}$, $\mathcal{R}$ stand, respectively, for the natural numbers, the integers, the rational numbers, and the real numbers.

EXAMPLE 6: Rewrite $\{x \mid x^2 + 5x + 6 = 0 \wedge x \text{ is a real number}\}$ in roster notation.

Solution: $\{-2, -3\}$. This is the solution set of the open sentence $x^2 + 5x + 6 = 0$. The set $\mathcal{R}$ of real numbers is the scope of x.

Some people find that the following diagram is helpful in visualizing relationships among these number sets.

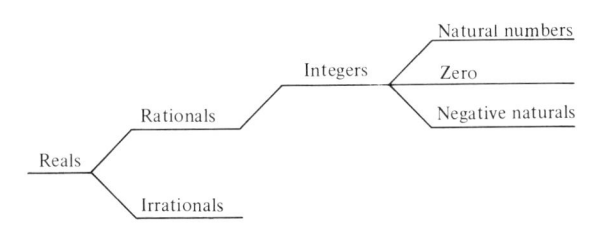

```
                                              Natural numbers
                              Integers    <   Zero
               Rationals  <                   Negative naturals
  Reals    <
               Irrationals
```

Figure 3.1

EXAMPLE 7: List some solution sets determined by typical open sentences $p(x)$ where x has the two given scopes $\mathcal{R}$, $\mathcal{N}$.

Solution:

$p(x)$	solution set on $\mathcal{R}$	solution set on $\mathcal{N}$
$(x-3)(x+2) = 0$	$\{3, -2\}$	$\{3\}$
$x^2 = x$	$\{1, 0\}$	$\{1\}$
$x = x$	$\mathcal{R}$	$\mathcal{N}$
$x = x + 1$	$\varnothing$	$\varnothing$

The examples thus far have specified the rule $p(x)$ which determines the elements of a set. Often in mathematics the reverse is true: that is, we know the elements and need to find the rule.

EXAMPLE 8: Find an open sentence which determines membership in the set $A = \{2, 4, 8, 16\}$.

Solution: One open sentence which will determine this set is $x = 2^n$ and $n \in \mathcal{N}$ and $n < 5$.

EXERCISE SET

1. List some of the everyday synonyms for *set* and for each give an example of a thing usually considered to belong to the set designated by that word. Example: bunch—grapes.

2. Let P be the set of presidents of the U.S. Answer true or false.
   a) Harry S. Truman $\in P$
   b) Abraham Lincoln $\in P$
   c) General Eisenhower $\notin P$
   d) General Patton $\in P$

3. List the elements of each set described (i.e., use roster notation).
   a) the set of vowels in the English language
   b) the set of positive even numbers less than 2
   c) the set of New England states in the U.S.
   d) the set of natural numbers less than 7

4. State a property or condition which determines membership in each set. The sets are not meant to be exhaustive in all cases.
   a) {air, water, fire, earth}
   b) {1, 4, 9, 16}
   c) $\{\frac{1}{1}, \frac{1}{2}, \frac{1}{3}, \frac{1}{4}, \frac{1}{5}\}$
   d) {2, 4, 6, 8, 10}
   e) {pencil, pen, crayon chalk, ... }
   f) {Puerto Rico, Samoa, Virgin Islands, Guam, ... }
   g) {Shell, Socony, Gulf, Sinclair, ... }
   h) {spoon, fork, knife, ... }.

5. Make a plausible sentence containing these terms: like, maple, tree, sunshine, dog.

6. Using roster notation, indicate at least four elements of each set. If there are five or fewer elements in the set, make a complete roster.
   a) the presidents of the U.S.
   b) the playing positions on a baseball field
   c) states bordering on the Pacific

§3b Relations Between Sets

d) mountains higher than Mt. Everest
e) national emblems which have colors red, white, and blue
f) $\{x \mid x$ is the name of a state in the U.S. beginning with the letter $C\}$

7. Which of the following are empty sets?
   a) voters in the U.S. under 21 years of age
   b) all odd numbers ending in 6
   c) all integers which are perfect squares and also have the unit digit 2
   d) the months of the year which begin with the letter L
   e) presidents of the U.S. under 35 years of age

8. Make a roster of each solution set.
   a) $\{x \mid x + 5 = 7 \wedge x \in \mathcal{N}\}$
   b) $\{z \mid 15 - z = 9 \wedge z \in \mathcal{Z}\}$
   c) $\{x \mid -3 < x < 8 \wedge x \in \mathcal{N}\}$
   d) $\{x \mid -3 < x < 8 \wedge x \in \mathcal{Z}\}$
   e) $\{x \mid x^2 + 10x + 25 = 0 \wedge x \in \mathcal{N}\}$
   f) $\{x \mid x - 3 = 7 \wedge x \in \mathcal{Z}\}$
   g) $\{x \mid 3x + 28 = \pi \wedge x \in \mathcal{R}\}$
   h) $\left\{x \mid \dfrac{x}{5} - 7 = 2 \wedge x \in \mathcal{Z}\right\}$
   i) $\{x \mid x^2 - 9 = 0 \wedge x \in \mathcal{N}\}$

9. Which of the following are ways of writing the empty set? Explain.
   a) $\{x \mid x$ is negative $\wedge x \in \mathcal{N}\}$
   b) $\{0\}$
   c) $\{\varnothing\}$

§3b RELATIONS BETWEEN SETS

Given any two sets A, B in a specific universe, what are all the possible ways that they could be arranged with relation to each other? Calling upon a geometric *model* can be helpful in answering this question. The idea of using circles to represent sets containing specific elements was first used by Leonhard Euler (1707–1783). In 1876 the method was extended by John Venn to cover any general case.

In using such set diagrams, let a box or "square" labeled I denote the universe of the discourse and "circles" inside the square represent the sets under discussion. Each such circle is considered to *enclose* the elements of the set it represents.

Now let sets A, B be represented by circles and the set I of the axioms be represented by the Euclidean plane. Our question becomes "What are all the possible ways two circles may be related in the plane?"

First: The two circles could be separate. Now keep B fixed and move A toward B.

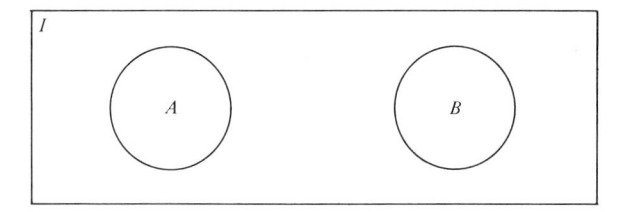

Figure 3.2

Second: The two circles enclose a common region. Now continue moving A.

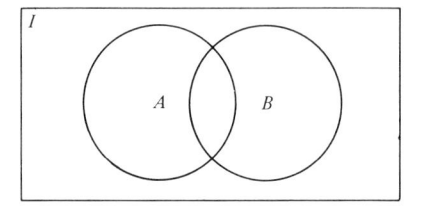

Figure 3.3

Third: The A circle could be entirely within the B circle. This can occur when the A circle has a smaller radius than the B circle, or

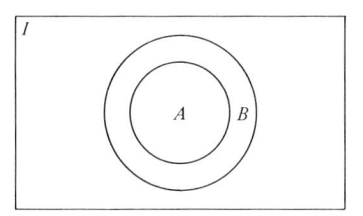

Figure 3.4

Fourth: The A circle and the B circle may coincide. This can occur when the radii of the A circle and the B circle are congruent.

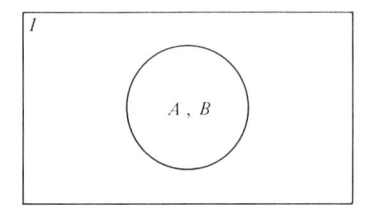

Figure 3.5

§3b   **Relations Between Sets**   121

Now if you continue moving A you will see that the next possible situation is the same as the second, and if you continue further you will repeat the first. Thus you should conclude that there are four possible ways two circles may be related in the plane. Now carrying this analogy back to

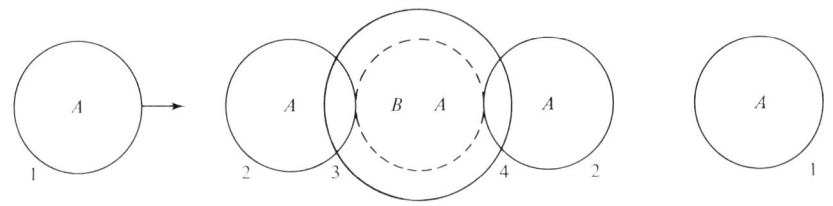

Figure 3.6

the original question, there are four possible ways that the two sets could be arranged in relation to each other. This prompts several definitions. The first two are given informally.

(1)   Two sets A, B are **disjoint** iff they have no elements in common.

Figure 3.2 illustrates disjoint sets. In §3f we shall formally define this concept using notions introduced in the intervening material.

(2)   Two sets A, B **overlap** iff *some but not all* of the elements of A are elements of B and *some but not all* elements of B are elements of A.

Figure 3.3 illustrates overlapping sets. Note that according to this statement the sets A, B of Figures 3.4, 3.5 do not overlap.

EXAMPLE 1:   Represent the sets $\{5, 2, 1\}$ and $\{1, 3, 5, 7, 9\}$ in a single set diagram.

Solution: These sets are overlapping because they have some elements in common, and each set contains at least one element which does not belong to the other. See (2). They may be drawn as shown in Figure 3.7.

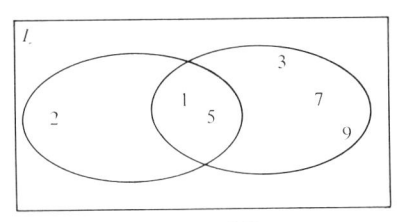

Figure 3.7

The relationship shown in Figure 3.4 where all elements of A are also elements of B is formalized as follows.

Definition (3-3):   For any two sets A, B, A is a **subset** of B (denoted $A \subset B$) iff all elements of A are elements of B; i.e.,

$$A \subset B \text{ iff } (x \in A \Rightarrow x \in B)$$

When $A \subset B$ we may also say "A is *included in B.*" Sometimes it is convenient to write $B \supset A$. Then we say "B is a **superset** of A" or "B *includes A.*" In Figure 3.4, for example, B is a superset of A.

EXAMPLE 2: Is $\{\alpha, \beta\}$ a subset of $\{\alpha, \pi, \beta\}$?

Solution: (i) $\alpha \in \{\alpha, \beta\} \Rightarrow \alpha \in \{\alpha, \pi, \beta\}$ is true;
(ii) $\beta \in \{\alpha, \beta\} \Rightarrow \beta \in \{\alpha, \pi, \beta\}$ is true.
Thus $\{\alpha, \beta\}$ is a subset of $\{\alpha, \pi, \beta\}$ by the definition.

The relationship illustrated in Figure 3.5 is formalized as follows:

Definition (3-4): For any two sets A, B, A **equals** B (denoted $A = B$) iff all ellements of A are elements of B and vice versa; i.e.,

$$A = B \text{ iff } (x \in A \Longleftrightarrow x \in B).$$

The symbol "$=$" in combination with the empty set is verbally a little awkward. The sentence "$S = \varnothing$" is read "S is the null set" or "S is empty."

EXAMPLE 3: Are $\{1, 2\}$, $\{2, 1\}$ equal sets?

Solution: $1 \in \{1, 2\} \Rightarrow 1 \in \{2, 1\}$ is true;
$2 \in \{1, 2\} \Rightarrow 2 \in \{2, 1\}$ is true;
$1 \in \{2, 1\} \Rightarrow 1 \in \{1, 2\}$ is true;
and $2 \in \{2, 1\} \Rightarrow 2 \in \{1, 2\}$ is true.
Thus $\{1, 2\} = \{2, 1\}$ by (3-4).

It is possible that other relationships could be specified based on Figures 3.2 through 3.5, but the four we have mentioned will serve our purpose well for the time being.

Now recall that by our axiom S-1

$$x \in I$$

is true for all meaningful x in the discourse. Thus the implication

$$x \in A \Rightarrow x \in I$$

is true for each set A, whence

$$A \subset I$$

is also true by Definition (3-3). We have proved the following proposition.

Theorem (3b-1): For each A, $A \subset I$.

Returning to Axiom S-2, let us recall that for a particular set A determined by some $p(x)$ we have

$$p(x) \text{ is true} \Longleftrightarrow x \in A$$

Since the bicondition is true here, its components are actually logically equivalent. Thus either may be substituted for the other at any time. In

particular we may write
$$A = \{x|\, p(x)\}$$
as

(3) $$A = \{x \mid x \in A\}$$

EXAMPLE 4: Using (3), we can rewrite the result of Example 6, §3a, as
$$\{x \mid x^2 + 5x + 6 = 0 \wedge x \in \mathcal{R}\} = \{-2, -3\}$$

EXAMPLE 5: Let $I = \{1, 2, 3, 4, 5\}$, $A = \{x \mid x \in I \wedge x < 1\}$, $B = \{x \mid x \in I \wedge x \notin A\}$. How are sets A, B, I related?

Solution: Since $x < 1$ is false for each element of I, $A = \{\ \}$. Thus $x \notin A$ is always true for which reason $B = \{x \mid x \in I\}$; by (3) we then have $B = I$. (See Figure 3.8.) Also, since
$$x \in A \Rightarrow x \in B$$
is true, we have $A \subset B$ and $A \subset I$.

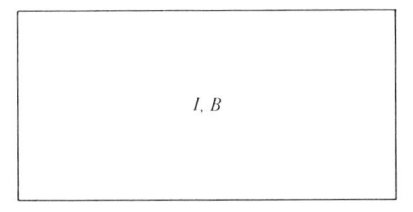

I, B

Figure 3.8

Still another useful method of comparing sets involves the notion of one-to-one correspondence.

Definition (3-5): The elements of two sets are in **one-to-one correspondence** iff each element of one set is paired with exactly one (i.e., at least one and at most one) element of the other and vice versa.

Definition (3-6): Two sets are **matched** (denoted "$\leftrightarrows$") or **equivalent** iff their *elements* are in one-to-one correspondence.

EXAMPLE 6: Are the sets $\{\Delta, \beta, \alpha\}$ and $\{1, 2, 3\}$ matched?

Solution: We can pair Δ to 2, β to 3, and α to 1, thus producing a one-to-one correspondence by Definition (3-5). Therefore, $\{\Delta, \beta, a\} \leftrightarrows \{1, 2, 3\}$ by Definition (3-6).

Intuitively we see that *finite* sets having the same number of elements are matched.

EXAMPLE 7: Compare the sets $\{x, y\}$, $\{a, x, y\}$.

Solution: $\{x, y\} \neq \{a, x, y\}$; $\{a, x, y\} \neq \{x, y\}$; $\{x, y\} \subset \{a, x, y\}$; $\{a, x, y\} \not\subset \{x, y\}$; the two sets are not disjoint; the two sets are nonoverlapping; also $\{x, y\} \not\leftrightarrow \{a, x, y\}$.

Using the concepts developed in this section, there is a somewhat different way of depicting the number relationships illustrated in Figure 3.1. Note that $\mathscr{Q}$, $\mathscr{Q}'$ are disjoint. As shown here the important number sets can be considered as having the subset relationships $\mathscr{N} \subset \mathscr{W}$, $\mathscr{W} \subset \mathscr{I}$, $\mathscr{I} \subset \mathscr{Q}$, $\mathscr{Q} \subset \mathscr{R}$.

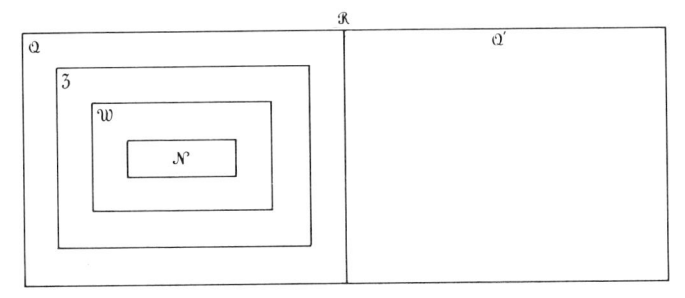

Figure 3.9

EXERCISE SET

1. Using logical or mathematical symbols for as many terms as possible, write definitions of the following concepts which were specified informally.
    a) disjoint
    b) overlap

2. State whether each pair of sets is equal.
    a) the set of letters $\{a, b, c, d\}$; the set $\{b, c, a, d\}$
    b) the set of cars in the United States; the set of licensed drivers in the United States
    c) the members of Mrs. Jones's family; the members of Mr. Jones's (Mrs. Jones's husband) family
    d) the set of fingers on the left hand; the set of fingers on the right hand (barring accidents and birth defects)

3. Let $S = \{1, 4, 9, 16, 17\}$. Write each solution set in roster notation.
    a) $\{x \mid x \in S \wedge x^2 \neq 16\}$
    b) $\{x \mid x \in S \wedge \sim (x \text{ is odd})\}$
    c) $\{x \mid x \in S \wedge x \text{ is odd}\}$
    d) $\{x \mid x \in S \wedge x + 1 \in S\}$
    e) $\{x \mid x \in S \wedge x + 5 = 8\}$
    f) $\{x \mid x \in S \wedge x^2 - 5x + 4 = 0\}$

4. If P is the set of baseball players on a team, make a roster of each set.

a) $\{x \mid x \in P \wedge x \text{ is an outfielder}\}$
  b) $\{x \mid x \in P \wedge x \text{ is in the battery}\}$
  c) $\{x \mid x \in P \wedge x \text{ is a baseman}\}$
  d) $\{x \mid x \in P \wedge x \text{ is an infielder}\}$
5. Answer true or false:
   a) Each set of a pair of equal sets is a subset of the other.
   b) All susets of a given set are equal.
   c) $263 \in \{5, 6, 7, \dots\}$
   d) $\{1, 2, 3, \dots\} \leftrightharpoons \{101, 102, 103, \dots\}$
   e) $\{0\} = \{\ \}$
   f) $\{\{\varnothing\}\} = \{\{\ \}\}$
   g) $\{\varnothing\} = \varnothing$
   h) $\{x \mid x \in \mathscr{L} \wedge -1 \leq x \leq 5\} \leftrightharpoons \{0, 1, 2, 3, 4\}$
   i) $0 \subset \{0, 1\}$
   j) $\{\varnothing\} \subset \{\{\varnothing\}\}$
   k) $\varnothing \subset \{\{\varnothing\}\}$
   l) $a = \{a\}$
   m) $\varnothing \in \{\varnothing\}$
6. Which pairs of sets are matched?
   a) $\{1, 2, 3, 4\}, \{101, 102, 103, 104\}$
   b) $\{5, 7, 9, 11\}, \{4, 6, 8, 10\}$
   c) $\varnothing, \{\text{George, Tom, Carl}\}$
   d) $\{\text{red, white, blue}\}, \{U, S, A\}$
7. Which of the following correctly relate the sets $S = \{a, b, y, z\}$, $M = \{w, x, y, z\}$?
   a) $S \subset M$ b) S overlaps M
   c) $S \leftrightharpoons M$ d) S, M are disjoint
   e) $S \supset M$ f) $S = M$
8. Use the choices in Exercise 7 to relate sets $S = \{2, 5, 7\}$, $M = \{2, 5\}$ correctly.
9. Use the choices in Exercise 7 to relate the sets $S = \{\&, \S, \Delta\}$ and $M = \{\S, \Delta, \&\}$.

§3c PROPERTIES OF EQUALITY AND INCLUSION

At this point we can develop some of the formal properties of set inclusion and set equality. In the process the reader will gain some experience with application of principles of logic to mathematical arguments.

In the preceding section we made use of diagrams to illustrate set relationships. In Figure 3.10, the elements of S, if any, are represented by the region enclosed by the circle. I must be nonempty, but the diagram is not meant to exclude the possibility that S may be empty. Also, all the elements of I might belong to S. That is, a set diagram is drawn to represent some general case, and we must keep in mind that it may be subject to various special interpretations.

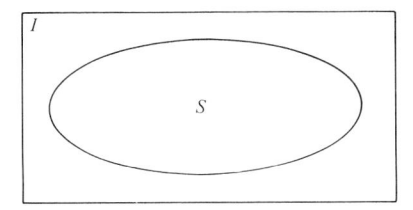

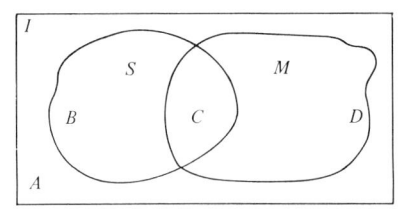

Figure 3.10

Figure 3.11

With these possibilities in mind let us consider Figure 3.11. There, two sets S, M are illustrated within the Universe of Discourse I. The set A is the region outside both S, M. The set B is the region within S but outside M. The set C is the region within both S, M. The set D is the region within M but outside S.

It may seem at first as though Figure 3.11 simply illustrates a case in which S, M overlap, but we must remember that every region but I itself could be empty. The following interpretations, then, are possible.

(i) A is empty $\Rightarrow$ S or M or both fill up I.
(ii) B is empty $\Rightarrow$ S is within M, i.e., $S \subset M$.
(iii) C is empty $\Rightarrow$ S, M are disjoint.
(iv) D is empty $\Rightarrow$ M is within S, i.e., $M \subset S$.
(v) B, D are empty $\Rightarrow$ M is within S and S is within M, i.e., both sets are the same as C.
(vi) B, C, D are empty $\Rightarrow$ A includes all of I.

From consideration of these cases, you can see that we must be very careful when we try to draw conclusions from a set diagram. For example, we cannot be *certain* that a diagram of two sets actually represents an overlapping case unless we know that regions B, C, D in Figure 3.11 are *each* nonempty. Figure 3.12 shows how we can illustrate sets S, M which actually are known to overlap. According to the statement of (2), §3b, we mean

$$a, b \in S; b, c \in M.$$

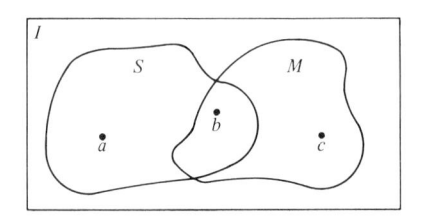

Figure 3.12

Properties of Equality and Inclusion

From consideration of Case (v) above, it can be seen that the following proposition provides a useful way to establish equality of sets as an alternative to that of the definition.

Theorem (3c-1): $A = B$ iff $(A \subset B) \wedge (B \subset A)$.

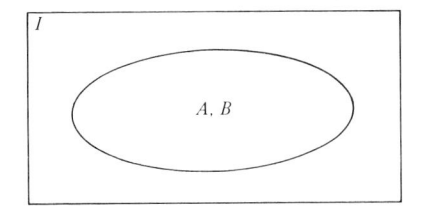

Proof of "only if" part:

1.	$A = B$	1.	Hypothesis (C.P.)
2.	If $A = B$, then $(x \in A \Rightarrow x \in B)$	2.	Definition of $=$, (3-4)
3.	$x \in A \Longleftrightarrow x \in B$	3.	R.D.
4.	$A \subset B \wedge B \subset A$	4.	Definition of $\subset$, (3-3)
5.	If $A = B$, then $A \subset B \wedge B \subset A$.	5.	C.I.

Done

Proof of the "if" part is left to the exercises.

Thus, to prove two sets equal it is sufficient to show that each is a subset of the other.

Use of the term "hypothesis" as a reason in Step 1 is common practice. In mathematical proofs it is a synonym for C.P. and serves just as well as the latter to stress the fact that this premise is not one of the given definitions, axioms, or previously proved theorems of the discourse, but is actually the hypothesis of the implication to be proved. Thus, either label indicates that the Rule of Conditional Inference is being used.

Before going on to the next theorem, it may be instructive to list the subsets of $S = \{\alpha, \beta\}$. Is $\{\alpha\} \subset S$? Since

$$\alpha \in \{\alpha\} \Rightarrow \alpha \in S$$

is true, we have

$$\{\alpha\} \subset S$$

by Definition (3-3) of subset. Similarly

$$\{\beta\} \subset S$$

Now note that both

$$\alpha \in S \Rightarrow \alpha \in S, \; \beta \in S \Rightarrow \beta \in S$$

are true, whence $\{\alpha, \beta\} \subset S$; that is

128   Sets

(1) $$S \subset S$$

Now let us look carefully at the proposition

(2) $$x \in \varnothing \Rightarrow x \in S$$

where x represents either α or β. Now recall that $x \notin \varnothing$ is true by Axiom S-3. Hence

$$x \in \varnothing$$

is false. Thus the implication of (2) has a false hypothesis, and the implication itself is true. We then have

(3) $$\varnothing \subset S$$

by the definition. Hence $S = \{\alpha, \beta\}$ has the four subsets $\{\alpha\}$, $\{\beta\}$, S, $\varnothing$.

We shall now prove general theorems concerning the principles of (1) and (3). You will see that some portions of the preceding argument will appear in the proofs.

Theorem (3c-2): If A is any set, then $A \subset A$.

Proof:

1. A is a set   1. ?
2. $x \in A$ is a proposition   2. S-2
3. $x \in A \Rightarrow x \in A$ 3. $p \Rightarrow p$ is a tautology (2e-1)
4. $(x \in A \Rightarrow x \in A) \Rightarrow A \subset A$ 4. Definition of Subset (3-3)
5. $A \subset A$ 5. ?
6. If A is a set, then $A \subset A$ 6. ?

                                                                Q.E.D.

If the "reason" for the crucial Step 3 of this proof seems out of place, recall that a "statement" in an axiomatic discourse may be a *true* proposition either from logic or from the particular discourse. The corresponding "reason" simply tells us *why* the corresponding statement *is* actually true. Now, by Axiom S-2

$$x \in A$$

is a *proposition* for elements of a given universe I. Thus,

(4) $$x \in A \Rightarrow x \in A$$

is a compound proposition of the form

$$p \Rightarrow p$$

Since $p \Rightarrow p$ is a tautology for any p, (4) is true and may be used as a statement in a proof.

Proof of the following corollary is left to the exercises.

Corollary (3c-2a): If A is a set, then $A = A$.

§3c Properties of Equality and Inclusion

This proposition "$A = A$ for all sets A" is called the *reflexive* property of set equality; correspondingly, the proposition "$A \subset A$ for all sets A" is called the *reflexive* property of set inclusion. Compare with E-1 of §1f.

The principle of this corollary shows us why we should be careful not to confuse either verbally or in mathematical symbols the notions of set belonging and set inclusion. We see that $A \subset A$ is *always* true. However, we do not want an element of a set to be equal to the set, e.g.,

$$3 \neq \{3\}$$

That is, the number 3 is not the same as the set containing 3. Also, the set of all cats is not a cat; a set of dishes is not a dish. To prevent such a situation, it is agreed that a set A is not to be an element of itself.

To avoid any tendency to confuse the two ideas of belonging and inclusion, some books never use the terms "in" for either "$\in$" or "$\subset$." Further, some even avoid both of the corresponding terms "contained in" and "included in."

The following proposition establishes the principle of (3) for all sets.

Theorem (3c-3): If A is a set and $\varnothing$ is the empty set, then $\varnothing \subset A$.

Proof:

1. $\varnothing$ is the empty set, A is any set
2. $x \notin \varnothing$
3. $x \in \varnothing \Leftrightarrow f$
4. $x \in \varnothing \Rightarrow x \in A$
5. $(x \in \varnothing \Rightarrow x \in A) \Rightarrow \varnothing \subset A$
6. $\varnothing \subset A$
7. A is a set and $\varnothing$ is the empty set $\Rightarrow \varnothing \subset A$

1. Hypothesis (C.P.)
2. ?
3. Definition of $\sim$
4. $f \Rightarrow q$ is a tautology (2e-5)
5. ?
6. ?
7. ?

Q.E.D.

The next theorem states the so-called *symmetric* property of set equality. Compare with E-2, §1f.

Theorem (3c-4): For any sets A, B, $A = B \Rightarrow B = A$.

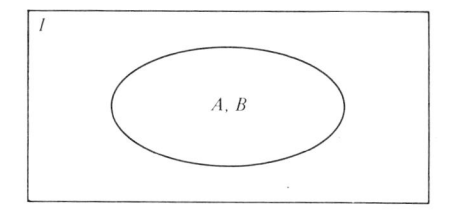

Proof:

1. $A = B$ 1. Hypothesis
2. $A \subset B \wedge B \subset A$ 2. (3c-1)
3. $B \subset A \wedge A \subset B$ 3. Commutativity of $\wedge$ (2e-14)
4. $B = A$ 4. (3c-1)
5. $A = B \Rightarrow B = A$ 5. C.I.

                                                                           Q.E.D.

This proof has been shortened by omission of the statement

(5) $A = B \Rightarrow (A \subset B) \wedge (B \subset A)$

which is how part of Theorem (3c-1) actually reads. Had this proposition been included as a step in the proof, then the statement following (5) would be that given in Step 2 by reason of R.D. Combining two steps into one in this way is a labor-saving device frequently used. When this is done, the "reason" quoted is usually the one which would go with the *missing* "statement," as in this case. Such combining was also done going from Step 3 to Step 4 of this proof.

When a person is sure of both his argument mechanics and the basic principles of the discourse being developed, many other steps are often left out. As experience is gained, the column arrangement may be discarded and the essential points of the argument simply given in paragraph form, leaving to the reader the task of filling in details. Ultimately, a terse sentence or two may be all that would be given in a proof.

The following two theorems state, respectively, the *transitive* property of set inclusion and set equality. The proofs are left to the exercises.

Theorem (3c-5):   For all sets S, M, L,

$$(S \subset M \wedge M \subset L) \Rightarrow S \subset L$$

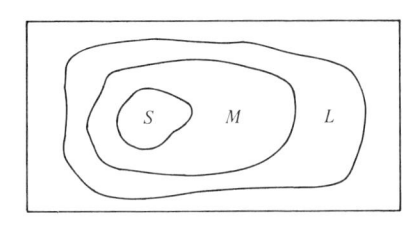

Theorem (3c-6):   For all sets S, M, L,

$$(S = M \wedge M = L) \Rightarrow S = L$$

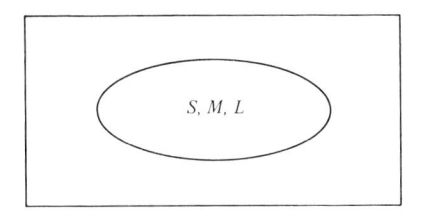

Another property of the empty set is worth investigating. As we saw in Example 2, §3a, $x^2 < 0$ determines $\varnothing$. In Example 7, §3a, we saw that $x = x + 1$ also determines $\varnothing$. Axiom S-3 guarantees the existence of an empty set, but are the sets

$$\{x \mid x^2 < 0 \wedge x \in \mathscr{R}\}, \{x \mid x = x + 1\}$$

in some way different even though both are empty? Can we show that these two sets and any other set A having the property that $x \notin A$ for all $x \in I$ are actually the same, i.e., equal? If we can show this fact, we would say that $\varnothing$ is *unique*.

Theorem (3c-7): $\varnothing$ is unique.

Proof ("at least one"):

1. $\varnothing$ is a set. 1. Why?

Done

Proof ("at most one"): As stated above, this part requires us to prove: "For any set $A \cdot \ni \cdot \forall\, x, x \notin A$, then $A = \varnothing$." This is because the determining property of $\varnothing$ is "$\forall\, x, x \notin \varnothing$." See Axiom S-3.

1. A is a set $\cdot \ni \cdot \forall\, x, x \notin A$ 1. Hypothesis
2. $x \in A$ is false   2. Definition of $\sim$
3. $x \in A \Rightarrow x \in \varnothing$ 3. $f \Rightarrow p$ is a tautology
4. $A \subset \varnothing$ 4. ?
5. $\varnothing \subset A$ 5. ?
6. $A \subset \varnothing \wedge \varnothing \subset A$ 6. C. A.
7. $A = \varnothing$ 7. ?
8. A is a set $\cdot \ni \cdot \forall\, x, x \notin A \Rightarrow A = \varnothing$ 8. C. I.

Q.E.D.

We often wish to show the uniqueness of some mathematical entity as was done in this theorem. All of the terms "exactly one," "one and only one," "at least one and at most one" are used as synonyms for each other and for "unique." Thus, when we state that a thing is unique, the statement is actually a conjunction of two propositions. Hence, a proof of uniqueness must be in two parts: one, to show that the thing exists; the other, to show that it is the *only* such thing which does exist in that discourse.

Proof of the following corollary is left to the exercises.

Corollary (3c-7a): For any set A,
$$A \neq \emptyset \Rightarrow \exists\, x \cdot \ni \cdot x \in A$$

EXERCISE SET I

1. Make a set diagram to illustrate each situation.
   a) the students in your room; the boys in your room
   b) the students in your room; the students in the next room
   c) the set of natural numbers from 1 to 12; the set of odd numbers; the set of numbers divisible by 3
   d) the students in your school; the students named John

2. Which of the following specify the empty set?
   a) men who can high jump 8 feet from a standing start
   b) odd numbers exactly divisible by 2
   c) squares of odd numbers that are even

3. List the subsets of the given set.
   a) $\{x, y\}$ b) $\{\Delta\}$
   c) $\{\ \}$ d) $\{G, T, C\}$

4. Which propositions correctly relate the sets $S = \{\ \}$, $M = \{\Delta, \S, \&\}$?
   a) $S \subset M$ b) S overlaps M
   c) $S \leftrightarrows M$ d) S and M are disjoint
   e) $S \supset M$ f) $S = M$

5. Use the choices in Exercise 4 to relate the sets $S = \emptyset$, $M = \emptyset$.

6. Supply the missing reasons in the proof of Theorem (3c-2).

7. Supply the missing reasons in the proof of Theorem (3c-3).

8. Supply the missing reasons in the proof of Theorem (3c-7).

9. Prove or disprove: $(A = B) \wedge (x \in A) \Rightarrow x \in B$.

10. Prove or disprove: If $A = B \wedge x \notin A$, then $x \notin B$.

11. Prove or disprove: $(A \leftrightarrows B) \Rightarrow (A = B)$.

12. Prove: The "if" part of Theorem (3c-1).

13. Prove: $\emptyset \subset \emptyset$
    a) as a corollary of (3c-2);
    b) as a corollary of (3c-3).
14. a) Prove Corollary (3c-2a).
    b) Prove the principle of this corollary independently of Theorem (3c-2).
15. Prove: (3c-5).
16. Prove: (3c-6).
17. Prove: (3c-7a).
18. See (3c-4) for the symmetric property of set equality. State a corresponding symmetric property of set inclusion. Prove or disprove your statement.

Definition (3-4) introduced the concept of equality of *sets*. We have made no other use of equality up to this point. In talking about sets of things, we obviously need some way of relating the elements themselves.

It is common practice to introduce the notion of equality of set elements by stating desired properties as axioms. This was done, for example, in §1f. Note particularly E-1, E-2, and E-3.

It is, however, easy enough to define the concept in terms of set equality and then show that the requisite properties follow as theorems.

Definition (3-7): For set elements a, b, $\mathbf{a} = \mathbf{b}$ iff $\{a\} = \{b\}$.

The following three theorems establish respectively the requisite *reflexive*, *symmetric*, and *transitive* properties of element equality.

Theorem (3c-8): If x is a set element, $x = x$.

Proof:

1. x is a set element	1. Hypothesis.
2. A set $\{x\}$ exists	2. Axiom S-2.
3. $\{x\} = \{x\}$	3. $A = A$ for any set A (3c-2a).
4. $x = x$	4. Definition (3-7).
5. If x is a set element, then $x = x$	5. C.I.

Q.E.D.

Theorem (3c-9): $\forall a, b$, if $a = b$, then $b = a$.

Theorem (3c-10): $\forall a, b, c$, if $a = b \wedge b = c$, then $a = c$.

Proof:

1. a, b, c are set elements	1. ?
2. Sets $\{a\}, \{b\}, \{c\}$ exist	2. ?
3. $a = b \wedge b = c$	3. ?
4. $\{a\} = \{b\} \wedge \{b\} = \{c\}$	4. Definition (3-7).
5. $\{a\} = \{c\}$	5. Theorem (3c-6).
6. $a = c$	6. ?
7. $a = b \wedge b = c \Rightarrow a = c$.	7. ?

Q.E.D.

The following propositions give further properties of the "=" relation. They can all be proved without direct use of Definition (3-7).

Theorem (3c-11): If $a = b$, $b = c$, and $c = d$, then $a = d$.

Theorem (3c-12): If $a \neq b$, then $b \neq a$.

Theorem (3c-13): If $a \neq b$ and $b = c$, then $a \neq c$.

EXERCISE SET II

19. In each case what can we conclude about the elements of sets
$$A = \{3\}, B = \{2x + 3\}, C?$$
   a) $A = B$ b) $A = C$ c) $B = C$

The usual definitions for subtraction and division are given in terms of addition and multiplication, i.e., $a - b = x$ iff $b + x = a$ and $a \div b = x$ iff $b \cdot x = a$.

20. If $a, b \in \mathcal{N}$, is $(a - a) \in \mathcal{N}$? Is $(a - b) \in \mathcal{N}$?
21. If $a, b \in \mathcal{W}$, is $(a - a) \in \mathcal{W}$? Is $(a - b) \in \mathcal{W}$?
22. If $a, b \in \mathcal{Z}$, is $(a - a) \in \mathcal{Z}$? Is $(a - b) \in \mathcal{Z}$?
23. Now answer the questions in Exercises 20, 21, and 22 in terms of division.
24. Supply the missing reasons in the proof of Theorem (3c-10).

Prove or disprove the proposition given in each of Exercises 25 through 30.

25. (3c-9)
26. (3c-11)
27. (3c-12)
28. (3c-13)

29. If $a \neq b$ and $b \neq c$, then $a \neq c$.
30. If $a \neq b$, $a = c$, and $b = d$, then $c \neq d$.

§3d COMPLEMENTATION

Axiom S-2 specifies a set A as containing particular elements, if any, of a nonempty set I. It should seem reasonable to have a designation for the set of all those elements of I which do not belong to A.

Definition (3-8): For any nonempty I and $A \subset I$, the set A', called the **complement of A** *in I*, is the set for which $x \notin A \longleftrightarrow x \in A'$.

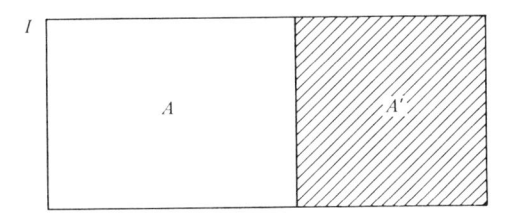

Figure 3.13

Figure 3.13 shows a way of drawing the relationship specified in this definition. The hatched region represents A', the complement of A.

To write the requirement for the complement of a set A in set builder notation we have only to recall that $A = \{x \mid x \in A\}$. See (3) of §3b. For the set A' this becomes

$$A' = \{x \mid x \in A'\}$$

Now recall from §2i that when the sentence $p \longleftrightarrow q$ is true in a mathematical discourse we may substitute either proposition for the other. Substituting $x \notin A$ for $x \in A'$ from the definition yields

(1) $$A' = \{x \mid x \notin A\}$$

If this proposition were to list all the requirements of the definition, it would be stated as

$$A' = \{x \mid x \in I \land x \notin A \land I \neq \varnothing\}$$

but since $x \in I$ and $I \neq \varnothing$ are always true and since $p \land t \equiv p$, the statement of (1) is sufficient. If we know that some $p(x)$ determines a set A, i.e., $A = \{x \mid p(x)\}$, then we have

(2) $$A' = \{x \mid \sim p(x)\}$$

as an alternative form of (1).

EXAMPLE 1: If $I = \{0, 1, 2, 3, 4, 5, 6, 7, 8, 9\}$, $A = \{0, 1, 2, 3, 4\}$, $B = \{4, 5, 6\}$, find A' and B'.

Solution:
$$A' = \{x \mid x \in I \wedge x \notin A\}$$
thus
$$A' = \{5, 6, 7, 8, 9\}$$
$$B' = \{x \mid x \in I \wedge x \notin B\}$$
thus
$$B' = \{0, 1, 2, 3, 7, 8, 9\}$$

Can we be certain that any particular set A has a complement A'? Axiom S-1 specifies that the elements of any set A must be elements of a universal set I. Since the set A of Definition (3-8) is a subset of I, "$x \in A$" is either true or false for the elements of I, i.e., "$x \in A$" is a proposition determinable on I. We know that the negation $\sim p$ of any proposition p is also a proposition relative to the same truth standard. That is, $x \notin A$ is determinable on I. Since

$$x \notin A \longleftrightarrow x \in A'$$

by the definition, the latter is also determinable on I. Thus

$$A' \text{ exists for all } A \cdot \ni \cdot A \subset I$$

Since the complement of a set always exists, then I' and $\varnothing'$ must exist. What can we discover about these two complements? Recall that by Axiom S-3, $x \notin \varnothing$ is always true. Also $x \in I$ is always true by S-1. Hence, we have

$$x \notin \varnothing \longleftrightarrow x \in I$$

is true. According to Definition (3-8), this sentence becomes

(3) $$x \in \varnothing' \longleftrightarrow x \in I$$

whence
$$\varnothing' = I$$

by the Definition of Set Equality.

The proposition of (3) must also be true for the negation of each side, and, using D.N. on the left, we obtain

$$x \in \varnothing \longleftrightarrow x \notin I$$

In turn, this proposition can be written as

$$x \in \varnothing \longleftrightarrow x \in I'$$

whence
$$\varnothing = I'$$

§3d Complementation

The main facts presented in the preceding discussion are restated in the following theorem and in its corollaries.

Theorem (3d-1): For any set A, its complement A' exists.

Corollary (3d-1a): If $\varnothing$ is the empty set and I is the Universe of Discourse, then $\varnothing' = I$.

Corollary (3d-1b): If I is the Universe of Discourse and $\varnothing$ is the empty set, then $I' = \varnothing$.

Since any set A has a complement, it follows that the set A' must also have a complement $(A')'$. Consider Figure 3.14(i). Here the circular region represents A, and its complement A' is represented by the hatched region

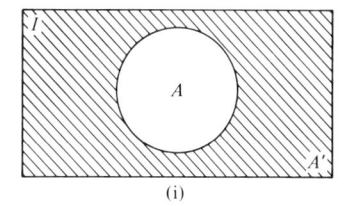

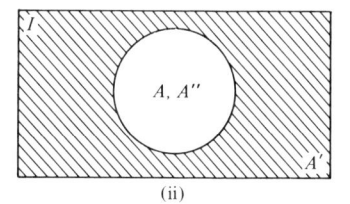

Figure 3.14

outside the circle. Now everything in I but not within the hatched region should represent the complement of A', that is, $(A')'$. However, this is just the region enclosed by the circle again! See Figure 3.14(ii), where $(A')'$ is written as A'' for brevity. It seems that

(4) $$(A')' = A$$

Can we prove this conjecture? According to Definition (3-8) the set $(A')'$ has the property

(5) $$x \in A'' \iff x \notin A'$$

For the same reason we know that

$$x \in A' \iff x \notin A$$

By Theorem (2e-15) and D.N., this bicondition becomes

$$x \notin A' \iff x \in A$$

Using this and (5) with the Chain Rule, we obtain

$$x \in A'' \iff x \in A$$

whence

$$A'' = A$$

We state this principle as a theorem.

Theorem (3d-2): For any set A, $(A')' = A$.

The set $(A')'$ is often called the *Double Complement* of A.

The following theorem and its corollaries should seem plausible from a study of the accompanying set diagram.

Theorem (3d-3): For any sets A, B, $A \subset B \Longleftrightarrow B' \subset A'$.

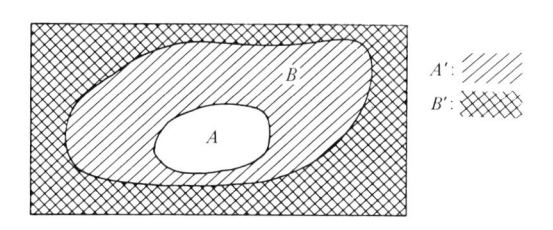

Proof:

1. A, B are sets.   1. ?
2. $A \subset B \Longleftrightarrow (x \in A \Rightarrow x \in B)$ 2. ?
3. $A \subset B \Longleftrightarrow (x \notin B \Rightarrow x \notin A)$ 3. ?
4. $A \subset B \Longleftrightarrow (x \in B' \Rightarrow x \in A')$ 4. ?
5. $A \subset B \Longleftrightarrow B' \subset A'$.   5. ?

Q.E.D.

Corollary (3d-3a): For any sets A, B, $A = B \Longleftrightarrow A' = B'$.

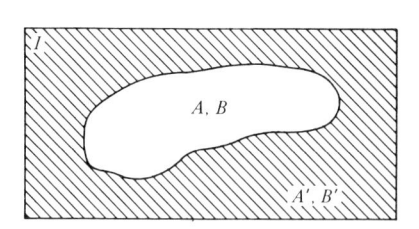

Corollary (3d-3b): If A, B are sets $\cdot \ni \cdot A = B'$, then $A' = B$.

EXERCISE SET

1. If A is a subset of the universal set $I = \{1, 3, 6, 10, 15, 21, 28, 36, 45\}$, write the complement of each A.
    a) $A = \{3, 6, 21, 45\}$

b) $A = \{1, 3, 6, 10\}$
   c) $A = \{\ \}$
   d) $A = \{1, 3, 6, 10, 15, 28, 36, 45\}$
2. If $I = \{3, 8, 9, 14, 27\}$, write the complement of each set in roster notation.
   a) $A = \{x \mid x \in I \land x \text{ is even}\}$
   b) $B = \{x \mid x \in I \land 3x \in I\}$
   c) $C = \{x \mid x \in I \land (x + 5) \in I\}$
   d) $D = \{x \mid x \in I \land x^2 \in I\}$
3. Write complements of each set.
   a) $S = \{x \mid x \leq 0 \land x \in \mathscr{I}\}$
   b) $S = \{x \mid x \leq 0 \land x \in \mathscr{N}\}$
4. If $I = \{0, 1\}$, write complements of:
   a) $S = \{x \mid 0 \leq x \leq 1 \land x \in I\}$
   b) $S = \{x \mid \sim (0 < x < 1) \land x \in I\}$
5. Why do you think I is specified as nonempty in Definition (3-8)? Carefully consider the sentence $x \notin A \leftrightarrow x \in A'$.
6. Supply the missing reasons in the proof of Theorem (3d-3).
7. Prove:
   a) Corollary (3d-3a)
   b) Corollary (3d-3b)
8. Prove Corollary (3d-3b) independently of any other theorem in this section.
*9. Using Corollary (3d-3b) and Corollary (3d-1a), prove the following principles independently of other theorems in this section.
   a) Corollary (3d-1b)
   b) Theorem (3d-2).

§3e OPERATIONS ON SETS

The discussion of logic in Chapter II began with single propositions, and it quickly became necessary to combine and modify them in various ways. This combining and modifying could be described as operating on propositions. An operation which changes a single entity to a new entity is called a *unary operation*. For example, negating p gave us a new proposition $\sim p$. A *binary operation* combines two entities to produce one. For example, combining p, q by disjunction gave us a new proposition $p \lor q$.

We now turn our attention to operations on sets. We have already discussed complementation, a unary operation on sets. Given two sets A, B, an interesting new set would be a combination of all the elements from both.

Definition (3-9): For any two sets A, B, the set

$$A \cup B = \{x \mid x \in A \vee x \in B\}$$

is called the **union** of A, B.

Although this definition is given using set builder notation, an equally acceptable way of specifying a union of two sets would be

(1) $\qquad x \in A \cup B \Longleftrightarrow x \in A \vee x \in B$

Thus the union of two sets is a set which contains all the elements found in either set.

It has been shown that two sets may be disjoint, overlapping, or equal, or that one may be a subset of the other (see Figures 3.2 through 3.6). In Fig. 3.15 the new set $E \cup S$ is hatched in each of four cases shown.

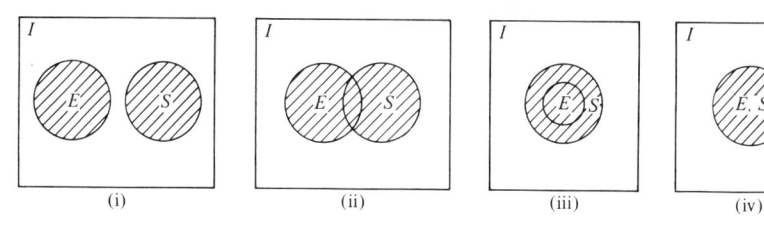

Figure 3.15

EXAMPLE 1: Find the union of $\{1\}$ and $\{\pi\}$.

Solution: $1 \in \{1\}$ is true. Therefore, the disjunction $1 \in \{1\} \vee 1 \in \{\pi\}$ is also true. Since $\pi \in \{\pi\}$ is true, the required disjunction is true again. Hence, $\{1\} \cup \{\pi\} = \{1, \pi\}$.

EXAMPLE 2. Find the union of $\{b, f\}$ and $\{b, k, n\}$.

Solution: $\{b, f, k, n\}$. See Figure 3.16.

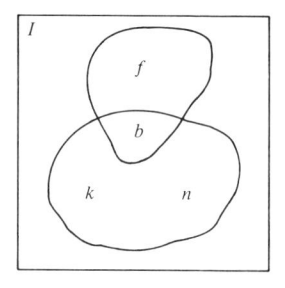

Figure 3.16

Another possible new set would be the set of elements common to the two sets. We shall call this binary operation the intersection of two sets.

Definition (3-10): For any two sets A, B, the set
$$A \cap B = \{x \mid x \in A \wedge x \in B\}$$
is called the **intersection** of A, B.

An alternative way of specifying the intersection of two sets is
(2) $\qquad x \in A \cap B \Longleftrightarrow x \in A \wedge x \in B$

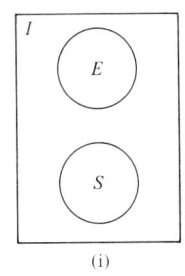

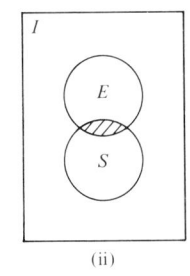

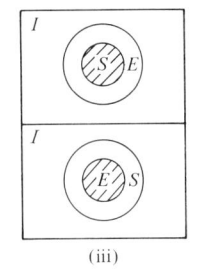

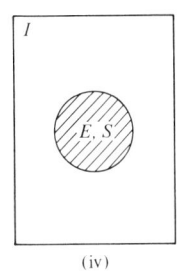

Figure 3.17

In Figure 3.17 the new set $E \cap S$ is hatched in each of the four cases shown. Both E, S are assumed to be nonempty. Note that $E \cap S = \emptyset$ in Diagram (i). This suggests an effective way to state a formal definition of disjoint sets, a concept first introduced informally in (1) of §3b.

Definition (3-11): Two sets A, B are **disjoint** iff $A \cap B = \emptyset$.

EXAMPLE 3: Find the intersection of sets $\{1\}, \{\pi\}$.

Solution: $1 \in \{1\}$ is true, but $1 \in \{\pi\}$ is false. Thus $1 \in \{1\} \wedge 1 \in \{\pi\}$ is also false, and 1 is not an element of the intersection. Similarly, π does not belong to the intersection. Since neither of the two elements under discussion belongs to the intersection, $\{1\} \cap \{\pi\} = \emptyset$ by Theorem (3c-7). That is, the sets $\{1\}, \{\pi\}$ are disjoint. See Diagram (i), Figure 3.17.

EXAMPLE 4: Find the intersection of $\{b, f\}$ and $\{b, k, n\}$.

Solution: $\{b\}$. See Figure 3.16.

EXAMPLE 5: From Figure 3.18 express $S \cap M'$ in roster notation.

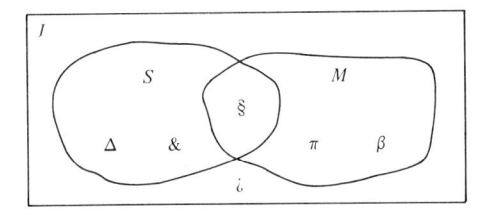

Figure 3.18

Solution:

$$I = \{\Delta, \&, \S, \pi, \pi, \beta, \mathit{l}\}$$
$$S = \{\Delta, \&, \S\}$$

and

$$M = \{\S, \pi, \beta\}$$

whence

$$M' = \{\Delta, \&, \mathit{l}\}$$

Therefore,

$$S \cap M' = \{\Delta, \&\}$$

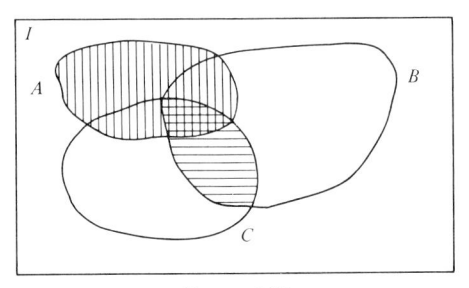

Figure 3.19

EXAMPLE 6: Hatch the region of Figure 3.19 which illustrates $A \cup (B \cap C)$.

Solution: $B \cap C$ is hatched horizontally; A is hatched vertically. The entire hatched region represents $A \cup (B \cap C)$.

Definitions (3-9) and (3-10) define binary operations. That is, they provide us with a way of combining only *two* sets. As in the following example, however, we often do not wish to restrict ourselves to combinations of only two sets. The following definition gives meaning to unions and intersections of three sets.

Definition (3-12): $\mathbf{A} \cup \mathbf{B} \cup \mathbf{C} = (A \cup B) \cup C$

and

$$\mathbf{A} \cap \mathbf{B} \cap \mathbf{C} = (A \cap B) \cap C$$

EXAMPLE 7: Hatch $A \cup B \cup C$ in Figure 3.20.

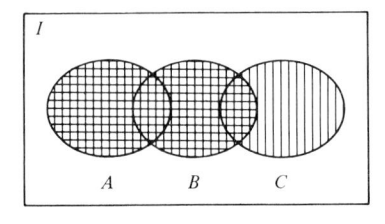

Figure 3.20

§3e   *Operations on Sets*   143

Solution: By Definition (3-12) we do $(A \cup B) \cup C$. $A \cup B$ is hatched horizontally. The union of $(A \cup B)$ and C is hatched vertically. The entire hatched region represents $(A \cup B) \cup C$.

EXAMPLE 8: Hatch the region of Figure 3.21 which illustrates $(A \cap B) \cup (B \cap C)$.

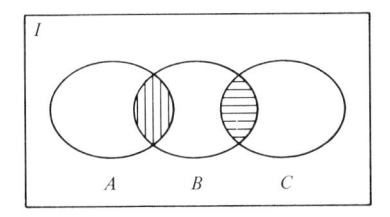

Figure 3.21

Solution: $A \cap B$ is hatched vertically; $B \cap C$ is hatched horizontally. The entire hatched region represents $(A \cap B) \cup (B \cap C)$.

EXAMPLE 9: Hatch the region of Figure 3.22 which illustrates $A \cap B'$.

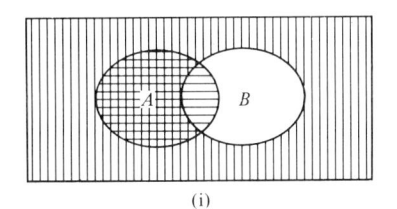

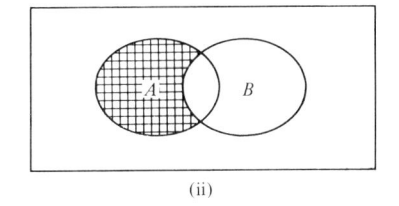

Figure 3.22

Solution: B' is hatched vertically and A is hatched horizontally in Diagram (i) of the figure. The required intersection is cross-hatched as shown in (ii).

EXAMPLE 10: Make rosters to represent the specified regions in Figure 3.23.

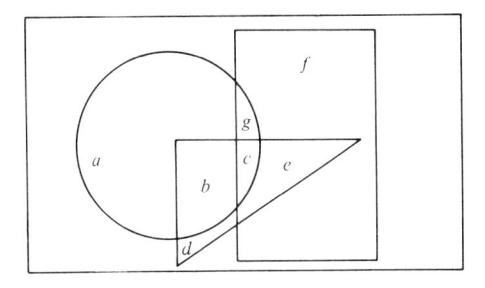

Figure 3.23

Solution:

1. The elements in the circle only. Answer: $\{a\}$.
2. The elements common to the circle and the triangle. Answer: $\{b, c\}$.
3. The elements in the triangle only. Answer: $\{d\}$.
4. The elements common to the triangle and rectangle. Answer: $\{c, e\}$.
5. The elements common to all three. Answer: $\{c\}$.
6. The elements in the circle or the rectangle. Answer: $\{a, b, c, g, e, f\}$.
7. The elements in the circle or the triangle. Answer: $\{a, b, c, d, e, g\}$.

EXAMPLE 11: In a certain region, 69% of the residents receive television Channel 4 (A); 75% receive Channel 5 (B); and 70% receive Channel 7 (C). Of these, 40% receive 4 and 7, 45% receive 4 and 5, and 59% receive 5 and 7. 30% receive all 3 channels. What percentage can receive only one of the channels? See Figure 3.24.

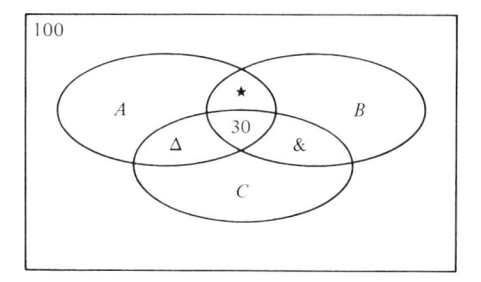

Figure 3.24

Solution: We use a set diagram with three overlapping sets. It is convenient to introduce the symbol $n(S)$ which is read "the number of elements in the set S." We are given that $n(A \cap B \cap C)$ is 30. Since $n(A \cap B)$ is 45, $*$ has 15 elements. Since $n(A \cap C)$ is 40, Δ has 10 elements. Since, $n(B \cap C)$ is 59, & has 29 elements. Since A has 69 elements, 14 receive Channel 4 alone. Since B has 75 elements, 1 receives Channel 5 alone. Since C has 70 elements, 1 receives Channel 7 alone. Hence, 16% receive only one of the channels.

Using the notation $n(A)$ as in Example 11 to mean "the number of elements in set A," it can be shown that for sets A, B, C not necessarily disjoint,

(3) $\quad n(A \cup B \cup C) =$
$\quad\quad n(A) + n(B) + n(C) - n(A \cap B) - n(A \cap C) -$
$\quad\quad n(B \cap C) + n(A \cap B \cap C)$

EXERCISE SET I

1. a) Name at least two unary operations from arithmetic.
   b) Name at least two binary operations from arithmetic other than addition and multiplication.

2. As a means of giving further examples of binary operations, consider the operation "∘" operating on numbers $\{a, b, \ldots\}$. $(a \circ b)$ is read "a bubble b.") What is the standard name for ∘ in each of the following?
   a) $2 \circ 3 = 5$
   b) $2 \circ 3 = 6$
   c) $2 \circ 3 = \frac{2}{3}$
   d) $2 \circ 3 = -1$
   e) $2 \circ 3 = 8$
   f) $8 \circ 3 = 2$
   g) $2 \circ 3 = 9$ and $3 \circ 3 = 12$

3. Use the concept of intersection to simplify the definition of overlapping sets produced in Exercise 1b of §3b.

4. Express the set $\mathscr{W}$ of whole numbers as a union of $\mathscr{N}$ and another set.

5. Fill in the blanks. $\mathscr{Q}$ is the set of rational numbers.
   a) $\mathscr{Q} \cup \mathscr{Q}' = $ _____
   b) $\mathscr{Q} \cap \mathscr{Q}' = $ _____

6. Punctuate the expression $A \cup B \cup C \cup D$ with parentheses and brackets to make it read correctly according to Definition (3-12).

7. Using Figure 3.25, express each of the following sets in roster notation.

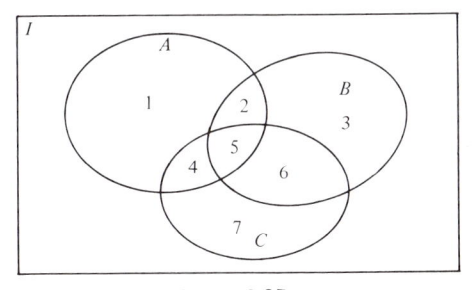

Figure 3.25

   a) $A \cap B$
   b) $A \cap B'$
   c) $A' \cap B'$
   d) $(A \cap B)'$
   e) $A \cap (B \cup C)$
   f) $(A \cap C') \cap (B \cup C)$

8. Find an algebraic expression for the hatched region in each diagram of Figure 3.26.

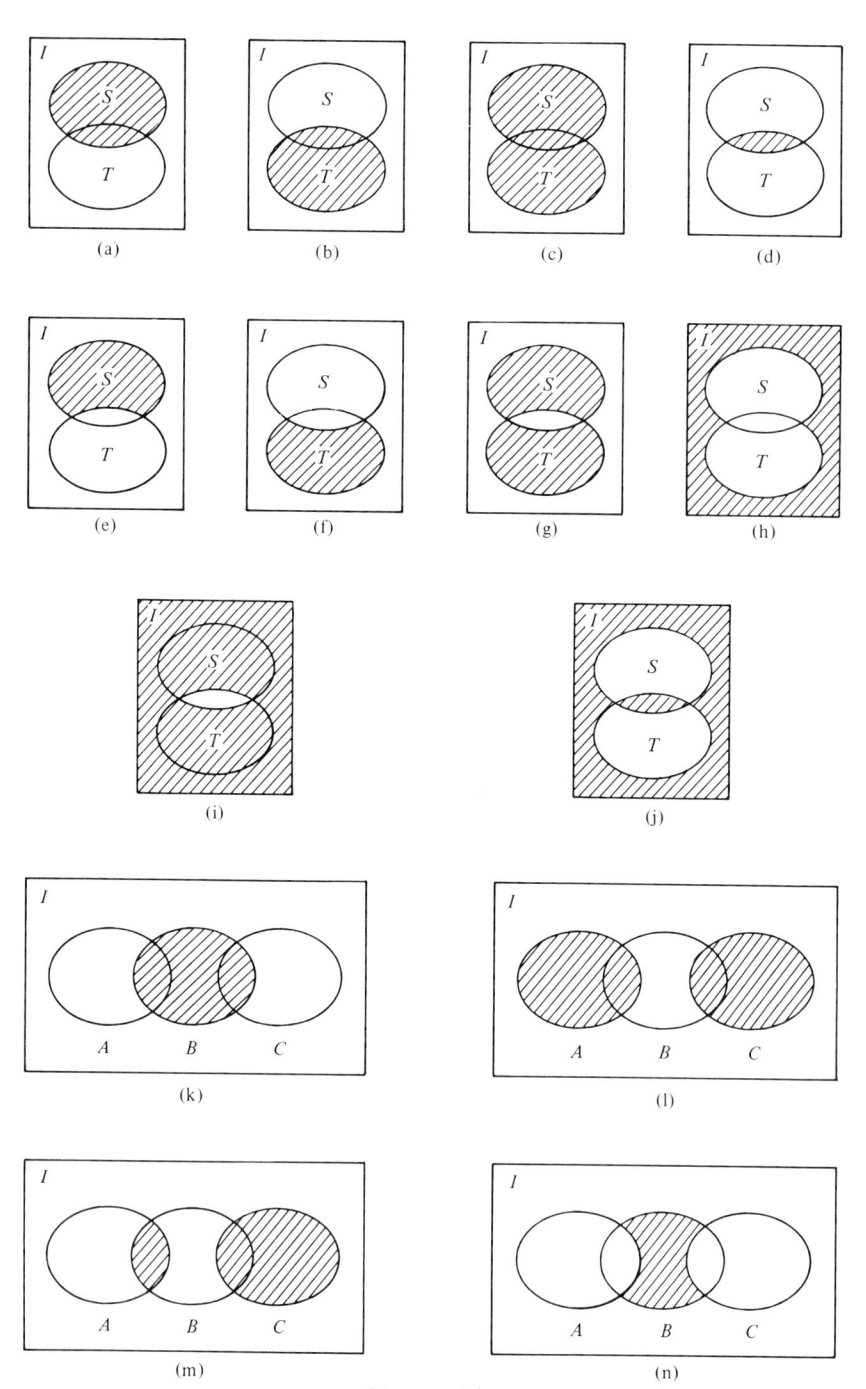

Figure 3.26

§3e   **Operations on Sets**   147

9. On a set diagram similar to that of Fig. 3.27 hatch the region which represents the given expression. I is the entire circular region; S the region bounded by the inner circle; A, B, C, D each a wedge-shaped region reaching to the center of I.

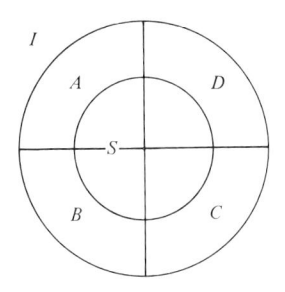

Figure 3.27

a) S'
b) $(A \cap S) \cup (C \cap S)$
c) $(A \cup C) \cap S$
d) $(A \cup B \cup C) \cap S$
e) $I \cap S'$
f) $S \cap S'$
g) $(A \cap S') \cup (B \cap S') \cup (C \cap S') \cup (D \cap S')$

10) On a set diagram similar to the one of Figure 3.28, hatch the region which represents the given expression. Set C is the circular region; Set S is the square region; and Set T is the triangular region.

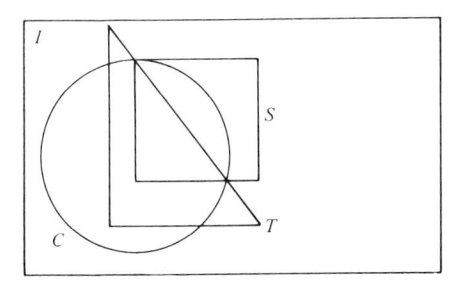

Figure 3.28

a) $S \cap T$
b) $T \cap C$
c) $T \cap C'$
d) $(C \cap S) \cap T'$
e) $(C \cap T) \cap S'$
f) $[(C \cap S) \cap T'] \cup (C \cap T')$

11. Given the sets $A = \{-, ¿, \Delta\}, B = \{\#, \theta\}$,
    a) find $A \cup B$
    b) find $B \cap A$
    c) Are the sets A, B disjoint?

12. Perform the indicated operations on the sets $A = \{a, b, c, d, e, f\}$;

$B = \{a, e, i, o, u\}$; $C = \{m, n, o, p, q, r, s, t, u\}$.
- a) $A \cup B$
- b) $A \cap B$
- c) $A \cup B \cup C$
- d) $A \cup C$
- e) $A \cap C$
- f) $B \cup C$
- g) $B \cap C$
- h) $A \cap B \cap C$

13. Given $A = \{a, b, c, d\}$, $B = \{\Delta, ¿, \theta\}$, $C = \{\#, \$\}$, find $(A \cup B) \cup C$.

14. The following are subsets of the universal set $I = \{x \mid x \text{ is a natural number less than } 20\}$: $A = \{1, 2, 3, 4, 5\}$; $B = \{2, 4, 6, 8, 10\}$; $C = \{3, 6, 9, 12, 15\}$. Write rosters of the following sets.
- a) $(A \cup B) \cap C$
- b) $A \cup (B \cap C)$
- c) $(A \cap C) \cup B$
- d) $A \cup B'$
- e) $B \cap A'$

15. If $A = \{a, c, e, g, i, k, m\}$ and $B = \{a, b, c, d, e, f, g\}$ write the following, if possible:
- a) $(A \cap B)'$ if $I = A \cup B$
- b) A' if $I = A \cup B$
- c) A' if $I = B \cup \{i, k, m\}$
- d) B' if $I = A \cup \{d, f, b\}$
- e) $(A \cup B)'$ if $I = \emptyset$

16. Use a set diagram to illustrate the principle of (3).

17. There are 57 students in a certain academic group. Three study French and German only. Six study French and Russian. Thirteen study German and Russian. Six take German only. Four take French, German, and Russian. Seven study Russian only. There are nineteen in all who study French.
- a) How many take no languages?
- b) How many take only French?

18. In a store fifteen people buy rulers and pens. Three buy only rulers. Ten buy rulers, pens, and pencils. Seventeen buy pens and pencils. No people buy only rulers and pencils. Four buy only pencils, six buy only pens.
- a) How many bought pens and pencils only?
- b) How many people are there?

19. Human blood is often classified according to presence or absence of antigens A, B, and Rh. Presence of A or B is indicated by this label, and absence of both by the label O. Presence of the Rh antigen is indicated by calling the blood positive and its absence by calling the blood negative. Suppose that 47 out of 120 different blood samples had antigen A, 35 had B, 102 had Rh, 12 had A and B, 39 had A and Rh, 30 had B and Rh, and 10 had all three antigens. How many of the samples are of each of these types?

§3e   Operations on Sets   149

- a) A-positive
- b) A-negative
- c) AB-negative
- d) B-positive
- e) B-negative
- f) O-positive
- g) O-negative

20. Given $X = \{a, b, c, d\}$, $Y = \{a, c, f, g\}$, $Z = \{c, d, e, f\}$, state whether each of the following propositions is true or false.
    - a) $(X \cup Y) \cap Z = X \cup (Y \cap Z)$
    - b) $(X \cup Z) \cap Y = X \cup (Z \cap Y)$
    - c) $(Z \cup X) \cap Y = Z \cup (X \cap Y)$

21. Given $A = \{\text{red, blue, green}\}$ and $B = \{\ \}$,
    - a) find $A \cup B$
    - b) find $A \cap B$
    - c) What plausible conjecture, if any, can be stated about the union of any set with the empty set?
    - d) What plausible conjecture, if any, can be stated about the intersection of any set with the empty set?

22. Given the set $X = \{a, b, c, d\}$ and $Y = \{c, d, e, f\}$, find
    - a) $X \cap Y$
    - b) $Y \cap X$
    - c) Using a set diagram, show $X \cap Y$ and $Y \cap X$.
    - d) Can you make a plausible conjecture regarding $X \cap Y$ and $Y \cap X$?

23. Make a conjecture about the relationships shown in (iii) of
    - a) Figure 3.15
    - b) Figure 3.17

24. For the sets X, Y, Z of Exercise 20, find
    - a) $(X \cap Y) \cap Z$
    - b) $X \cap (Y \cap Z)$
    - c) Make a conjecture about how $(A \cap B) \cap C$ and $A \cap (B \cap C)$ are related in general.

25. For the sets X, Y of Exercise 22 and for $I = \{a, b, c, d, e, f, g\}$, find
    - a) $X' \cap Y'$
    - b) $(X \cap Y)'$
    - c) $(X \cup Y)'$
    - d) Make a conjecture about how the expressions $X' \cap Y'$, $(X \cap Y)'$ $(X \cup Y)'$ are related.

The following propositions state many of the formal properties of union and intersection of sets. You should note that these theorems occur in pairs—

Part (a) and Part (b). One part makes a statement about union, the other about intersection.

For the most part, the theorems presented here will not be proved formally. Appeal will be made to set diagrams to see that the principle in question at least seems plausible. In certain cases for which it is difficult to draw a diagram which really illustrates the relationship, an appeal will be made to a basic principle upon which a proof could rest. The quantification in each theorem is universal unless specifically stated otherwise.

Now recall from Chapter 1 that a logically correct definition does not state existence of an entity. It merely describes something which may exist. Logically, then, a proof of existence should accompany any definition which does more than rename an entity. For the sake of brevity, such proofs are often omitted from mathematical developments.

The following statement of the existence of unions and intersection is included here basically for the sake of making a complete listing of properties of sets in order to facilitate comparison with properties of numbers (§1f and Chapter 6) and properties of propositions (§2j).

Theorem (3e-1): (a) $A \cup B$ is a set; (b) $A \cap B$ is a set.

The proof of (a) would depend on the facts that $x \in A$ and $x \in B$ are both propositions; thus, $x \in A \lor x \in B$ is also a proposition which then determines a set according to S-2.

Consideration of the figure for Theorem (3e-2) should make the stated propositions seem plausible. See also Figures 3.15 and 3.17.

Theorem (3e-2): (a) $A \subset (A \cup B)$; (b) $(A \cap B) \subset A$.

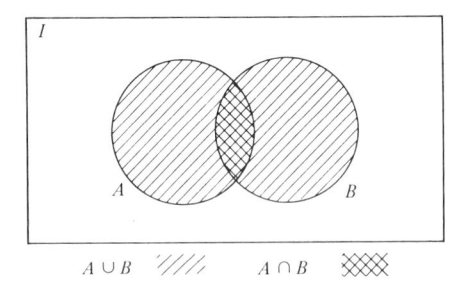

Proof of these principles would depend on the facts that $p \Rightarrow p \lor q$ and $p \land q \Rightarrow p$ are both tautologies.

EXAMPLE 12: Relate $A = \{3, 4\}$ to $A \cup B$ if $B = \{0, 1, 2\}$.

Solution: $A \cup B = \{3, 4\} \cup \{0, 1, 2\} = \{0, 1, 2, 3, 4\}$. Since
$$3 \in \{3, 4\} \Rightarrow 3 \in \{0, 1, 2, 3, 4\}$$

and
$$4 \in \{3, 4\} \Rightarrow 4 \in \{0, 1, 2, 3, 4\}$$
both hold true,
$$\{3, 4\} \subset \{0, 1, 2, 3, 4\}$$
i.e., $A \subset A \cup B$ as predicted in Theorem (3e-2), Part (a).

In Exercise 23 you were asked to form conjectures about Figures 3.15(iii) and 3.17(iii). If you saw the important relationships, the conjectures would have been as follows.

Theorem (3e-3): (a) $A \subset B$ iff $A \cup B = B$;
(b) $A \subset B$ iff $A \cap B = A$.

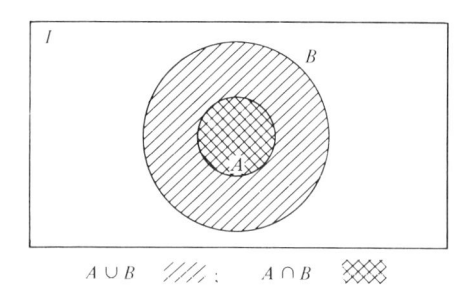

$A \cup B$ ////: $A \cap B$ ✗✗✗

Proofs of these principles can be constructed fairly easily if we realize that $(p \Rightarrow q) \subseteq (p \vee q \Leftrightarrow q)$ and $(q \Rightarrow p) \subseteq (p \wedge q \Leftrightarrow q)$. Using the basic theorem, the following corollaries can be quickly established.

Corollary (3e-3a): (a) $A \cup A = A$; (b) $A \cap A = A$.

EXAMPLE 13: Find $A \cap A$ if $A = \{s, t, u\}$.

Solution: $A \cap A = \{s, t, u\}$ by Part (b) of Corollary (3e-3a).

Corollary (3e-3b): (a) $A \cup \emptyset = A$: (b) $A \cap I = A$.
In words, the things which belong to A or to $\emptyset$ are simply the elements of A; the things which are common to A and I are precisely the elements of A.

Corollary (3e-3c): (a) $A \cup I = I$; (b) $A \cap \emptyset = \emptyset$.

In words, the things which belong to A or to I are precisely the elements of I; no elements are common to A and $\emptyset$.

Now consider the diagram for Theorem (3e-4) which shows a set and its complement. The things which belong to A or to A' are all the elements of I. From the definition of A' we can see that A and A' have no elements in common. The theorem restates these observations.

Theorem (3e-4): (a) $A \cup A' = I$; (b) $A \cap A' = \varnothing$.

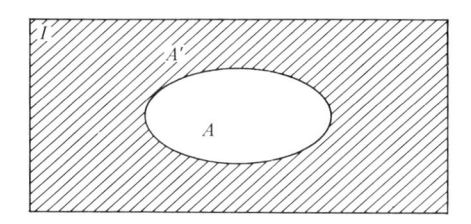

EXAMPLE 14. Find $A \cap A'$ if $A = \{1, 2, 3, 4\}$ and $I = \{0, 1, 2, 3, 4, 5, 6\}$.

Solution:
$$A' = \{x \mid x \notin A \wedge x \in I\}$$
$$= \{0, 5, 6\}$$

Now
$$A \cap A' = \{1, 2, 3, 4\} \cap \{0, 5, 6\}$$
$$= \varnothing$$

That is, A, A', have nothing in common.

In many "modern" mathematics textbooks written for the elementary grades, some of the properties of addition and multiplication of numbers are explained by appealing to the corresponding properties of sets. Some of these properties appear in the following three theorems.

Theorem (3e-5): (a) $A \cup B = B \cup A$; (b) $A \cap B = B \cap A$.

In Exercise 22 you were asked to form the conjecture that Part (b) of this proposition should hold. The theorem states that the result of combining sets under union or intersection alone is independent of order.

A conjecture formed in Exercise 24 indicates that the following principles should hold. They state that the result of combining sets by either union or intersection alone is independent of the grouping used.

Theorem (3e-6): (a) $(A \cup B) \cup C = A \cup (B \cup C)$
(b) $(A \cap B) \cap C = A \cap (B \cap C)$

The fact that these principles hold does *not* mean that parentheses, brackets, or other signs of grouping are unnecessary in set expressions. On the contrary, the result of Exercise 20 shows clearly that with a mixture of operations the signs of grouping are very necessary.

Even though all of the propositions of Exercise 20 are false, there are important true equations which relate the two operations. These are stated in the following theorem.

Theorem (3e-7): (a) $A \cap (B \cup C) = (A \cap B) \cup (A \cap C)$
(b) $A \cup (B \cap C) = (A \cup B) \cap (A \cup C)$

EXAMPLE 15: Using $X = \{a, b, c, d\}$, $Y = \{a, c, f, g\}$, $Z = \{c, d, e, f\}$ as in Exercise 20, verify that Part (a) of Theorem (3e-7) holds.

Solution:
$$X \cap (Y \cap Z) = \{a, b, c, d\} \cap [\{a, c, f, g\} \cup \{c, d, e, f\}]$$
$$= \{a, b, c, d\} \cap \{a, c, d, e, f, g\}$$
$$= \{a, c, d\}$$
$$(X \cap Y) \cup (X \cap Z) = [\{a, b, c, d\} \cap \{a, c, f, g\}] \cup [\{a, b, d, f\} \cup \{c, d, e, f\}]$$
$$= \{a, c\} \cup \{c, d\}$$
$$= \{a, c, d\}$$

As the theorem predicts, $\{a, c, d\} = \{a, c, d\}$.

Now recall Exercise 25. The conjecture formed there indicates that the following propositions hold.

Theorem (3e-8): (a) $(A \cup B)' = A' \cap B'$
(b) $(A \cap B)' = A' \cup B'$

EXAMPLE 16: Show that Part (a) of Theorem (3e-8) holds for the sets $I, \varnothing$.

Solution: $(I \cup \varnothing)' = I'$ by (3e-3b) (a)
           $= \varnothing$ by (3d-1b)

Also $I' \cap \varnothing' = \varnothing \cap \varnothing'$ by (3d-1b)
           $= \varnothing' \cap \varnothing$ by (3e-5)
           $= \varnothing$ by (3e-3c) (b)

We see that $(I \cup \varnothing)' = I' \cap \varnothing$

Using the set properties we have developed, many set expressions can be simplified algebraically. The following example illustrates one such case.

EXAMPLE 17: Simplify $(A' \cap B) \cup A$.

Solution:

$(A' \cap B) \cup A = A \cup (A' \cap B)$	(3e-5)
$= (A \cup A') \cap (A \cup B)$	(3e-7)
$= I \cap (A \cup B)$	(3e-4)
$= (A \cup B) \cap I$	(3e-5)
$= A \cup B$	(3e-3)

See Figure 3.29.

154 Sets

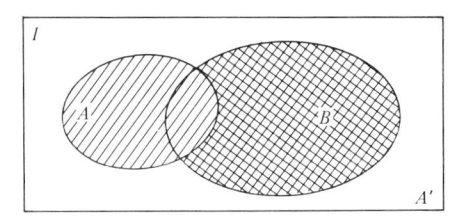

Figure 3.29

EXERCISE SET II

26. Compare the following theorems and corollaries with properties in the lists in §1f and §2j. Give each theorem the name which seems appropriate according to the comparison.
    a) (3e-1) Part (a)   b) (3e-1) Part (b)
    c) (3e-3a)   d) (3e-3b)
    e) (3e-4)   f) (3e-5)
    g) (3e-6)   h) (3e-7)
    i) (3e-8)

27. Using set diagrams or theorems, simplify the following expressions.
    a) $(\varnothing \cap M) \cup S$ b) $(A \cup I) \cup \varnothing$
    c) $M \cap (M \cup S)$ d) $M \cup (M \cap S)$

28. Reduce each of the following expressions to a single set, if possible, or make the simplest roster. See page 117 for description of number sets.
    a) $\{x \mid 0 < x < 3 \wedge x \in \mathscr{L}\} \cup \{y \mid y^2 + 5y + 6 = 0 \wedge y \in \mathscr{R}\}$
    b) $\{t \mid 0 < t < 5 \wedge t \in \mathscr{L}\}$
    c) $\{c \mid c^2 = -3 \wedge c \in \mathscr{R}\}$
    d) $\{1, 2, 3, 4\} \cup \{4, 5, 6\} \cap \{x \mid x^2 < 0 \wedge x \in \mathscr{L}\}$
    e) $\mathscr{N} \cup \mathscr{R}$
    f) $\mathscr{N} \cap \mathscr{Q}$
    g) $\mathscr{L} \cap \mathscr{W}$
    h) $\mathscr{R} \cap \varnothing$

29. Identify by *name* the principle which justifies the given statement. (See Exercise 26).
    a) $A \cup \varnothing = \varnothing \cup A$
    b) $(A \cap \varnothing) \cap A' = A \cap (\varnothing \cap A')$
    c) $\{\ \} \cap (S \cup M) = (\{\ \} \cap S) \cup (\{\ \} \cap M)$
    d) $(A' \cup B) \cap C = C \cap (A' \cup B)$
    e) $A' \cup \varnothing = A'$
    f) $S' \cap I = S'$
    g) $I \cap \varnothing = \varnothing$
    h) $(S \cup A')' = S' \cap A''$

§3e Operations on Sets

30. Rewrite the given expressions by using the commutativity principles (3e-5).
    a) $S \cup M'$
    b) $(S \cap M) \cup I$

31. Use the associativity principles (3e-6) to rewrite the following expressions.
    a) $(A \cap B) \cap B'$
    b) $A \cup [B \cup (M \cap S)]$

32. Use the distributive principles (3e-7) to find a set equal to each of the following expressions.
    a) $R \cap (S \cup M)$
    b) $S \cup (R \cap M)$
    c) $(Z \cup Y) \cap X$
    d) $(B \cap A) \cup (B \cap C)$
    e) $(M \cup R) \cap (M \cup S)$
    f) $(W \cap P) \cup (H \cap P)$

33. Simplify the following expressions, justifying each step by stating the theorem used.
    a) $(C \cup B) \cup C'$
    b) $(C \cap I) \cup C'$
    c) $(C \cup B') \cap (C \cup B')$
    d) $(A' \cap I)' \cup A'$
    e) $[(C \cup \emptyset)' \cup C]'$
    f) $(V \cap W') \cup W'$
    g) $S \cap (S \cap B)$
    h) $M \cap (M' \cap B)$
    i) $(S' \cap M')'$
    j) $(A \cup \emptyset)' \cap A$

34. Prove or disprove:
    a) $A \cap (B \cup C) = (A \cap B) \cup C$
    b) $M' \cup M' = M$
    c) $A \subset B \Rightarrow A \cap B = A$
    d) $(A \cap B)' = A' \cap B'$
    e) $(A \cup B)' = A' \cup B'$
    f) $(A \cap \emptyset)' = I$
    g) $A \cap B = A \cap C \Rightarrow B = C$
    h) $A \cap B = \emptyset \Rightarrow (A = \emptyset \vee B = \emptyset)$

35. Prove the *uniqueness* theorems: for all sets A, B, C,
    a) $A = B \Rightarrow A \cup C = B \cup C$
    b) $A = B \Rightarrow A \cap C = B \cap C$
    c) Are the converses of these propositions true? Give examples.

*36. Prove:
    a) Part (a) of (3e-1)
    b) Part (b) of (3e-1)
    c) Part (a) of (3e-2)
    d) Part (b) of (3e-2)
    e) Part (a) of (3e-3)
    f) Part (b) of (3e-3)
    g) Part (a) of (3e-3a)
    h) Part (b) of (3e-3a)
    i) Part (a) of (3e-3b)
    j) Part (b) of (3e-3b)
    k) Part (a) of (3e-3c)
    l) Part (b) of (3e-3c)
    m) Part (a) of (3e-4)
    n) Part (b) of (3e-4)
    o) Part (a) of (3e-5)
    p) Part (b) of (3e-5)
    q) Part (a) of (3e-6)
    r) Part (b) of (3e-6)
    s) Part (a) of (3e-7)
    t) Part (b) of (3e-7)
    u) Part (a) of (3e-8)
    v) Part (a) of (3e-8)

156 Sets

*37. Prove the following propositions without using any theorems of this section as premises.
    a) Part (a) of (3e-3a)   b) Part (b) of (3e-3a)
    c) Part (a) of (3e-3b)   d) Part (b) of (3e-3b)
    e) Part (a) of (3e-3c)   f) Part (b) of (3e-3c)

*38. Prove or disprove the following propositions.
    a) $A \cap B = A \iff A \cap B' = \emptyset$
    b) $A \cup B = B \iff A \cap B = A$
    c) $A \cap B' = A \iff A' \cup B' = I$

§3f   THE ALGEBRA OF SETS

The basic properties of sets under the operations of complementation, union, and intersection are called an algebra, in this case the algebra of sets. The ones regarded as most important are listed below with their usual names. Basic properties of set equality are listed first. Propositions stating actual set properties follow and are numbered A-1 through A-22. Note that set properties of $\in$ and $\subset$ are not considered a part of this algebra of sets.

    E-1: For any set S, $S = S$.   (3c-2a)

    E-2: For any sets S, M, $S = M \Rightarrow M = S$.   (3c-4)

    E-3: For any sets S, M, R, $S = M \wedge M = R \Rightarrow S = R$.   (3c-6)

These are called respectively the *reflexive*, *symmetric*, and *transitive* properties of set equality.

    A-1: For all S, R, $S \cap R$ is a set.   (3e-1)

    A-2: For all S, R, $S \cup R$ is a set.   (3e-1)

    A-3: For all S, R, M, $S = R \Rightarrow S \cap M = R \cap M$.   Ex. 35, §3e

    A-4: For all S, R, M, $S = R \Rightarrow S \cup M = R \cup M$.   Ex. 35, §3e

    A-5: For all S, $S \cap I = S$.   (3e-3b)

    A-6: For all S, $S \cup \emptyset = S$.   (3e-3b)

    A-7: For all S, $S \cap S' = \emptyset$.   (3e-4)

    A-8: For all S, $S \cup S' = I$.   (3e-4)

    A-9: For all S, R, $S \cap R = R \cap S$.   (3e-5)

    A-10: For all S, R, $S \cup R = R \cup S$.   (3e-5)

    A-11: For all S, R, M, $(S \cap R) \cap M = S \cap (R \cap M)$.   (3e-6)

§3f The Algebra of Sets

A-12: For all S, R, M, $(S \cup R) \cup M = S \cup (R \cup M)$. (3e-6)

A-13: For all S, R, M,
$$S \cap (R \cup M) = (S \cap R) \cup (S \cap M).$$ (3e-7)

A-14: For all S, R, M,
$$S \cup (R \cap M) = (S \cup R) \cap (S \cup M).$$ (3e-7)

A-15: For all S, R, $(S \cap R)' = S' \cup R'$. (3e-8)

A-16: For all S, R, $(S \cup R)' = S' \cap R'$. (3e-8)

A-17: For all S, $S \cap S = S$. (3e-3a)

A-18: For all S, $S \cup S = S$. (3e-3a)

A-19: For all S, $S'' = S$. (3d-2)

A-1 and A-2 state that sets are *closed* under each operation. A-3 and A-4 state that intersection and union are *uniquely defined* (well defined) operations. A-5 and A-6 state that I and $\varnothing$ are *neutral (identity) elements* for the operations of $\cap$ and $\cup$ respectively. A-7 and A-8 state that the complement of a set is its *inverse* relative to $\cap$ and its *inverse* relative to $\cup$. A-9 and A-10 state that each operation is *commutative*. A-11 and A-12 state that each operation is *associative*. A-13 and A-14 state that each operation is *distributive over* the other. A-15 and A-16 are called *DeMorgan's Laws* for sets. A-17 and A-18 are called *idempotent* properties. A-19 is the *Rule of Double Complementation*.

The last three properties are separated from the rest only because they have no standard names.

A-20: For all S, $S \cap \varnothing = \varnothing$. (3e-3c)

A-21: For all S, $S \cup I = I$. (3e-3c)

A-22: For all S, R, $S = R \Rightarrow S' = R'$. (3d-3a)

Now compare these statements about sets with properties A-1 through A-17 of propositions as given in §2j, using the correspondence.

(1)
disjunction to union	$\vee \longleftrightarrow \cup$
conjunction to intersection	$\wedge \longleftrightarrow \cap$
negation to complement	$\sim \longleftrightarrow '$
true proposition to universe	$t \longleftrightarrow I$
false proposition to null set	$f \longleftrightarrow \varnothing$

We see that the properties of the two algebras are basically the same in form.

In some developments of the algebra of sets alone, i.e., without use of $\in$ and $\subset$, such properties as A-1, A-2, A-3, A-4, A-5, A-6, A-7, A-8, A-9,

A-10, A-13, and A-14 are restated as axioms. The remaining properties are then developed from these. An algebra having these basic properties is called a Boolean algebra in honor of George Boole (1815–1864) who first developed a system of symbolic logic.

New principles of this algebra may be established by using A-1 through A-22 as shown in the following examples.

EXAMPLE 1: Prove or disprove: $A \cap B = \varnothing \Rightarrow A \cap B' = A$.

Solution: Although a diagram is not a proof, it can be used to see whether the conjecture seems plausible. Figure 3.30 is drawn so that the hypothesis is satisfied. The intersection of A and B is obviously empty. Since B' is represented by the hatched region, the crosshatched intersection of A and B' represents A. The conjecture does seem plausible. If it did not, we would attempt a *disproof*.

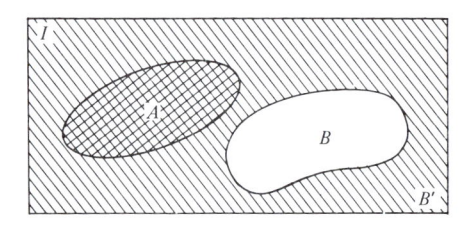

Figure 3.30

Proof:

1.	$A = A \cap I$	1. Neutral element for $\cap$	A-5
2.	$= A \cap (B \cup B')$	2. Inverse property for $\cup$	A-8
3.	$= (A \cap B) \cup (A \cap B')$	3. Distributivity of $\cap$ over $\cup$	A-13
4.	$= \varnothing \cup (A \cap B')$	4. Hypothesis	
5.	$= (A \cap B') \cup \varnothing$	5. Commutativity of $\cup$	A-10
6.	$= A \cap B'$	6. Neutral element for $\cup$	A-6
7.	$A \cap B' = A$	7. Symmetric property of $=$	E-2

Q.E.D.

Although use of E-2 is illustrated in Step 7 of this proof, arguments of this kind are traditionally shortened by making almost no explicit use of E-1 through E-3, A-1 through A-4, and the inference rules. For example, Step 4 states that

$$A = \varnothing \cup (A \cap B')$$

and "Hypothesis" is given as the reason. Without abbreviating one would have

$$A \cap B = \varnothing$$

by hypothesis,
$$A \cap B'$$
is a set by A-1,
$$A \cap B' = A \cap B'$$
by E-1, whence
$$(A \cap B) \cup (A \cap B') = \varnothing \cup (A \cap B')$$
by A-4.

Since $A = (A \cap B) \cup (A \cap B')$ from Step 3, we would now have
$$A = \varnothing \cup (A \cap B')$$
by E-3.

Thus, going from Step 3 to Step 4 with all this detail included would be more laborious but might be more understandable to the beginner.

EXAMPLE 2: Prove or disprove: $A \cap B' = \varnothing \Rightarrow A \cup B' = I$.

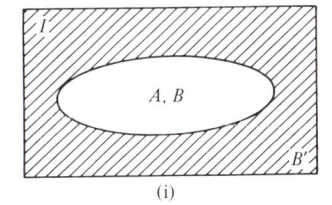

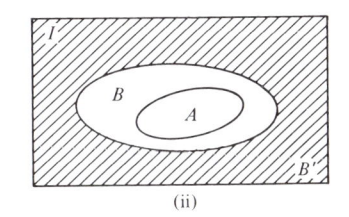

Figure 3.31

Solution: Again we use set diagrams to test plausibility. In Diagram (i) of Figure 3.31, $A \cap B' = \varnothing$ is true and $A \cup B' = I$ is also true. Thus the given proposition would be true. However, a look at Diagram (ii) shows a case for which $A \cap B' = \varnothing$ is true but $A \cup B' = I$ is false. In this case the given proposition is false. Thus, a disproof is required. To find a concrete counterexample, let $A = \varnothing$. Then
$$\begin{aligned} A \cap B' &= \varnothing \cap B' \\ &= B' \cap \varnothing \\ &= \varnothing \end{aligned}$$
as required to have the hypothesis true. However,
$$\begin{aligned} A \cup B' &= \varnothing \cup B' \\ &= B' \cup \varnothing \\ &= B' \\ &\neq I \text{ for all possible } B \end{aligned}$$

Thus, the conclusion is false, therefore, the given proposition is false.

EXERCISE SET

1. Using the correspondences of (1), write the logical proposition which corresponds to the given set proposition.
   a) $(\emptyset \cap M')' = \emptyset' \cup M$
   b) $M' \cap (I \cup M') = M'$

2. Simplify the following expressions using only E-1 through E-3 or A-1 through A-22 to justify steps.
   a) $(C \cup B) \cup C'$
   b) $(C \cap I) \cup C'$
   c) $(C \cup B') \cap (C \cup B')$
   d) $(A \cap I)' \cup A'$
   e) $[(C \cup \emptyset)' \cup C]'$
   f) $(V \cap W') \cup W'$
   g) $S \cap (S \cap B)$
   h) $M \cap (M' \cap B)$
   i) $(S' \cap M')'$
   j) $(A \cup \emptyset)' \cap A$

3. The **relative complement of B in A** is defined as $A - B = A \cap B'$, i.e., the relative complement of B in A is the set of elements in A but not in B. For each relative complement write a simplified expression without the minus sign.
   a) $I - A$
   b) $A - \emptyset$
   c) $\emptyset - A$
   d) $A - I$
   e) $A - A$
   f) $A - A'$
   g) $A - (B \cap C)$
   h) $(A \cup B) - C$

4. The **symmetric difference** $A \Delta B$ of two sets A, B is defined as $A \Delta B = (A \cup B) \cap (A \cap B)'$. For each symmetric difference, write a simplified expression without the "Δ" sign.
   a) $S \Delta \emptyset$
   b) $S \Delta S$
   c) $S \Delta I$
   d) $S \Delta S'$

5. Prove or disprove the following propositions using only E-1 through E-3 or A-1 through A-22 to support your argument.
   a) $(I \cup M) \cap S = S$
   b) $(A \cap \emptyset) \cup I = \emptyset$
   c) $A \cap B = A \iff A \cap B' = \emptyset$
   d) $A \cup B = B \iff A \cap B = A$
   e) $A \cap B' = A \iff A' \cup B' = I$
   f) $A' \cap B' \neq \emptyset \iff A' \cap B \neq \emptyset$
   g) $A \cap B' = \emptyset \iff A' \cap B = \emptyset$
   h) $A \cap B = \emptyset \iff A' \cup B' = I$
   i) $A - B = B - A$
   j) $A \Delta B = B \Delta A$

§3g QUANTIFIED ARGUMENTS

Axiom S-2 states that a set A such that $A \subset I$ has elements for which some open sentence $p(x)$ is true. Recall that A is termed the solution set or truth set

§3g   **Quantified Arguments**   161

of $p(x)$. Under certain circumstances the relationships among such solution sets can be used to represent arguments and these representations can in turn indicate whether the arguments are valid or invalid.

This is especially useful in the case where premises and conclusions are quantified. In §2h we saw that any quantified open sentence in one variable is actually a proposition. Now by Axiom S-2 each such proposition determines a set.

In analyzing quantified propositions by set diagrams, the concepts of subsets, equality of sets, and the complement of a set as specified in Definitions (3-3), (3-4), and (3-8), respectively, are often employed. For example, let us interpret this sentence.

(1) All s's are m's.

This sentence states that each element s of a set S is also an element of set M, i.e.,

$$s \in S \Rightarrow S \in M$$

By Definition (3-3) this means

(2) $$S \subset M$$

In attempting to illustrate (2), recall the discussion of set diagrams in §3c.

For nonempty S, M, there are two possible ways of representing the relationship of (2). Part (i) of Figure 3.32 illustrates the case for which $S \neq M$ and Part (ii) the case for which $S = M$. Both cases are possible since by Theorem (3c-2) every set is a subset of itself. Thus, S may be all of M as shown in (ii).

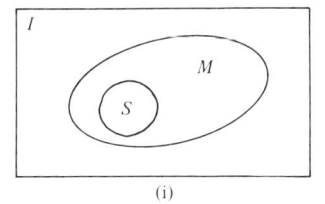

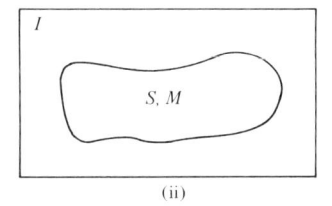

Figure 3.32

Now that we know how to use sets to diagram a universally quantified proposition, how do we apply this to analysis of arguments? Recall that an argument is valid when there is no case in which the premises are each true and the conclusion false.

(3) In analyzing an argument which contains quantified propositions, our aim is to produce diagrams which show the premises to be simultaneously true. We look at *all* such diagrams to see if the conclusion *must* be true in each one.

EXAMPLE 1: Use a set diagram to determine whether the following argument is valid or invalid.

162 *Sets*

P(a) All zoys are glachos.
P(b) <u>All glachos are ootsters.</u>
        All zoys are ootsters.

Solution: Let

$$Z = \{x \mid x \text{ is a zoy}\}$$
$$G = \{x \mid x \text{ is a glacho}\}$$
$$O = \{x \mid x \text{ is an ootster}\}$$

We interpret P(a) as $Z \subset G$, P(b) as $G \subset O$. Each of the four diagrams in Figure 3.33 shows the premises simultaneously true. In each of these possible diagrams the conclusion

$$Z \subset O \text{ is true.}$$

Hence the argument is valid.

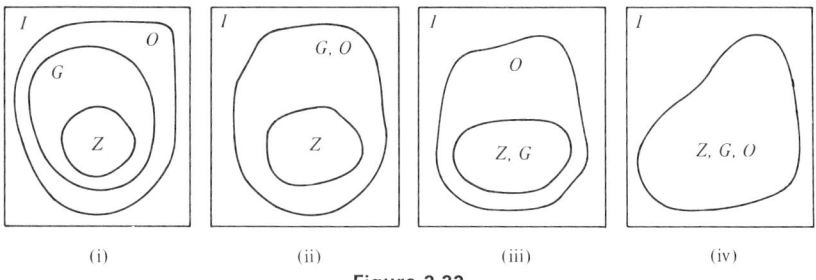

Figure 3.33

Note: In this example, Diagram (i) is the diagram that is minimally satisfied and usually we use it alone to represent all four cases without drawing the others but remembering to check mentally whether it is possible to draw a set diagram which would invalidate the argument.

EXAMPLE 2: Use a set diagram to determine whether the following argument is valid or invalid.

P(a) All zoys are glachos.
P(b) No shfnexi is a glacho.
∴, no zoy is a shfnexi.

Solution: Let

$$Z = \{x \mid x \text{ is a zoy}\}$$
$$G = \{x \mid x \text{ is a glacho}\}$$
$$S = \{x \mid x \text{ is a shfnexi}\}$$

We interpret P(a) as $Z \subset G$, P(b) as S and G are disjoint. Diagramming, we obtain (i) and (ii) of Fig. 3.34. In each of the possible diagrams the conclusion "Z and S are disjoint" is true. Hence the argument is valid.

§*3g*   *Quantified Arguments*

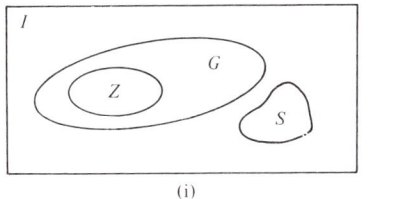

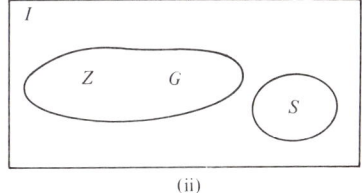

Figure 3.34

Note: In this example, Diagram (i) is the diagram that is minimally satisfied, and usually we use it alone to represent all cases.

EXAMPLE 3: Use a set diagram to determine whether the argument of Exercise 8, §2a, is valid or invalid.

Solution: We are given:

         P(a)   All tired tigers are ticklish.
         P(b)   All men are ticklish.
         ∴,   all tired tigers are men.

Letting $S = \{x \mid x$ is a tired tiger$\}$, $T = \{x \mid x$ is ticklish$\}$, and $M = \{x \mid x$ is a man$\}$, we interpret P(a) as $S \subset T$ and P(b) as $M \subset T$. These relationships are illustrated in Figure 3.35. Note in particular that we can conclude nothing about the relationship of S and M to each other, except that they are both subsets of T. They do not need to overlap as in (ii). We *might* have $S \subset M$ (iii), $M \subset S$ (iv), $M = S$ (v) and (vi), or M and S completely disjoint (i) as far as the premises are concerned; (iii), (v), and (vi) show that the stated conclusion $S \subset M$ is true. However, (i), (ii), and (iv) are also all possible and represent cases for which the premises are each true but the conclusion is false. Therefore, the argument is invalid.

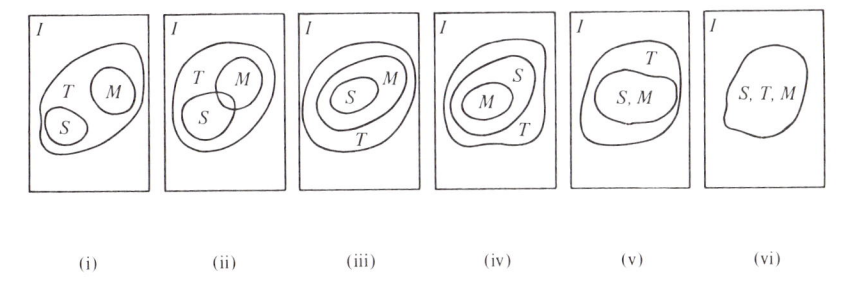

Figure 3.35

In (1) we diagrammed a universally quantified proposition. Now let us consider the existential case.

(4) Some *s*'s are *m*'s.

This sentence states that *at least one* element x of a set S is also an element of another set M, i.e., the two sets S and M cannot be disjoint. Hence, any of the diagrams in Figure 3.36 illustrates this relationship. In each diagram

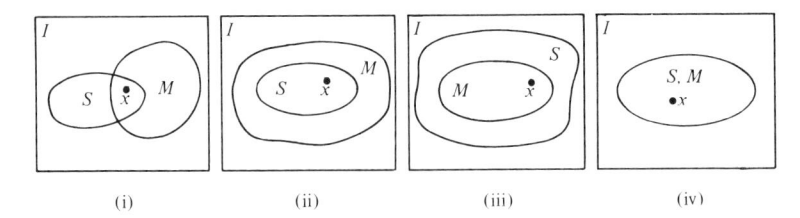

Figure 3.36

the dot labeled x represents the "at least one x" which is an element of both S, M. If any of these diagrams seems unreasonable, the reader should review Definition (2-15) of the existential quantifier.

You must keep in mind that "some *s*'s are *m*'s" means "at least one *s* is an *m*." The statement says nothing about the rest of the elements of set *S*.

Note: Diagram (i) of Fig. 3.36 is the diagram that is minimally satisfied, and we usually use it by itself to represent all cases.

Arguments that use both quantifiers can be diagrammed by combining the methods which have been discussed.

EXAMPLE 4: Use set diagrams to determine whether the following argument is valid or invalid.

P(*a*) Some *m*'s are *s*'s. ∴, some *s*'s are not *m*'s.

Solution: Let $S = \{s \mid s \in S\}$; $M = \{m \mid m \in M\}$. Any of the diagrams in Figure 3.36 shows the premise true. However, diagrams (ii), (iv) show the conclusion false. Hence, this argument is *invalid*.

EXAMPLE 5: Analyze the following argument by set diagrams.
(a) All integers are rational numbers.
(b) Some integers are even.
∴, some rational numbers are even.

Solution: Let $\mathscr{I} = \{x \mid x \text{ is an integer}\}$; $\mathscr{Q} = \{x \mid x \text{ is a rational number}\}$, and $E = \{x \mid x \text{ is even}\}$. Figure 3.37 shows some of the possible ways of diagramming the argument. The dot labeled x represents the "at least one" integer which belongs to the set of even numbers. The conclusion can be stated as

(5) $$\exists x \in \mathscr{Q} \cdot \ni \cdot x \in E$$

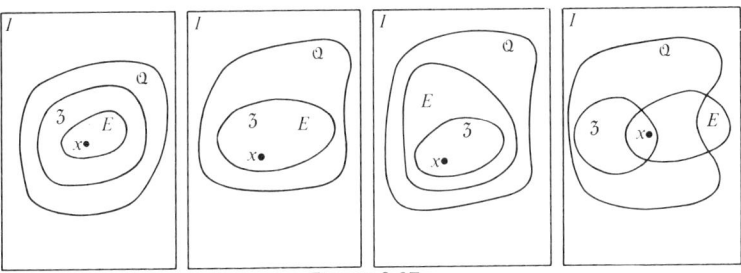

Figure 3.37

Each of the diagrams in Figure 3.37 shows this conclusion true, but we have not exhausted all the possible diagrams and cannot yet decide whether the argument is valid.

(6) Rather than attempting to draw all possible cases to show such an argument valid, let us look instead for a case that would make it invalid. An invalid case would have the premises true and the conclusion false.

The negation of the conclusion in this argument would be

(7) $\sim [\,\exists\,(x) \in \mathscr{Q}, x \in E\,] \equiv \forall (x) \in \mathscr{Q}, x \notin E$

i.e., no rational numbers are even. Thus, conclusion (7) requires that $\mathscr{Q}$ and E be disjoint. Hence, to show the invalid case we must have:

    (a) $\mathscr{X} \subset \mathscr{Q}$;
    (b) $x \in \mathscr{X} \wedge x \in E$;
    (c) $\mathscr{Q}$ and E disjoint.

However, this is impossible because $x \in \mathscr{X}$ requires $x \in \mathscr{Q}$ by definition of subset. Then $\mathscr{Q}$ and E have at least one element in common and cannot be disjoint. We have shown that it is impossible to have the invalid case; hence, this argument is *valid*.

Note: The perceptive reader will see a similarity in the approach outlined in (6) and the Rule of Indirect Inference (I.I.).

These examples illustrate that in order to test a quantified argument by Venn or Euler diagram, it is necessary to consider all the possible interpretations of the premises. If any *one* interpretation will lead to the negation of the conclusion, then the argument is invalid. *To show validity then, all possible diagrams must lead to the same conclusion.* It should be sufficient in this regard to show on a minimally satisfied diagram that the conclusion must be true. That is, we use one that is overlapping rather than showing a subset or equal sets.

166 Sets

Examples 6 through 12 illustrate *invalid* forms.

EXAMPLE 6: All A is in B.
$\underline{x \text{ is in } B.}$
x is in A.

See Figure 3.38.

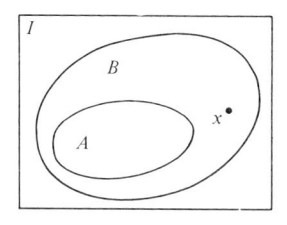

Figure 3.38

EXAMPLE 7: All A is in B.
$\underline{x \text{ is not in } A.}$
x is not in B.

See Figure 3.38.

EXAMPLE 8: All A is in B.
$\underline{\text{All } C \text{ is in } B.}$
All C is in A.

See Figure 3.39.

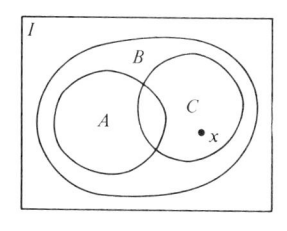

Figure 3.39

EXAMPLE 9: All A is in B.
$\underline{\text{It is false that all } C \text{ is in } A.}$
It is false that all C is in B.

See Figure 3.39.

EXAMPLE 10: Some A is in B.
$\underline{z \text{ is in } A.}$
z is in B.

See Figure 3.40.

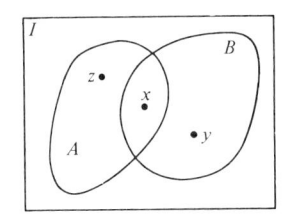

Figure 3.40

EXAMPLE 11: Some A is in B.
y is not in A.
───────────
y is not in B.

See Figure 3.40.

EXAMPLE 12: All A is in B.
───────────
Some A is in B.

See Figure 3.41. Since A may be empty, we have no right to indicate an element x as shown in Diagram (ii).

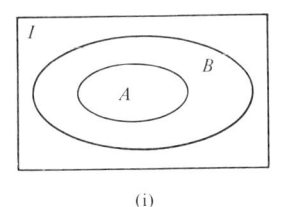

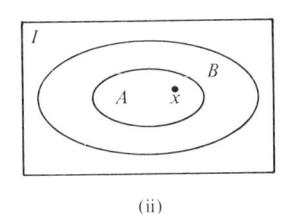

Figure 3.41

EXERCISE SET

1. Use set diagrams to determine the validity or invalidity of the following arguments from §2a.
   a) The argument of Example 2 leading to conclusion (ii).
   b) Exercise 5.
   c) The argument of Exercise 7.

In Exercises 2 through 35, use set diagrams to analyze the given quantified arguments for validity or invalidity.

2. Some cats (C) are dogs (D). No lizards (L) are cats. Therefore, no cats are lizards.

3. a) Normal human beings have no arms.

b) All snails are normal human beings.
∴, no snail has arms.

4. Use the premises of Exercise 2 leading to the conclusion: "Therefore, no lizards are dogs."

5. (a) Roses are red and violets are blue.
   (b) All blue things are red.
   (c) Some red things are blue.
   ∴, some violets are roses.

6. Use the argument of Exercise 5 leading to: "Therefore, some roses are blue."

7. All pumas (P) are cats (C). All angoras (A) are pets (Q). Therefore, some pets are pumas.

8. No irrational numbers ($\mathscr{Q}'$) are rational numbers ($\mathscr{Q}$). All rational numbers are real numbers ($\mathscr{R}$). Therefore, some real numbers are not rationals.

9. No bobolinks (B) are robins (R). Some robins are people (P). Therefore, some people are not bobolinks.

10. All sea gulls (S) eat garbage (G). No garbage eaters fly (F). Therefore, no fliers are sea gulls.

11. Some Salem students (S) are mathematicians (M). All mathematicians are teachers (T). Therefore, some teachers are Salem students and some Salem students are teachers.

12. No even numbers (E) are divisible by nine (N). All numbers divisible by nine are divisible by three (T). Therefore, some numbers divisible by three are not even.

13. All differentiable functions (F) are continuous (C). Some relations (R) are continuous. Therefore, some relations are not differentiable functions.

14. All rhombi (R) are kites (K). All squares (S) are rhombi. Therefore, all squares are kites.

15. All equilateral triangles (E) are isosceles (S). Some acute triangles (A) are equilateral. Therefore, some acute triangles are isosceles.

16. All equiangular triangles (A) are equilateral (L). All equilateral triangles are isosceles (S). Some right triangles (R) are not equiangular. Therefore, some equilateral triangles are not right triangles.

17. No prime numbers ($\mathscr{P}$) are irrationals ($\mathscr{Q}'$). Some integers ($\mathscr{I}$) are primes. Therefore, some integers are not irrationals.

§3g   **Quantified Arguments**   169

18. All A is in B. All B is in A. All A is in C. Therefore, some B is not in C.
19. All R is in M. No S is in M. Therefore, no S is in R.
20. All M is in S. All T is in S. Some M is in T. Therefore, not all M is in T.
21. Some M is not in R. All M is in S. Therefore, some S is not in R.
22. All R is in M. Some S is not in M. Therefore, some S is not in R.
23. All R is in M. All M is in S. Therefore, some S is in R.
24. $(\forall x \in R, x \notin M) \subseteq (\forall y \in M, y \notin R)$. $\exists y \in S, y \in M$. Hence, $\exists y \in S, y \notin M$.
25. $\forall x \in M, x \in R$. $(\exists x \in M, x \in S) \subseteq (\exists y \in S, y \in M)$. Hence, $\exists y \in S, y \in R$.
26. $\forall x \in M, x \in S$. $\forall x \in T, x \in S$. $\exists x \in M, x \in T$. Hence, $\forall x \in M, x \in T$.
27. $\exists x \in M, x \in T$. $\forall x \in M, x \in A$. $\forall x \in T, x \in B$. Hence, $\exists x \in B, x \in A$.
28. Some rational numbers have a factor of 5. All integers are rational numbers. Therefore, some integers have a factor of 5.
29. O_2 helps everyone. Someone wants air. Therefore, there is someone who is helped by O_2 and who wants air.
30. Little lambs are quadrupeds. Quadrupeds eat ivy. Therefore, little lambs eat ivy.
31. Wombats are marsupials. Opossums are marsupials. Therefore, some opossums are wombats.
32. All people who like baseball (B) like ice cream (C). No people who play the violin (V) like ice cream. Some Eldens play the violin. Therefore, no people who like baseball are Eldens.
*33. Persons of average intelligence can do calisthenics. A person without average intelligence is not physically fit. Your children cannot do calisthenics. Therefore, your children are not physically fit.
*34. No trustworthy person who is a lawyer is untruthful. Some people are untruthful. Therefore, no trustworthy people are lawyers.
*35. Some egotists are egoists. Some egotists like all egoists. Therefore, all egoists are not disliked by all egotists and some egoists are liked by some egotists.

§3h MISCELLANEOUS EXERCISES

1. a) Describe the set of eight-foot-tall teachers in your school in set builder notation. What is the universal set? Let it be represented by I. Thus: $\{x \mid x \in I \wedge x \text{ is eight feet tall}\}$.

   In the following exercises select the universal set and use the set builder notation to specify the given sets.
   b) The set of mathematics teachers in your school.
   c) The set of male teachers whom you have during the day.
   d) The set of white houses on your street.

2. Let the Universe of Discourse be "Presidents of the U.S.A.," represented by P. Use both set builder and roster notation to specify the set described.
   a) The last five presidents of the U.S.
   b) The last five presidents who were Democrats.
   c) One of the last five presidents and came from (was born in) New England.
   d) All the presidents who were formerly generals in the U.S. Army.
   e) The set which would be the complement of the set in Exercise 2b above. Describe this set.

3. Describe the complement of the set in each part of Exercise 1.

4. In Figure 3.42, the circle represents the set A, the triangle represents the set B, and the rectangle represents the set C.

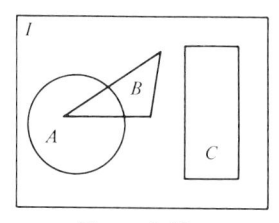

   Figure 3.42

   a) Draw horizontal lines through $(A \cup B)$.
   b) Draw vertical lines through $(A \cup C)$.
   c) What region represents $(A \cap B)$?
   d) What region represents $(A \cap C)$?
   e) What region represents $(A \cap B)' \cap C$?
   f) What region represents $(A \cup B) \cap (C \cup I)$?

5. If $I = \{a, b, c, d, e\}$, $A = \{a, c\}$, $B = \{a, b, d\}$, and $C = \{d, e\}$, find:
   a) $A \cup B$
   b) $A \cap B$
   c) $A \cap C$

d) $A \cup B'$
e) $(A \cap B) \cap B'$
f) How are A and B related?
g) How are A and C related?
h) How are $B \cap C$ and B related?

6. If $I = \{0, 1\}$, state the complement of each of the given sets.
   a) $\{x \mid x \cdot a = x \wedge a \in \mathscr{L} \wedge x \in \mathscr{L}\}$
   b) $\{x \mid a \cdot x = a \text{ for any number } a\}$
   c) $\{x \mid a + x = a \text{ for any number } a\}$
   d) $\{x \mid x + a = x \wedge a \in \mathscr{Q} \wedge x \in \mathscr{Q}\}$

7. Let A be the set of letters of the word $EVIL$ and let B be the set of letters in the spelling $LIVE$. Is $A = B$? Why?

8. Under what circumstances are the following statements true?
   a) $A \cup B = A \cap B$ b) $A \cap B = B'$
   c) $A \subset \emptyset$ d) $A \cap B = B$
   e) $(A \cup B) \cap B' = B$ f) $(A \cap B') \cup B = A \cup B$

9. We know that one possible subset of a set A is A itself. Sometimes one wishes to distinguish between this general subset concept and a concept which does not allow a set to be a subset of itself. For this purpose the concept of proper subset is introduced. Then the symbol $\subset$ is used to indicate a proper subset and the symbol $\subseteq$ to indicate any subset in general. Then one defines "A is a **proper subset** of B, denoted $A \subset B$, iff $A \subseteq B \wedge A \neq B$, where $A \subseteq B$ indicates that A is any subset of B in general." Using this notation, state which of the diagrams in Figure 3.43 correctly describes the relationship specified. The correct response may include more than one of the diagrams.

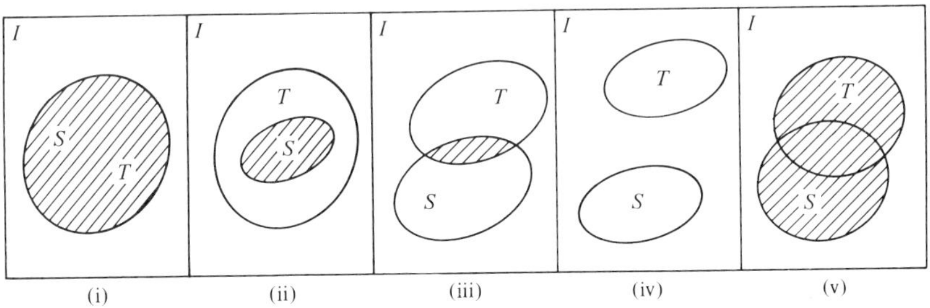

Figure 3.43

a) $S \cup T \wedge S$ overlaps T b) $S \cap T \wedge S$ overlaps T
c) $S \subset T$ d) $S = T$
e) $S \subseteq T$ f) $S \cup T = T$
g) $S \cap T = \emptyset$ h) S, T are disjoint
i) $S \neq T$ j) S, T are not disjoint
k) $S \not\subset T$ l) $S \not\subset T \wedge S \subseteq T$

172   Sets

10. Fill in the following statements:
    a) If $A = B$, then $A \cup B$ _____ A and $A \cup B$ _____ B.
    b) If A and B are overlapping sets, then $A \neq$ _____ and $A \cap B \neq$ _____.
    c) If $A \supset B$, then $A \cup B =$ _____.
    d) If $A = B$, then $A \cup B =$ _____ and $A \cap B =$ _____.
    e) If $A \subset B$, then $A \cap B =$ _____.
    f) $A = B$ if and only if A _____ B and B _____ A.
    g) If $A \cap B = \emptyset$, $A \neq \emptyset \lor B \neq \emptyset$, then set A' and set B' are _____.
    h) If $A' \cup B' = I$, then A, B are _____.

11. a) Is $\{\emptyset\}$ a permissible set? Explain.
    b) Is $\{\emptyset\} = \emptyset$? Explain.
    c) Is $\emptyset \subset \{\{\emptyset\}\}$? Explain.

12. a) If set S is a subset of the set T, to what set is $S \cup T$ equal?
    b) Is it possible for a set A to overlap a set B without having set B be a subset of set A or vice versa?

13. State whether $A = B$.
    a) $A = \{0, \text{I}, \text{II}\}$, $B = \{x \mid x + 1 = 1 \lor x^2 - 3x + 2 = 0\}$
    b) $A = \emptyset$, $B = \{0\}$
    c) $A = \emptyset$, $B = \{x \mid x = x + 4\}$
    d) $A = \emptyset$, $B = \{\emptyset\}$

14. State whether true or false.
    a) $\{1, 1, 1\} \neq \{1\}$
    b) $\{\text{Earth}\} = \{\text{Earth, solar planet between Venus and Mars}\}$
    c) Every element of $\emptyset$ is an even number.
    d) Every element of $\emptyset$ is an empty set.
    e) $\emptyset \in \{1, 2\}$

15. Find open sentences as simple as possible which determine each solution set over $\mathscr{L}$.
    a) $\{0, 2, 4, 6, 8\}$ b) $\{-2, -1, 0, 1, 2\}$
    c) $\{1, 4, 9, 16\}$ d) $\{1, 2, 3, 5, 8, 13\}$
    e) $\{0, 1\}$

16. Make a roster of the following sets, if possible.
    a) Set of all x such that
    $$x \in \{0, 1, 2, 3\} \land x \in \{1, 2, 3, 4\}$$
    b) Set of all x such that
    $$x \in \{0, 1, 2, 3\} \lor x \in \{1, 2, 3, 4\}$$
    c) Set of all x such that
    $$x \in \{0, 1, 2, 3\} \land x \notin \{1, 2, 3, 4\}$$

d) Set of all x such that
$$x \in \{0, 1, 2, 3\} \lor x \notin \{1, 2, 3, 4\}$$

17. Suppose a class of 30 students has the following statistics: 19 take mathematics, 17 take music, 11 take history, 12 take math and music, 7 take history and math, 5 take music and history, 2 take all three. Answer the following questions about the class.
    a) What number take history but not mathematics?
    b) How many study none of these subjects?
    c) How many study only math?
    d) How many study math and music but not history?

18. There are $\frac{1}{3}$ as many people in a group who swim and play tennis only as there are people who participate in all three sports (tennis, swimming, golf). There are 24 people who swim and play tennis. There are 20 who swim and play golf. Six people swim only. Twice as many people play golf only as those who swim only. Twenty-two people play tennis and golf. No one plays only tennis. How many people are there in all?

19. If "∘" (bubble) and "§" (squiggle) are binary operations on elements of a set $S = \{a, b, c, \ldots\}$, what would each of the following properties be called?
    a) $\forall a, b \in S, (a \,\S\, b) \in S$
    b) $\forall a, b, c \in S, a \circ (b \,\S\, c) = (a \circ b) \,\S\, (a \circ c)$
    c) $\exists \pi \in S \cdot \ni \cdot \forall a \in S, a \,\S\, \pi = a$
    d) $\forall a, b, c \in S, a \circ (b \circ c) = (a \circ b) \circ c$
    e) $\forall a, b \in S, a \,\S\, b = b \,\S\, a$
    f) $\forall a \in S, \exists B \in S \cdot \ni \cdot a \,\S\, B = \pi$, where π is the element specified in Part (c)
    g) $\forall a, b, c \in S, a = b \Rightarrow a \circ c = b \circ c$
    h) $\forall a, b \in S, a \circ b = b \circ a$

In Exercises 20 through 26 determine whether the stated argument is valid or invalid.

20. $\forall x \in R, x \notin M$. $\forall x \in S, x \in M$. Hence, $\forall x \in S, x \notin R$.

21. $\forall x \in M, x \in R$. $\forall x \in M, x \in S$. Hence, $\exists x \in S, x \in R$.

22. No one (P) likes radishes (R). Elden likes *die Zauberflöte* (F). Therefore, someone who likes *die Zauberflöte* does not like radishes.

23. No mice (M) are rats (R). All squirrels (S) are mice. Therefore, no squirrel is a rat.

24. If someone has fed the dog, then all of your cats are hungry. One of your cats is not hungry. Therefore, no one has fed the dog.

25. Some *M* is in *R*. Some *M* is in *S*. Therefore, some *S* is in *R*.
26. (a) Some apples (*A*) are McIntosh apples (*T*).
    (b) All McIntosh apples are Maine apples (*M*).
    ∴, some apples are Maine apples.

4
RELATIONS AND FUNCTIONS

§4a RELATIONS

The concepts of relation and function are of utmost importance in mathematics. The function concept will be used over and over in all your future work in mathematics. People who use mathematics to answer questions about the physical world depend heavily on the function concept.

Since a function is a particular kind of relation, we shall consider this latter topic before working with functions. In turn, the concept of relation depends on the notion of ordered pair. Thus our ultimate starting point for a discussion of relations is the concept of ordered pair.

As the name "ordered pair" would indicate, we are talking about two things, say a, b, for which a, b is not necessarily the same as b, a.

EXAMPLE 1: The pair of acts "take a shower" and "go to bed" can be thought of as ordered. Obviously the end result is different depending on which act is followed by the other. That is, "Take a shower, then go to bed" is not the same as "Go to bed, then take a shower."

Now how can we formalize this concept for mathematical use? First we must recognize that *any pair of things is a set*. That is, it is meaningful to write

$$a \in \{a, b\}, \quad b \in \{a, b\}$$

In this light, what might we mean by saying that a, b and b, a are not necessarily the same? They are certainly equal sets according to Definition (3-4) of set equality. For any pairs $\{a, b\}, \{c, d\}$, the requirements of (3-4) can ultimately be reduced to

(1) $\qquad \{a, b\} = \{c, d\}$ iff $(a = c \wedge b = d) \vee (a = d \wedge b = c)$

Since, for example, we can have either $a = c$ or $a = d$ here, no order of elements is specified. What we need is a kind of pairing for which only the first component of this disjunction would hold. As we shall see, the pair defined as follows has this desired property.

175

Definition (4-1): For any things a, b, the set $(\mathbf{a}, \mathbf{b}) = \{\{a, b\}, \{a\}\}$ is called an *ordered pair*.

Note that the *elements* of the ordered pair (a, b) are the sets
(2) $$\{a, b\}, \{a\}$$
We call a, b the **components** of the ordered pair (a, b).

The set $\{\{a, b\}, \{a\}\}$ may look like a strange thing to define as an ordered pair. If we consider this set carefully, however, we see that it possesses the desired attributes. It contains a pair $\{a, b\}$ of things and singles out one of them, a, for special treatment. This is surely what an ordered pair should be like. The following proposition states that the pair $\{a, b\}, \{a\}$ has the desired equality property.

Theorem (4a-1): For any things a, b, c, d,
$$(a, b) = (c, d) \text{ iff } a = c \wedge b = d$$

Proof: (Outline of "only if" part)

$$(a, b) = (c, d) \quad \text{Hypothesis}$$
$$\{\{a, b\}, \{a\}\} = \{\{c, d\}, \{c\}\} \quad \text{by (4-1)}$$

Using (1), we obtain

(3) $$\{a, b\} = \{c, d\} \wedge \{a\} = \{c\}$$
or
(4) $$\{a, b\} = \{c\} \wedge \{a\} = \{c, d\}$$

Now (3) reduces to

$$a = c \quad \text{by (3-8)}$$

and

$$\{a, b\} = \{c, d\}$$

whence

$$b = d$$

Thus

$$a = c \wedge b = d$$

From (4) we obtain

$$a = c = b \wedge a = c = d$$

whence

$$a = c \wedge b = d$$

as in the previous case.

The "if" part is even simpler to establish.

Done

It will be noted that this theorem tells us how ordered pairs and their *components* are related by the equality concept. The principle that

(5) $\qquad (a, b) = (c, d)$ iff $a = c \wedge b = d$

is a fact we must always keep in mind. The parentheses notation is particularly important to distinguish ordered pairs from other sets to which we apply the equality principle of Definition (3-4) to the *elements*.

EXAMPLE 2: $(x + 1, y - 3) = (5, 7)$. Find values for x and y.

Solution: $(x + 1, y - 3) = (5, 7)$ iff

$$x + 1 = 5 \wedge y - 3 = 7 \qquad \text{by (4a-1)}$$

Then $x + 1 = 5$ iff $x = 4 \wedge y - 3 = 7$ iff $y = 10$. Therefore, $x = 4$ and $y = 10$.

EXAMPLE 3: Is $\{(\text{one, hot}), (\text{dog, please})\} = \{(\text{hot, one}), (\text{please, dog})\}$?

Solution: Two sets are equal iff the elements are the same. Our question then depends on whether (one, hot) is the same as (hot, one), and whether (dog, please) is the same as (please, dog). By Theorem (4a-1) these ordered pairs are not the same. Hence the sets are unequal.

This example indicates that a set of ordered pairs is a reasonable notion. It is here that the relation concept enters the picture.

Definition (4-2): A **relation** is a set of ordered pairs.

EXAMPLE 4: Is $R = \{(0, 0)\}$ a relation?

Solution: Yes. $\{(0, 0)\}$ is obviously a set of ordered pairs and thus a relation. It contains one element $(0, 0)$ both components of which are 0.

It is desirable to have a general method of representing a relation in set builder notation. Ordered pairs (x, y) can be considered as being determined by values assigned to the two variables x, y. Thus the truth set of an open sentence $p(x, y)$ in two variables is a relation. Using this notation, a relation R of Definition (4-2) can be written

(6) $\qquad R = \{(x, y) | x \in A \wedge y \in B \wedge p(x, y)\}$

In this context we often call $p(x, y)$ a *rule of correspondence* or *condition on* x, y. The scope of x is A; the scope of y is B. If no explicit statement for $p(x, y)$ is known, R is simply given in roster notation.

EXAMPLE 5: Make a roster of the relation

$$R = \{(x, y) | x \in \mathcal{N} \wedge y \in \{1, 2\} \wedge x = y^2\}$$

Solution: $R = \{(1, 1), (4, 2)\}$ because $1 = 1^2$ and $4 = 2^2$.

EXAMPLE 6: Make a roster of
$$R = \{(x, y) | x \in \mathscr{I} \wedge y \in \mathscr{I} \wedge y = \sqrt{x}\}$$
Solution: $R = \{(0, 0), (1, 1), (4, 2), (9, 3), \ldots\}$.

It is often convenient in our work to have a way of talking about all the first components and all the second components of a relation.

Definition (4-3): For any relation
$$R = \{(x, y) | x \in A \wedge y \in B \wedge p(x, y)\}$$
the set $\mathscr{D}_R = \{x | (x, y) \in R\}$ is called the **domain** of R.

Definition (4-4): For any relation
$$R = \{(x, y) | x \in A \wedge y \in B \wedge p(x, y)\}$$
the set $\mathscr{E}_R = \{y | (x, y) \in R\}$ is called the **range** of R.

Thus the domain of a relation R is a subset of the scope A of x, and the range of R is a subset of the scope B of y.

EXAMPLE 7: Find the domain and range of the relation R in Example 5.
Solution: $\mathscr{D}_R = \{1, 4\}$; $\mathscr{E}_R = \{1, 2\}$.

EXAMPLE 8: Find the domain and range of the relation R in Example 6.
Solution: $\mathscr{D}_R = \{0, 1, 4, 9, 16, \ldots\}$
$\mathscr{E}_R = \{0, 1, 2, 3, 4, \ldots\}$

EXAMPLE 9: Consider
$$R = \{(x, y) | x^2 = x \wedge x \in \{-1, 0, 1\} \wedge y \in \{0, 1, 2\}\}$$
Write R, $\mathscr{D}_R$, $\mathscr{E}_R$ in roster notation.
Solution: $R = \{(0, 0), (0, 1), (0, 2), (1, 0), (1, 1), (1, 2)\}$. $\mathscr{D}_R = \{0, 1\}$. $\mathscr{E}_R = \{0, 1, 2\}$. Note that $\mathscr{D}_R \subset \{-1, 0, 1\}$, $\mathscr{E}_R \subset \{0, 1, 2\}$.

EXERCISE SET

1. State whether the proposition is true or false and explain.
    a) $\{1, 4\} = \{4, 1\}$
    b) $(1, 4) = (4, 1)$
    c) $\varnothing$ is a set of ordered pairs
2. Make a roster of the relation $\{(a, b) | a \in A \wedge b \in B\}$ for which
    a) $A = \{1, 2\}$, $B = \{0, 3\}$ b) $A = \{1, 2\}$, $B = \{1, 2\}$
3. If $(a, b) = (b, c) \wedge (b, c) = (c, d)$, what can be said about a, b, c, d?

4. Find any real number values of x, y for which the proposition is true.
   a) $(2x, -3y) = (-y, 7x)$
   b) $(x^2, -2y) = (2, -2)$
   c) $(x + y, 3) = (y, 4y)$
   d) $(x + 1, y) = (x, x)$
   e) $(2x - 3, 8) = (7, 5y - 2)$

5. Make a roster of
   a) $\{(x, y) \mid x \in \{0, 1, 2\} \land y = 2x\}$
   b) $\{(y, x) \mid x \in \{0, 1, 2\} \land y = 2x\}$

6. If $R = \{(3, 2), (2, 1), (1, 2), (2, 3), (3, 1), (1, 3)\}$:
   a) What is the domain of R?
   b) What is the range of R?

7. Is it possible that $\mathscr{D}_R = A$ in (4-3) and $\mathscr{E}_R = B$ in (4-4)? Explain.

8. Prove: $(a, b) \neq (c, d)$ iff $a \neq c \lor b \neq d$. This proposition states that two ordered pairs are distinct if either their first components are different or their second components are different.

§4b CARTESIAN PRODUCTS

Particular types of relations which can be regarded as important in mathematics are virtually legion. The relations called functions have already been mentioned and will be discussed in §4d. In this and the following section a few of the other important types will be considered.

To see how various types of relations are determined, let us concentrate on the open sentence $p(x, y)$ of the relation R in

(1) $\qquad R = \{x \mid p(x, y) \land x \in A \land y \in B\}$

We recall that an open sentence may be

(i) universally true;
(ii) true for some values of the variables and false for others;
(iii) universally false.

For the most part, particular relations are specified by imposing suitable restrictions in Case (ii). At this point, however, let us consider the special cases (i), (iii).

EXAMPLE 1: Make a roster of

$$R = \{(x, y) \mid x \in \mathscr{R} \land y \in \mathscr{R} \land x = x + 1\}$$

Solution: Since $x = x + 1$ is false for all possible real numbers, no ordered pairs are determined. That is, $R = \{\ \}$. Compare with Exercise 1c, §4a.

As shown in this example, the empty set is the relation determined by a universally false $p(x, y)$.

In Case (i) we have
$$p(x, y) \subseteq t$$
where t is a known true proposition. For this case, (1) becomes
$$R = \{(x, y) \mid x \in A \land y \in B \land t\}$$
and this reduces to
$$R = \{(x, y) \mid x \in A \land y \in B\}$$
by A-3, §2j.
This is a relation having the restriction that, for given A, B, *all* possible ordered pairs belong to it.

Definition (4-5): The **Cartesian product** of two sets A, B, is
$$A \times B = \{(x, y) \mid x \in A \land y \in B\}$$
i.e., it is the set of *all* ordered pairs $(x, y) \cdot \ni \cdot x \in A$ and $y \in B$.

The Cartesian product of two sets is sometimes called the *cross product*.

EXAMPLE 2: Let $A = \{1, 2, 3\}$ and $B = \{a, b\}$. Find $A \times B$.

Solution: Each element from A will be paired with each element from B. Thus
$$A \times B = \{(1, a), (1, b), (2, a), (2, b), (3, a)(3, b)\}$$

EXAMPLE 3: Let $A = \{1, 2\}$. Find $A \times A$.

Solution: Each element from A will be paired with each element from A. Hence,
$$A \times A = \{(1, 1), (1, 2), (2, 1), (2, 2)\}$$

It would now be instructive to count the elements in each solution set. The set of Example 2 has six elements; the set of Example 3 has four elements. What is striking here is that in Example 2, A has three elements, B has two elements, and $A \times B$ has six elements. In Example 3, A has two elements, and $A \times A$ has four elements. Perhaps that is why this operation is called a product! Whatever the reason, a good check on whether such a Cartesian product is correct or not is to see whether enough elements belong to it.

As these examples indicate, it can be fruitful to explore the way the number of elements in a combination of sets is related to the number of elements in the individual sets. Important use is made of the following principles in elementary school materials and elsewhere.

Definition (4-6): For any sets A, B,
$$\mathbf{n(A \cup B)} = n(A) + n(B) - n(A \cap B)$$
Compare this statement with that of (3), §3e.

Definition (4-7): For any sets A, B,
$$n(A \times B) = n(A) \cdot n(B)$$

The principle stated here is, of course, the one mentioned following Example 3.

EXAMPLE 4: If $A = \{2, 3, 4\}$, $B = \{3, 4, 5\}$, what is $n(A \cup B)$? What is $n(A \times B)$?

Solution: $n(A) = 3$, $n(B) = 3$, $n(A \cap B) = 2$. Thus
$$n(A \cup B) = 3 + 3 - 2 = 4$$
Check: $A \cup B = \{2, 3, 4, 5\}$ and $n(A \cup B) = 4$.
$$n(A \times B) = 3 \cdot 3 = 9$$

EXERCISE SET

1. Let $A = \{\text{cat, rat, bat}\}$, $B = \{\wedge, \vee, \Rightarrow\}$, $C = \{\text{big, bird}\}$, $D = \{1\}$. Find:
   a) $A \times B$
   b) $B \times C$
   c) $C \times B$
   d) $D \times D$

2. Let $I = \{1, 2, 3, 4, 5\}$, $A = \{1, 2\}$, and $B = \{2, 3, 4\}$. Find:
   a) $A \times B'$
   b) $A \times A$
   c) $(A \cap B) \times A'$
   d) $A \cap (B \times A')$

3. If $n(A) = 5$, $n(B) = 6$, $n(C) = 3$, and A, B, C are disjoint, find:
   a) $n(A \times B)$
   b) $n(A \times C)$
   c) $n(B \times C)$
   d) $n[(A \times B) \times C]$
   e) $n[A \times (B \times C)]$
   f) $n(A \cup B)$
   g) $n(A \cup C)$
   h) $n(B \cup C)$
   i) $n[(A \cup B) \cup C]$
   j) $n[A \cup (B \cup C)]$

4. If $A = \{a, b, c, d\}$, $B = \{d, e, f\}$, $C = \{b, c\}$, find a) through j) of Exercise 3 above.

5. What is the number of elements of $A \times B$ where $A = \{n \mid 1 < n < 9 \wedge n \in \mathcal{N}\}$, $B = \{x \mid x \text{ is a month of the year}\}$?

6. a) What is the domain of the Cartesian product in Example 2?
   b) What is the range of the Cartesian product in Example 2?
   c) What is a reasonable conjecture about the domain and range relative to their respective scopes?

7. If $R = \{(3, 2), (2, 1), (1, 2), (2, 3), (3, 1), (1, 3)\}$,
   a) What is the domain of R?
   b) What is the range of R?
   c) Is R a Cartesian product? Explain.

8. State whether the proposition is true or false and explain your answer.
$$R = \{(x, y) \mid x \in A \land y \in B \land p(x, y)\} \Rightarrow R \subset A \times B.$$
9. Make a roster of each Cartesian product.
    a) $\varnothing \times \varnothing$
    b) $\varnothing \times \{\varnothing\}$
    c) $\{\varnothing\} \times \{\varnothing\}$
10. Is $A \times B$ ever equal to $B \times A$? If you said yes, when? If you said no, you're wrong!
11. Find $A \cdot \ni \cdot A \times A \leftrightharpoons A$.
12. If we were to define $A \times \varnothing$, it would be $A \times \varnothing = \varnothing$. With this premise, find $B \cdot \ni \cdot B \times B = B$.

§4c EQUIVALENCE RELATIONS

Frequently we consider relations
$$R = \{(x, y) \mid x \in A \land y \in B \land p(x, y)\}$$
for which the scope of both variables is the same. If the scope is S, for example, R becomes

(1) $\qquad R = \{(x, y) \mid x \in S \land y \in S \land p(x, y)\}$

When the scope is clearly understood, we can simply write

(2) $\qquad R = \{(x, y) \mid p(x, y)\}$

This is often done when the scope is the set $\mathscr{R}$ of real numbers.

A common notational device for the open sentence $p(x, y)$ is to designate part of it by the Greek letter ρ (Rho). Then we write

$$x \, \rho \, y$$

for $p(x, y)$.
With this notation, (1) is written
$$R = \{(x, y) \mid x \in S \land y \in S \land x \, \rho \, y\}$$
and (2) becomes simply

(3) $\qquad R = \{(x, y) \mid x \, \rho \, y\}$

For example, ρ might stand for "weighs more than," and $A = B$ might be the set of people in your class.

EXAMPLE 1: Let $A = B = \{\text{cat, rat, gnat}\}$, and let ρ be "weighs more than." Find R.

Solution: $R = \{(\text{cat, rat}), (\text{cat, gnat}), (\text{rat, gnat})\}$. $\mathscr{D}_R = \{\text{cat, rat}\}$ and $\mathscr{E}_R = \{\text{rat, gnat}\}$.

§4c Equivalence Relations

We note that in this example domain and range are *not* the same and that neither is the same as the scope of the variables. That this is not always so is shown in the following example.

EXAMPLE 2: Let $A = B = \{1, 2, 3\}$, and let ρ be "is the same as." Find R.

Solution: $R = \{(1, 1), (2, 2), (3, 3)\}$. Hence
$$\mathscr{D}_R = \{1, 2, 3\} = \mathscr{E}_R = A = B$$

In §1f we listed the reflexive, symmetric, and transitive properties of number equality. Recall also from Chapter 3 that we proved that set equality is reflexive, symmetric, and transitive. We are now in a position to make formal definitions of these terms.

Definition (4-8): A relation R is **reflexive** iff
$$\forall x \in \mathscr{D}_R, (x, x) \in R \text{ (i.e., } x \rho x \text{ is true } \forall x \in \mathscr{D}_R)$$

EXAMPLE 3: $R_1 = \{(2, 3)\}, (3, 2), (2, 2), (3, 3)\}$; $R_2 = \{(1, 2), (2, 1), (1, 1)\}$. Are these relations reflexive?

Solution: $\mathscr{D}_{R_1} = \{2, 3\}$. Is $(2, 2) \in R_1$? Yes. Is $(3, 3) \in R_1$? Yes. Hence R_1 is reflexive. $\mathscr{D}_{R_2} = \{1, 2\}$. Is $(1, 1) \in R_2$? Yes. Is $(2, 2) \in R_2$? No. Hence R_2 is not reflexive.

Definition (4-9): A relation R is **symmetric** iff
$$\forall (x, y) \in R, (x, y) \in R \Rightarrow (y, x) \in R$$
(i.e., $x \rho y \Rightarrow y \rho x$ is true for each $(x, y) \in R$).

EXAMPLE 4: Consider $R_1 = \{(a, b), (b, a), (a, c)\}$; $R_2 = \{(c, d), (d, c), (c, c)\}$. Are they symmetric?

Solution:

$(a, b) \in R_1 \Rightarrow (b, a) \in R_1$ is true
$(b, a) \in R_1 \Rightarrow (a, b) \in R_1$ is true
$(a, c) \in R_1 \Rightarrow (c, a) \in R_1$ is false

Therefore, R_1 is not symmetric.

$(c, d) \in R_2 \Rightarrow (d, c) \in R_2$ is true
$(d, c) \in R_2 \Rightarrow (c, d) \in R_2$ is true
$(c, c) \in R_2 \Rightarrow (c, c) \in R_2$ is true

Thus, $\forall (x, y) \in R_2$,

$(x, y) \in R_2 \Rightarrow (y, x) \in R_2$ is a true implication.

Hence, the relation R_2 is symmetric.

Definition (4-10): A relation R is **transitive** iff

$$\forall (x, y), (y, z) \in R, \quad [(x, y) \in R \land (y, z) \in R] \Rightarrow (x, z) \in R$$

(i.e., $x \rho y \land y \rho z \Rightarrow x \rho z$ is true for each $(x, y), (y, z) \in R$).

EXAMPLE 5: Consider $R_1 = \{(1, 3), (3, 5), (1, 5), (2, 7)\}$; $R_2 = \{(1, 2), (2, 1), (1, 1)\}$. Are they transitive?

Solution: $\{(1, 3)\} \in R_1 \land (3, 5) \in R_1 \Rightarrow (1, 5) \in R_1$ is true. This is the only pair of elements of R_1 having first and second components in common. Hence the implication is true for all other possible combinations, and R_1 is transitive. For example, for R_2 we have

$(2, 7) \in R_1 \land (7, ?) \in R_1 \Rightarrow (2, ?) \in R_1$ is true because $(7, ?) \in R_1$ is false.
$(1, 2) \in R_2 \land (2, 1) \in R_2 \Rightarrow (1, 1) \in R_2$ is true,
$(2, 1) \in R_2 \land (1, 2) \in R_2 \Rightarrow (2, 2) \in R_2$ is false.

Hence R_2 is not transitive.

Definition (4-11): A relation R is called an **equivalence relation** iff R is reflexive, symmetric, and transitive.

EXAMPLE 6: Given: $R = \{(1, 2), (2, 3), (1, 3)\}$. Which of the properties of an equivalence relation does R have?

Solution: $\mathcal{D}_R = \{1, 2\}$. Therefore, we must test for the reflexive property by considering all ordered pairs having first component 1 or 2. Is it true that $(1, 1) \in R \land (2, 2) \in R$? No. Therefore, R is *not reflexive*.

We test for the symmetric property by considering each (a, b) and looking for its symmetric partner (b, a). Is "$(1, 2) \in R \Rightarrow (2, 1) \in R$" true? No; therefore, R is *not symmetric*.

To test for transitivity we select any ordered pair and look at its *second* component. Then we consider each other ordered pair having that component as its *first* component. Then we repeat for each ordered pair of the relation. Is $[(1, 2) \in R \land (2, 3) \in R] \Rightarrow (1, 3) \in R$" true? Yes. Are there any more ordered pairs having the second component of one the same as the first component of the other? No; therefore, *R is transitive*.

EXAMPLE 7: Given: $R = \{(1, 1), (2, 2)\}$. Which of the properties of an equivalence relation does R have?

Solution: R is an equivalence relation. (Try to prove this on your own.)

Note that R_1 of Example 3 is also an equivalence relation.

EXAMPLE 8: Which of the properties of an equivalence relation does $R = \{x, y\} \mid x \in \mathscr{Z} \land y \in \mathscr{Z} \land x \rho y\}$ possess where ρ stands for "is 2 times"? Explain.

Equivalence Relations

Solution: Does $x \rho x$ hold for $x \in \mathscr{L}$? No. For example, "2 equals 2 times 2" is false. R is not reflexive. Does $x \rho y = y \rho x$, $\forall x, y \in \mathscr{L}$? No. For example, $4 = 2 \cdot 2$ is true, but $2 = 2 \cdot 4$ is false; whence the implication is false, and R is not symmetric. Does $x \rho y \wedge y \rho z \Rightarrow x \rho z$, $\forall x, y, z \in \mathscr{L}$? No. For example, $12 = 2 \cdot 6 \wedge 6 = 2 \cdot 3$ is true but $12 = 2 \cdot 6 \wedge 6 = 2 \cdot 3 \Rightarrow 12 = 2 \cdot 3$ is false, and R is not transitive. Thus R has none of the three required properties of an equivalence relation.

EXERCISE SET

1. Given $A = B = \{1, 2, 3, 4, 5\}$, make a roster of R for each meaning of ρ.
   a) ρ stands for "is a factor of" (x is a factor of y)
   b) ρ stands for "equals" ($x = y$)
   c) ρ stands for "is less than" ($x < y$)
   d) ρ stands for "x is larger than $y \wedge y = 3$"
   e) ρ stands for "x is less than $y \wedge y = 1$"

2. Which of the properties of an equivalence relation are possessed by the relation
$$R = \{(3, 2), (2, 1), (1, 2), (2, 3), (3, 1), (1, 3)\}?$$

3. Decide whether each of the following rules determines an equivalence relation on the given set. If so, give an example to show how each of the three requirements is satisfied. If not, show by example why any of the requisite properties does not hold.
   a) "is longer than" for the set of measuring sticks
   b) "is farther away from Boston than" for the set of towns
   c) "is different from by 5" for the set of integers
   d) "is a factor of" for the set of integers
   e) "is equal to the square root of" for the set of natural numbers
   f) "is the ancestor of" for the set of people
   g) "is the sister of" for the set of people
   h) "is married to" for the set of people
   i) "weighs within two pounds of" for the set of people
   j) "is a contemporary of" for the set of people
   k) "is perpendicular to" for the set of nonconcurrent, coplanar straight lines
   l) "$\neq$" for $\mathscr{R}$

4. Write a rule of correspondence which will result in a relation having the stated property or properties
   a) transitive only
   b) reflexive and symmetric only

c) symmetric only
d) reflexive and transitive only
e) symmetric and transitive only
f) reflexive only

5. Let $A = \{$cat, mouse$\}$, There are 15 possible relations on A. Indicate some which are
    a) reflexive
    b) symmetric
    c) transitive
    d) equivalence relations

6. Which of the following are equivalence relations?
    a) $\perp, //$ for straight lines
    b) $\subset, =$ for sets

7. What other elements need to be in R to make the stated relation an equivalence relation?
    a) (cat, dog) $\in R$
    b) $(1, m) \in R \wedge (m, k) \in R$
    c) $(2, 8) \in R \wedge (3, 5) \in R$

§4d FUNCTIONS

This chapter began by stating that the function concept was important and then proceeded to discuss something called a relation, because a function is a particular kind of relation. However, not every relation is a function.

Consider, for example, the case of a weather station which gives hourly reports of temperature. Would the set of reports in Figure 4.1 be useful?

Time	6:00	7:00	7:00	8:00
Temperature	60°	62°	65°	60°

Figure 4.1

We note several things about the information presented in the table. First, each entry in the table is an ordered pair of the form

(time, temperature)

and thus the whole table is a relation. Second, the same temperature occurred at both 6 o'clock and 8 o'clock. Two different temperatures are reported at 7 o'clock.

The fact that the same temperature is reported for two different times of day should not be too disturbing. A shift in wind or other changes might easily cause such a duplication. However, listing two different temperatures at the same time for the same location does not seem too reasonable. In fact, the listing of two different temperatures at the same time of day makes the information given almost unusable. What really did happen at 7 o'clock?

For data to be meaningful in this and many other cases, we must have only one second component for each first component. When this condition holds, we call the relation a function.

(1) A relation F is a function iff no two distinct ordered pairs belonging to F have the same first component.

This description is satisfactory for some purposes, but we also need a definition of this concept which uses variables. To see how this can be accomplished, let us ask ourselves what we mean by *distinct* ordered pairs. For $(x_1, y_1), (x_2, y_2)$ to be distinct, we must have

$$x_1 \neq x_2 \lor y_1 \neq y_2$$

Now, if the x-components are the same, we must have

$$y_1 \neq y_2$$

by disjunctive simplification. We now have distinct ordered pairs having the same first component, but the relation $Q = \{(x_1, y_1), (x_1, y_2), \ldots\}$ is *not* a function by (1). In order to have a function, then, we must specify that if first components are the same, then the second components must also be the same. The following definition states this idea symbolically.

Definition (4-12): A relation F is a **function** iff $x_1 = x_2 \Rightarrow y_1 = y_2$ for each $(x_1, y_1), (x_2, y_2) \in F$.

The crucial point in Definition (4-12) is the extra restriction

(2) $$x_1 = x_2 \Rightarrow y_1 = y_2$$

which must be satisfied by a function in addition to being a set of ordered pairs. If the hypothesis of (2) is false for all ordered pairs in F, then all are distinct. If the hypothesis is true and the conclusion is false, then the implication is false and F is not a function.

EXAMPLE 1: Let $S = \{1, 2\}$. List all the possible relations using elements of S. Then label each as a function or not a function and explain your label.

Solution:

$R_0 = \{\quad\}$
$R_1 = \{(1, 1), (1, 2), (2, 1), (2, 2)\}$
$R_2 = \{(1, 1)\}$
$R_3 = \{(1, 2)\}$
$R_4 = \{(2, 1)\}$
$R_5 = \{(2, 2)\}$
$R_6 = \{(1, 1), (1, 2)\}$
$R_7 = \{(1, 1), (2, 1)\}$
$R_8 = \{(1, 1), (2, 2)\}$
$R_9 = \{(1, 2), (2, 1)\}$
$R_{10} = \{(1, 2), (2, 2)\}$
$R_{11} = \{(2, 1), (2, 2)\}$
$R_{12} = \{(1, 1), (1, 2), (2, 1)\}$
$R_{13} = \{(1, 1), (1, 2), (2, 2)\}$
$R_{14} = \{(1, 1), (2, 1), (2, 2)\}$
$R_{15} = \{(1, 2), (2, 1), (2, 2)\}$

The functions are $R_0, R_2, R_3, R_4, R_5, R_7, R_8, R_9$, and R_{10}. For a sample explanation, consider R_1. Let $(x_1, y_1) = (1, 1)$ and $(x_2, y_2) = (1, 2)$. Now $x_1 = x_2$ is true because $1 = 1$, but $y_1 = y_2$ is false because $1 = 2$ is false. Thus the implication

$$1 = 1 \Rightarrow 1 = 2$$

is false, and R_1 is not a function. Also, consider R_2. Here $(x_1, y_1) = (1, 1)$ and $(x_2, y_2) = (1, 1)$ because $(1, 1)$ is the only ordered pair we have. Certainly the implication

$$1 = 1 \Rightarrow 1 = 1$$

is true. Thus R_2 is a function. You provide the remaining explanations.

EXAMPLE 2: Note that the postage we pay for sending a letter by first class mail is a function of its weight. That is, each entry of a table such as the one of Figure 4.2 is an ordered pair. Fill in the remaining second components of this function.

Weight	0 oz.	$\frac{1}{2}$ oz.	1 oz.	$1\frac{1}{3}$ oz.	2 oz.	$2\frac{3}{4}$ oz.	3 oz.	4 oz.
Postage	0 ¢	10 ¢	10 ¢	20 ¢				

Figure 4.2

Solution: See Figure 4.3.

Weight	0 oz.	$\frac{1}{2}$ oz.	1 oz.	$1\frac{1}{3}$ oz.	2 oz.	$2\frac{3}{4}$ oz.	3 oz.	4 oz.
Postage	0 ¢	10 ¢	10 ¢	20 ¢	20 ¢	30 ¢	30 ¢	40 ¢

Figure 4.3

In addition to the letter F used for function in the definition, we shall use the capitals G, H, Φ for this purpose. Long-standing custom also sanctions the use of certain small letters such as f, g, h, ϕ for functions instead of the more usual capitals employed when discussing other kinds of sets. Thus any arbitrary function might be designated $f = \{(x, y) \mid x \, p \, y\}$.

Each second component of a function is called a *value* of the function. Hence the set of values of a function is its range. The following notational convention is convenient and frequently employed.

If a is a particular first component of a function f, we use $f(a)$ to stand for the corresponding value (second component) b. Then

(3) $$f(a) = b$$

means that $(a, b) \in f$. Similarly $f(x)$ stands for any of the values corresponding to a particular x, and we write

(4) $$f(x) = y$$

Thus $f(x)$ stands for an expression from which we can find a value y by substituting a particular thing for x. If $f = \{(x, y) \mid x \, p \, y\}$, we can then write

$$f = \{(x, y) \mid y = f(x)\}$$

using (4), or

$$f = \{(x, f(x)\}$$

EXAMPLE 3: Find the value of f which corresponds to the first component 3 if $f(x) = x^2 - 4x$.

Solution : Using form (3), by substitution we have

$$\begin{aligned} f(3) &= (3)^2 - 4(3) \\ &= 9 - 12 \\ &= -3 \end{aligned}$$

Thus $(3, -3) \in f$.

When using the writing conventions (3), (4), it is important to remember that f or another single letter stands for the *set* of ordered pairs, i. e., for the function itself. The corresponding $f(x)$ stands for a second component or components of f.

EXAMPLE 4: Write a few of the elements of

$$F = \{(x, y) \mid y = 5x \wedge x \in \mathscr{R}\}$$

Solution : Let $x = 3$, then $y = (5)(3)$ or 15; so $(3, 15) \in F$. Let $x = -2$, then $y = (5)(-2)$ or -10; so $(-2, -10) \in F$. Let $x = \pi$, then $y = (5)(\pi)$ or 5π; so $(\pi, 5\pi) \in F$. Hence

$$F = \{(x, 5x)\} \text{ or } F = \{(3, 15), (-2, -10), (\pi, 5\pi), \cdots\}$$

It is possible to define binary operations on functions if the domain is carefully specified.

Definition (4-13): For any functions f, g and $\forall x \in (\mathscr{D}_f \cap \mathscr{D}_g)$,

$$\mathbf{f+g} = \{(x, y) \mid y = (f + g)(x) = f(x) + g(x)\}$$

EXAMPLE 5: If $f = \{(2, 3), (3, 5), (4, 7), (5, 1)\}$ and $g = \{(0, -1), (1, 2), (2, 0), (3, -2)\}$, find $f + g$ and $g + f$.

Solution : Start with the first x in f, $x = 2$. Look in g to see if there is any ordered pair with first component 2. We find $(2, 0) \in g$. According to Definition (4-13), we add the second components of these ordered pairs, obtaining $(2, 3 + 0)$ or $(2, 3)$. Continuing in the same manner, we have

$$f + g = \{(2, 3), (3, 3)\}$$
$$g + f = \{(2, 3), (3, 3)\}$$

Note that $\mathscr{D}_{f+g} = \mathscr{D}_f \cap \mathscr{D}_g$ as required.

Definition (4-14): For any functions f, g and $\forall x \in \mathscr{D}_f \cap \mathscr{D}_g$,
$$\mathbf{f} - \mathbf{g} = \{(x, y) \mid y = (f - g)(x) = f(x) - g(x)\}$$

EXAMPLE 6: Using f, g from Example 5, find $f - g$ and $g - f$.

Solution:
$$f - g = \{(2, 3), (3, 7)\}$$
$$g - f = \{(2, -3), (3, -7)\}$$

Definition (4-15): For any functions f, g and $\forall x \in \mathscr{D}_f \cap \mathscr{D}_g$,
$$\mathbf{f} \cdot \mathbf{g} = \{(x, y) \mid y = (f \cdot g)(x) = f(x) \cdot g(x)\}$$

EXAMPLE 7: Using f, g from Example 5, find $f \cdot g$ and $g \cdot f$.

Solution:
$$f \cdot g = \{(2, 0), (3, -10)\}$$
$$g \cdot f = \{(2, 0), (3, -10)\}$$

Definition (4-16): For any functions f, g and $\forall x \in \mathscr{D}_f \cap \mathscr{D}_g$ where $g(x) \neq 0$,
$$\frac{\mathbf{f}}{\mathbf{g}} = \left\{(x, y) \mid y = \frac{f}{g}(x) = \frac{f(x)}{g(x)}\right\}$$

EXAMPLE 8: Using f, g from Example 5, find f/g and g/f.

Solution: $f/g = \{(3, \frac{5}{2})\}$. Note that $g(2) = 0$ and cannot be used; $g/f = \{(2, 0), (3, -\frac{2}{5})\}$.

EXAMPLE 9: If $f(x) = x - 2, g(x) = x - 3, x \in \mathscr{R}$, find: $f + g$, $f - g, f \cdot g, f/g$.

Solution:
$$f(x) = x - 2$$
$$g(x) = x - 3$$

so
$$f(x) + g(x) = 2x - 5$$
$$f(x) - g(x) = +1$$
$$f(x) \cdot g(x) = (x - 2)(x - 3)$$
$$= x^2 - 5x + 6$$
$$\frac{f(x)}{g(x)} = \frac{x - 2}{x - 3}, \quad g \neq 3$$

Thus,
$$f + g = \{(x, y) \mid y = 2x - 5\}$$
$$f - g = \{(x, y) \mid y = 1\}$$
$$f \cdot g = \{(x, y) \mid y = x^2 - 5x + 6\}$$
$$\frac{f}{g} = \{(x, y) \mid y = \frac{x-2}{x-3} \wedge x \neq 3\}$$

Still another binary operation on functions finds considerable use in later work. As the following definition indicates, it is produced by substitution and is called the *composition* function.

Definition (4-17): For any functions f, g and
$$\forall x \cdot \ni \cdot x \in \mathscr{D}_g \wedge g(x) \in \mathscr{D}_f,$$
$$\mathbf{f} \circ \mathbf{g} = \{(x, y) \mid y = (f \circ g)(x) = f(g(x))\}$$

A statement which is equivalent to the one given in the definition is the following:

(5) $\quad f \circ g = \{(x, z) \mid (x, y) \in g \wedge (y, z) \in f\}$

EXAMPLE 10: Using f and g in Example 5, find $f \circ g$ and $g \circ f$.

Solution: Look at the second components of the ordered pairs in the g function. Use only those which are also first components of the ordered pairs of the f function. Since $(1, 2) \in g$, $(2, 3) \in f$ are the only pairs which satisfy the requirement,
$$f \circ g = \{(1, 3)\}$$

In the same manner $g \circ f$ is found. This time begin with the second components of f and look at the first components of g. We have
$$g \circ f = \{(2, 2), (5, 2)\}$$

EXAMPLE 11: If $f(x) = 3x + 7$ and $g(x) = x^2 - 1$ for $x \in \mathscr{R}$, find $f(2), f(3), f(\pi), f(g(x)), g(f(x))$.

Solution: The open sentence for f tells us to multiply the variable by 3 and then add 7. Thus
$$f(2) = 3(2) + 7, \text{ i.e., } 13$$
$$f(3) = 3(3) + 7, \text{ i.e., } 16$$
$$f(\pi) = 3(\pi) + 7$$
$$f(g(x)) = 3(g(x)) + 7$$
$$= 3(x^2 - 1) + 7$$
$$= 3x^2 + 4$$
$$g(f(x)) = (f(x))^2 - 1$$
$$= (3x + 7)^2 - 1$$
$$= 9x^2 + 42x + 48$$

EXERCISE SET

1. Which of the following relations are functions?
   a) $\{(1, 2), (1, 3), (1, 4)\}$
   b) $\{(1, 1)\}$
   c) $\{(1, 2), (2, 2), (3, 2), (4, 2)\}$
   d) $\{(3, 1), (4, 2), (3, 2)\}$
   e) $\{(x, y) \mid y = 2x \land x, y \text{ are numbers}\}$
   f) $\{(x, y) \mid y^2 = x\}$
   g) $\{(x, y) \mid x^2 + y^2 = 1\}$

2. The basic salary for salesmen of a particular company is $10,000 a year. In addition, each gets 10% of his total dollar value of sales. Fill in the table of Fig. 4.4 for annual salary of salesmen a, b, c, d, e, f.

Salesman	a	b	c	d	e	f
Total sales	$5,000	$12,000	$15,000	$7,000	$70,000	$20,000
Annual Salaries						

Figure 4.4

3. If $f = \{(5, 12), (7, 0), (2, 2), (0, 3)\}$,
   $g = \{(5, -2), (7, 2), (2, 0), (0, 3), (4, 0), (-6, 2)\}$,
   $h = \{(0, 0), (-5, -5), (4, 4)\}$,

   make a roster of each specified function.
   a) $f + g$
   b) $f - g$
   c) $\dfrac{f}{g}$
   d) $f \cdot g$
   e) $f \circ g$
   f) $h \cdot h$
   g) $f \circ f$
   h) $g \circ f$
   i) $g - f$
   j) $h - f$
   k) $\dfrac{g}{f}$
   l) $\dfrac{f}{h}$
   m) $\dfrac{h}{f}$
   n) $\dfrac{g}{h}$
   o) $\dfrac{h}{g}$
   p) $h \circ f$
   q) $h \circ g$
   r) $g \circ h$
   s) $h \circ h$
   t) $f \circ h$

4. If $f = \{(x, x-2)\}$, i.e., $f(x) = x - 2$, and if $g(x) = 2x + 2$, find
   a) $f(0)$
   b) $f(3)$
   c) $f(\sqrt{3})$
   d) $\dfrac{f(\sqrt{3})}{f(3)}$

e) $f(g(x))$ f) $f(3) + f(0)$
g) $f(3) \cdot g(4)$ h) $f(x) \cdot g(x)$
i) $f(x) + g(x)$ j) $f(x) - g(x)$
k) $(f \circ g)(-1)$ l) $(f \circ g)(\pi)$
m) $(f \circ g)(x)$ n) $(g \circ f)(x)$

5. If $f = \{x, x^2 + 2)\}$ and $g = \{x, x - 1)\}$, find

a) $f(x) + g(x)$ b) $f(x) \cdot g(x)$
c) $f(g(3))$ d) $g(f(2))$
e) $f(3) + f(2)$ f) $\dfrac{f(3)}{g(3)}$
g) $(f \circ g)(x)$ h) $(g \circ f)(x)$

6. Find $\mathscr{D}_f$ and $\mathscr{E}_f$ if $f(x) = \sqrt{x^2 - 1}$ and $x \in \mathscr{L}$.

7. If $A = \{a, b, c\}$, find a function f such that

a) $\mathscr{D}_f \leftrightarrows \mathscr{E}_f$;
b) $\mathscr{D}_f = \mathscr{E}_f$;
c) f is symmetric;
d) f is transitive and not reflexive
e) f is an equivalence relation

8. Prove or disprove

a) $\emptyset$ is a function.
b) If f, g are functions, $f \cap g$ is a function.
c) If G, H are functions, $G \cup H$ is a function.

9. Prove: A relation which consists of a single ordered pair is a function.

10. If $*$ stands for any of the operations in Definitions (4-13) through (4-17), for which operations is $f * g = g * f$? That is, which operations are commutative?

§4e   REVERSE RELATIONS

A relation R was defined as a set of ordered pairs (x, y) such that $x \, \rho \, y$ is a true statement, where ρ is some rule of correspondence. Thus we could represent R as $\{(x, y) \mid x \, \rho \, y\}$. It is useful to introduce the concept of a reverse relation.

Definition (4-18): For any relation $R = \{(x, y) \mid x \, \rho \, y\}$, the **reverse relation** $\bar{R}$, read "R-bar", is

$$\bar{R} = \{(a, b) \mid (b, a) \in R\}$$

That is, each ordered pair of R becomes an ordered pair of $\bar{R}$ by *reversing* its components.

EXAMPLE 1: If $R = \{(1, 2), (2, 3), (2, 4), (5, 7)\}$, what is $\bar{R}$?

Solution: $\bar{R} = \{(2, 1), (3, 2), (4, 2), (7, 5)\}$.

If you let ρ be "is less than" in Example 1, you will see that the components of R satisfy ρ. Do the components of $\bar{R}$ also satisfy ρ? From the definition, $\bar{R} = \{(x, y) \mid y \, \rho \, x\}$. Now $(2, 1) \in \bar{R}$. Is 1 less than 2? Yes. The reader can test the remaining components of $\bar{R}$ to see that they also satisfy ρ. Only the components of the ordered pairs have been interchanged.

Now, $\bar{R}$ is also a relation by definition; hence it has a domain and range. The range of R is the domain of $\bar{R}$, and the domain of R is the range of $\bar{R}$.

EXAMPLE 2: If $R = \{(x, y) \mid y = 5x\}$, what is $\bar{R}$?

Solution: Here $x = f(y)$ is $x = 5y$. Thus $\bar{R} = \{(x, y) \mid x = 5y\}$ or $\bar{R} = \{(x, y) \mid y = x/5\}$. Note that $x = 5y \Longleftrightarrow y = x/5$.

EXAMPLE 3: If $R = \{(x, y) \mid y = \sqrt{x}\}$, find a few sample pairs of both R and $\bar{R}$ over $\mathscr{R}$. Note that

$$\mathscr{D}_R = \{x \mid x \geq 0\}, \quad \mathscr{E}_R = \{y \mid y \geq 0\}$$

Solution:

$$R = \{(1, 1), (4, 2), (9, 3), (16, 4), \cdots\}$$

Thus

$$\bar{R} = \{(1, 1), (2, 4), (3, 9), (4, 16), \cdots\}$$

Also

$$\bar{R} = \{(x, y) \mid x = \sqrt{y}\}$$

Since the reverse of a relation is a relation, does the following conjecture hold?

Conjecture: The reverse relation $\bar{F}$ of a function F is also a function.

We examine this conjecture by considering the following example.

EXAMPLE 4: If $F = \{(2, 1), (3, 1)\}$, is $\bar{F} = \{(1, 2), (1, 3)\}$ a function?

Solution: No, $\bar{F}$ does not satisfy the definition of function because the proposition

$$1 = 1 \Rightarrow 2 = 3$$

is false.

Thus the stated conjecture is false: the reverse of a function is *not* necessarily a function.

§4e   Reverse Relations

What kind of function will have a reverse relation which is a function? Recalling that ordered pairs of F are reversed to obtain $\bar{F}$, we can see that the requirement that the implication

(1) $$y_1 = y_2 \Rightarrow x_1 = x_2$$

must hold for components of F will guarantee that the reverse relation $\bar{F}$ will also be a function. Combining the statement of (1) with the defining statement of a function, we can state that "the reverse relation $\bar{F}$ of a function F is also a function iff

(2) $$x_1 = x_2 \Longleftrightarrow y_1 = y_2$$

for each $(x_1, y_1), (x_2, y_2) \in F$." Thus a function F has a reverse function iff no two distinct ordered pairs have the same *second* element. In §4d one designation of a function F was $F = \{(x, y) \mid y = f(x)\}$. When $\bar{F}$ is a function, it can then be written $\bar{F} = \{(x, y) \mid x = f(y)\}$.

The reverse $\bar{F}$ of a function F is denoted F^{-1} and is given the name "inverse function" in some texts. We have avoided this terminology to prevent possible confusion with the inverse of an implication.

EXERCISE SET

1. What is the reverse of each of the following relations?
    a) $\{(1, 1), (2, 2), (3, 0)\}$
    b) $\{(2, 3), (2, 4), (2, 5), (2, 6)\}$
    c) $\{(x, y) \mid x \text{ weighs less than } y\}$
    d) $\{(1, 1), (2, 1), (3, 1), (4, 1), (5, 1)\}$
    e) $\{(1, 1), (1, 2), (2, 2), (2, 1)\}$
    f) $\{(x, y) \mid y = 3x\}$
    g) $\{(x, y) \mid y = 2x + 1\}$
    h) $\{(x, y) \mid y = \sqrt{4 - x^2}\}$
    i) $\{(x, y) \mid y = x^2 + 5\}$
    j) $\{(x, y) \mid y = \sqrt{x^2 + 1}\}$
    k) $\{(x, y) \mid y = 0 \lor x = 0\}$
    l) $\{(x, y) \mid y = \sqrt{-x^2}\}$

2. Find the domain and range of the reverse relations for a, b, d, e in Exercise 1.

3. Which of the reverse relations for a, b, d, e are functions? Which are reverse functions?

4. Let x be 1, 2, 3 and write the corresponding 3 elements for relations f through l in Exercise 1; then follow directions in Exercises 2 and 3.

5. If $F = \{(2, 3), (4, 7), (9, \sqrt{8}), (\pi, \sin 12)\}$, find $\bar{F}$, $F \circ \bar{F}$, and $\bar{F} \circ F$.

§4f GRAPHING

Since any numerical relation is just a set of ordered pairs $\{(x, y)\}$, we can use the Cartesian coordinate system to picture the relation by assigning each ordered pair to a unique point in the system. Although any number set can be used, the scope of each variable x, y of the relation is, unless otherwise stated, the set $\mathscr{R}$ of real numbers.

EXAMPLE 1: Find the points assigned to the ordered pairs A: $(3, 7)$, B: $(4, -2)$, C: $(-5, -2)$, D: $(-1, 5)$.

Solution: See Figure 4.5. The origin O: $(0, 0)$ is also marked.

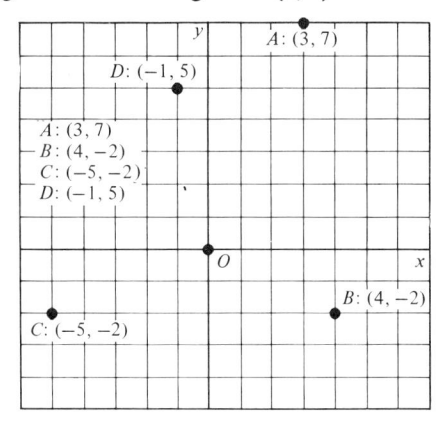

Figure 4.5

Definition (4–19): The **locus** of a relation is a set of points whose coordinates are elements of the relation.

The *graph* of a relation is a picture or sketch of the locus. Let us now consider a few examples.

EXAMPLE 2: Graph: $f = \{(x, y) \mid x = y \land x \in \mathscr{N}\}$.

Solution: In this function, the scope of x is restricted to the set $\mathscr{N}$ of natural numbers. Since $y = x$, the scope of y is also $\mathscr{N}$. See Fig. 4.6.

EXAMPLE 3: Graph: $f = \{(x, y) \mid x = y\}$.

Solution: Here no restriction on the scopes of the variables is stated. Thus we use $\mathscr{R}$, and ordered pairs such as $(3/2, 3/2), (-\sqrt{2}, -\sqrt{2})$ will determine points of the locus. See Fig. 4.7.

§4f *Graphing* 197

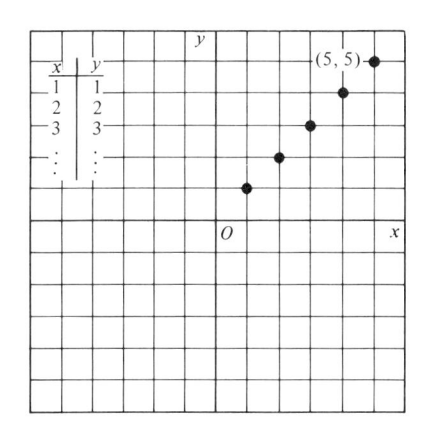

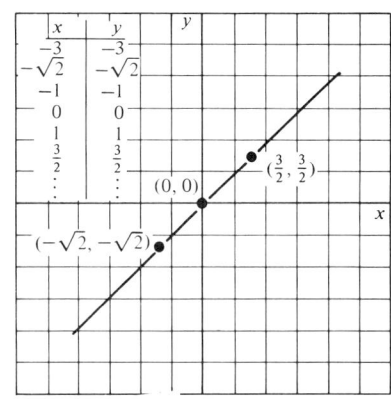

Figure 4.6

Figure 4.7

EXAMPLE 4: Graph: $f = \{(x, y) \mid x^2 = y \wedge x \in \mathscr{Z}\}$.

Solution: See Fig. 4.8.

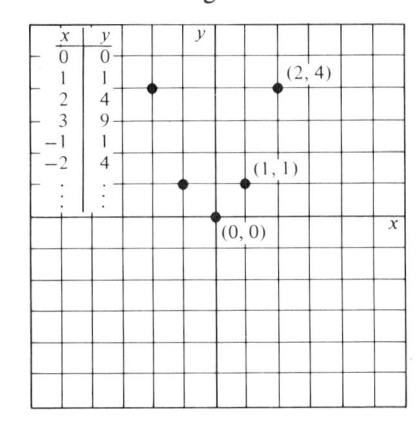

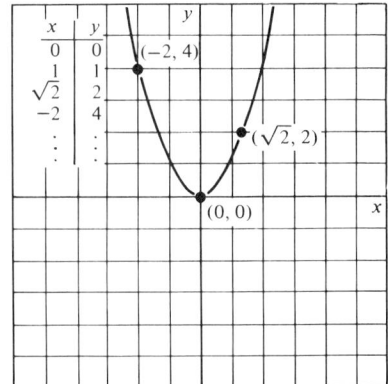

Figure 4.8

Figure 4.9

EXAMPLE 5: Graph: $f = \{(x, y) \mid x^2 = y\}$.

Solution: See Fig. 4.9.

EXAMPLE 6: Graph: $\bar{f}$ where f is the function specified in Example 5.

Solution: $\bar{f} = \{(x, y) \mid y^2 = x\}$. See Fig. 4.10.

A not so obvious shortcut in graphing reverse relations is to use the concept of symmetry.

198 *Relations and Functions*

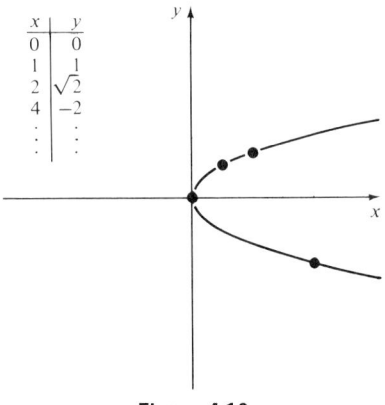

x	y
0	0
1	1
2	$\sqrt{2}$
4	-2
⋮	⋮

Figure 4.10

A graph such as that of $y = x^2$ (Example 5) is said to be symmetric with respect to y-axis since whenever $(x, f(x))$ is on the graph, $(-x, f(x))$ is also on the graph.

Closer inspection of the graph of $y = x^2$ indicates that the line of symmetry acts like a mirror, and the image of a point is the same distance from the line of symmetry as the point. The line represented by $x = y$ is the line of symmetry for a relation and its reverse relation.

To find the graph of a reverse relation, we can simply find the image points of the relation using the line of $x = y$ as our mirror.

EXAMPLE 7: Graph the reverse of the relation graphed in Example 5.

Solution: See Fig. 4.11.

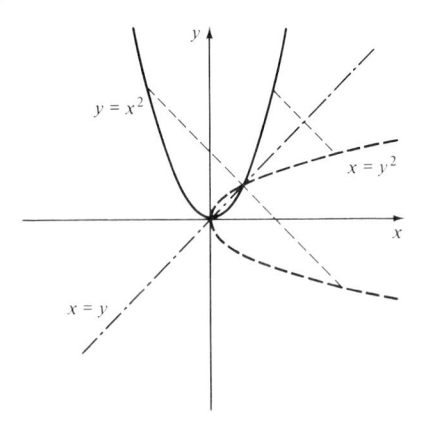

Figure 4.11

EXAMPLE 8: Graph: $R = \{(x, y) \mid x < 0 \wedge x \in \mathscr{X} \wedge y \in \mathscr{R}\}$

Solution: See Fig. 4.12.

§4f *Graphing* 199

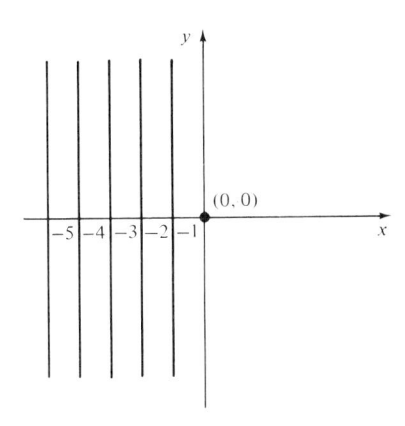

Figure 4.12

None of the open sentences which determine the relations graphed to this point have been compounds. Conjunctive and disjunctive open sentences are frequently encountered. The sentence $x \geq a$, for example, is a disjunction which stands for $x > a \lor x = a$.

EXAMPLE 9: Graph $R = \{(x, y) \mid x \geq 0\}$.

Solution: All the points for which either $x > 0$ or $x = 0$ are true will be points of the locus. See Fig. 4.13.

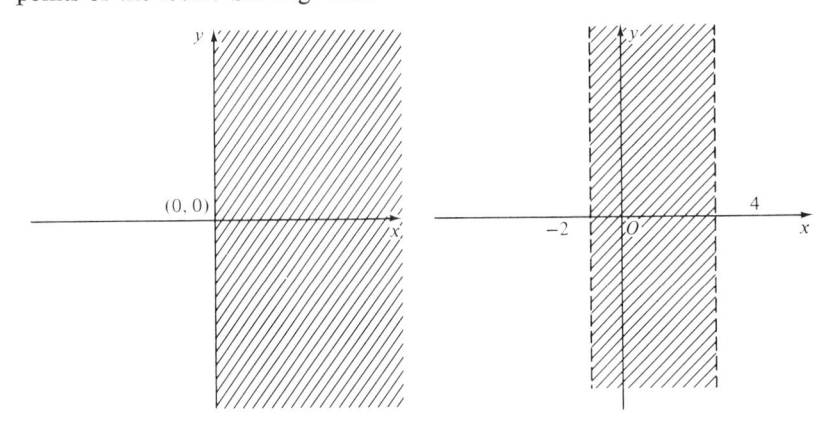

Figure 4.13   **Figure 4.14**

The sentence $a < x < b$ is a conjunction which stands for $x > a \land x < b$.

EXAMPLE 10: Graph the relation $\Phi = \{(x, y) \mid -1 < x < 3\}$.

Solution: We note that y is unrestricted but that x is any real number for which $x > -1$, $x < 3$ are simultaneously true. See Fig. 4.14.

EXAMPLE 11: Graph the relation $H = \{(x, y) \mid x = 1 \lor y = -2\}$.

Solution: The points of the locus are those for which either $x = 1$ or $y = -2$ are true. See Fig. 4.15.

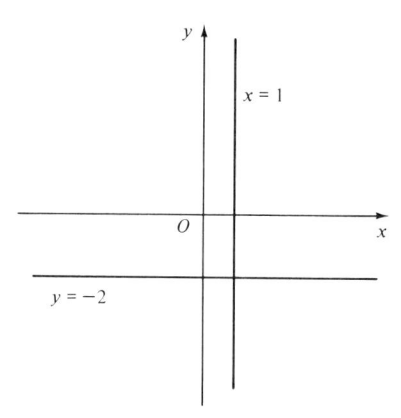

Figure 4.15

Certain special relations are useful in mathematics and in graphing. Two of these are now defined and illustrated.

Definition (4-20): The **absolute value** of any real number x, denoted $|x|$, is x if $x > 0$, 0 if $x = 0$, $-x$ if $x < 0$.

For example, $|3| = 3$ because $3 > 0$; $|x - y| = 0$ if $x = y$ because $x - y = 0$; $|-3| = -(-3) = 3$ because $-3 < 0$.

EXAMPLE 12: Graph $R = \{(x, y) \mid y = |x|\}$.

Solution: See Fig. 4.16.

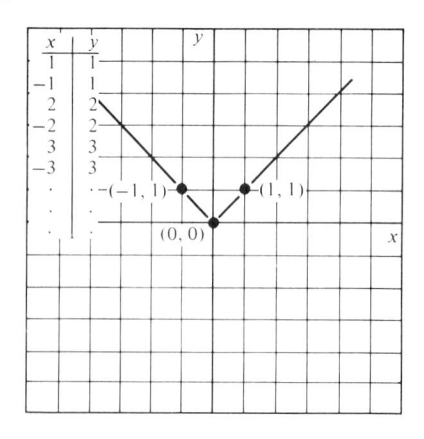

Figure 4.16

§4f Graphing

The second function is called the *greatest integer*, *step*, or *bracket function*. In addition to a variety of strictly mathematical uses, it is sometimes used by business men who know that some people read a price of $3.95 as three dollars rather than four dollars.

Definition (4-21): The **step function**, denoted $[x]$, is

$$[x] = \{(x, y) \mid n \in \mathscr{L} \wedge y = n \text{ for all } n \leq x < n + 1\}$$

EXAMPLE 13: Graph $[x]$.
Solution: See Fig. 4.17.

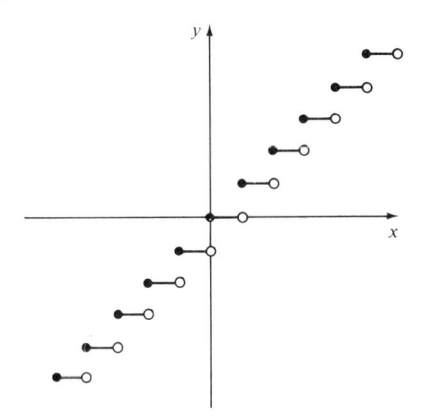

Figure 4.17

EXERCISE SET

1. Graph each relation and state whether it is a function.
    a) $F = \{(x, y) \mid y = 2x \wedge x \in \mathscr{L}\}$
    b) $G = \{(x, y) \mid 2y = x \wedge x \in \mathscr{L}\}$
    c) $R = \{(x, y) \mid y > 2x \wedge x \in \mathscr{R}\}$
    d) $F = \{(x, y) \mid y = \pi \wedge x \in \mathscr{L}\}$
    e) $R = \{(x, y) \mid x + y = y + x \wedge x, y \in \mathscr{L}\}$
    f) $R = \{(x, y) \mid n \leq x \leq n + 1 \wedge k \leq y \leq k + 1 \wedge n, k \in \mathscr{L} \wedge x, y \in \mathscr{R} \wedge n, k \text{ are even integers}\}$
    g) $F = \{(x, y) \mid y = x^3 \wedge x \in \mathscr{R}\}$

2. Graph each relation and state whether it is a function.
    a) $R = \{(x, y) \mid x = y \wedge x > -1\}$
    b) $F = \{(x, y) \mid y = x^2 \wedge y > 1\}$
    c) $G = \{(x, y) \mid y = \sqrt{x+1} \wedge x > 0\}$

d) $H = \{(x, y) \mid x = 4 \land |y| < 0\}$
e) $\Phi = \{(x, y) \mid x > -1 \land x < 5\}$
f) $R = \{(x, y) \mid x = y \lor x = -y\}$
g) $F = \{(x, y) \mid x = y^2 \land x \leq 0\}$
h) $G = \{(x, y) \mid x^2 \geq 0\}$

3. Graph the reverse of the relations found in:
   a) Exercise 1c   b) Exercise 1d   c) Exercise 1g

4. Graph each of the following.
   a) $F = \{(x, y) \mid y = [x]\}$ b) $F = \{(x, y) \mid y = [2x]\}$
   c) $F = \{(x, y) \mid y = [x + 3]\}$ d) $F = \{(x, y) \mid y = \left[\dfrac{x}{3}\right]\}$

5. Graph the reverse of each relation specified in Exercise 4.

6. Graph the reverse of the relations found in:
   a) Example 8   b) Example 9   c) Example 10

7. Graph the following relations and their reverse relations.
   a) $f = \{(x, y) \mid y = |x - 1| \land x \in \mathcal{R}\}$
   b) $\phi = \{(x, y) \mid y = |x + 1| \land x \in \mathcal{R}\}$
   c) $g = \{(x, y) \mid y = |2x + 2| \land x \in \mathcal{R}\}$
   d) $f = \{(x, y) \mid |x| + |y| = 0 \land x, y \in \mathcal{R}\}$
   e) $\phi = \{(x, y) \mid \dfrac{x}{|x|} = y \land x, y \in \mathcal{R}\}$
   f) $g = \{(x, y) \mid x > 0 \land x < 0 \land x \in \mathcal{R}\}$
   g) $h = \{(x, y) \mid |x| + |y| = 1\}$

§4g   CURVE SKETCHING

The loci sketched in the preceding section were of fairly simple nature. It was necessary to determine by substitution only a very few points in order to gain a reasonable idea of the general shape.

In more complicated cases it would often be necessary to find many points if this were the only method used to obtain a good sketch of a locus. Since point plotting can be both tedious and time consuming, we use various other techniques to determine important properties so as to reduce the number of actual points needed. Among these techniques are finding *intercepts*, determining various types of *symmetry*, and locating *excluded regions*.

Definition (4-21):

a) Two **points** M and N are said to be **symmetric about a line** iff the line is the perpendicular bisector of segment MN.

§4g Curve Sketching

b) A **locus** is **symmetric about a line** iff each of its points is one of a pair of points symmetric about the line.
c) Two **points** M, N are **symmetric about a point** O iff O is the midpoint of segment MN.
d) A **locus** is **symmetric about a point** O iff each of its points is one of a pair of points symmetric about O.

We are particularly interested in symmetry about the coordinate axes and the origin. For example, if $(-x, y)$ determines a point of the locus whenever (x, y) does, then the locus is symmetric about the y-axis. In addition, if whenever (x, y) determines a point of the locus, $(x, -y)$ and $(-x, -y)$ do also, then the locus is symmetric about the x-axis and the origin. From the definition of symmetry and these observations, we formulate the following tests:

(i) If an equation is unchanged when y is replaced by $-y$, then the locus of the equation is symmetric about the x-axis.
(ii) If an equation is unchanged when x is replaced by $-x$, then the locus of the equation is symmetric about the y-axis.
(iii) If an equation is unchanged when x is replaced by $-x$ and y by $-y$, then the locus of the equation is symmetric about the origin.

EXAMPLE 1: Test each equation for symmetry of the locus.

a) $x^2 + y^2 = 1$
b) $\dfrac{x^2}{5} + \dfrac{y^2}{8} = 1$
c) $y = 8x^2$
d) $\dfrac{x^2}{7} - \dfrac{y^2}{9} = 1$
e) $y = |x|$
f) $y = \sqrt{x}$
g) $y = x$

Solution:

a) The locus is symmetric about x-axis, y-axis, and origin. It is, in fact, a circle.
b) The locus is symmetric about x-axis, y-axis, and origin.
c) The locus is symmetric about the y-axis only.
d) The locus is symmetric about x-axis, y-axis, and origin.
e) The locus is symmetric about the y-axis only.
f) The locus has no usable symmetry.
g) The locus is symmetric about the origin only.

The second aid to sketching is called finding the intercepts.

Definition (4-22):

a) The abscissa or x-coordinate of any point of intersection of a locus and the x-axis is called an **x-intercept**.

204 Relations and Functions

b) The ordinate or y-coordinate of any point of intersection of a locus and the y-axis is called a **y-intercept**.

To find the y-intercepts of a locus, we set $y = 0$ and solve the resulting equation for x. Similarly, to find the x-intercepts of a locus, we set $x = 0$ and solve the resulting equation for y.

The third aid to sketching mentioned was location of excluded regions. The concept of excluded regions is closely related to the concepts of domain and range of relations. Thus it might be well to review Definitions (4-3) and (4-4) before going on to consider examples of finding excluded regions.

EXAMPLE 2: What region must be excluded in the graph of $y = 1/x$? For this purpose, we regard a *region* as any nonempty subset of the Cartesian plane.

Solution: Since y is a real number and equals $1/x$, then $1/x$ must also be a real number. Here x cannot be zero; thus $x = 0$ is excluded from the domain. Now solve the equation for x in terms of y, i.e., $x = 1/y$. Since x is a real number and equals $1/y$, then $1/y$ must also be a real number. Hence y cannot be zero, and $y = 0$ is excluded from the range. Thus the x- and y-axes are excluded regions. See Fig. 4.18.

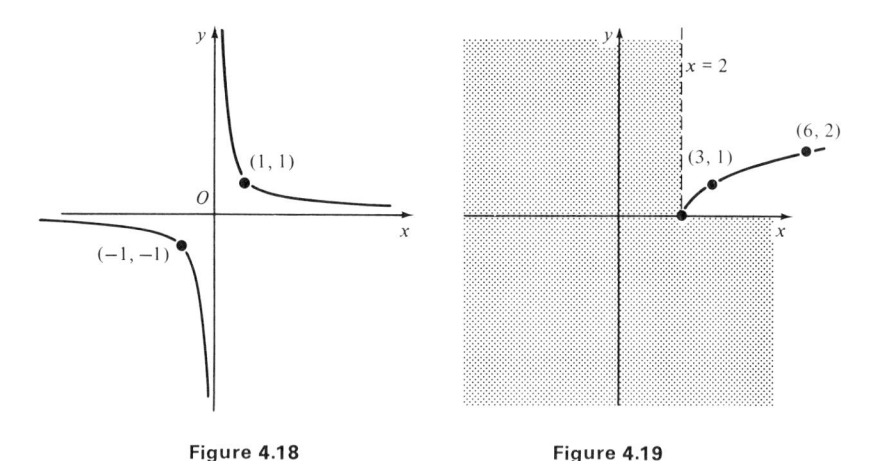

Figure 4.18   Figure 4.19

EXAMPLE 3: What region must be excluded in the graph of $y = \sqrt{x-2}$?

Solution: Since y is a real number and equals $\sqrt{x-2}$, then $\sqrt{x-2}$ must also be a real number. Hence $x - 2$ must be greater than or equal to zero; that is, $x - 2 \geq 0$ or $x \geq 2$. So we must exclude $x < 2$ from the domain. Since the principal square root of a number is nonnegative, we also have $y \geq 0$. That is, all values $y < 0$ are excluded from the range. Now solve for x in terms of y.

§4g   Curve Sketching   205

$$y^2 = x - 2$$
$$x = y^2 + 2$$

Since $y^2 + 2$ is always real, there is no further restriction on the range. Thus $\mathscr{D} = \{x \mid x \geq 2\}$ and $\mathscr{E} = \{y \mid y \geq 0\}$. Therefore, the region below the axis and to the left of the line $x = 2$ is excluded. See Figure 4.19.

These two examples hopefully make the following rule for finding excluded regions plausible.

Rule: To find excluded vertical regions (restriction on domain), solve the given equation for y in terms of x. If the result is a radical of even index or a fraction with $f(x)$ in the denominator, exclude all values of x which cause the radicand to be negative or which cause the denominator to be zero. Similarly, to find excluded horizontal regions (restrictions on range), solve the given equation in terms of y and exclude all values of y that make the resulting radicand negative or the resulting denominator zero. Also be sure to note that *radicals and absolute values stand for nonnegative numbers*. The important fact is that it can be as necessary in sketching to know where a locus is *not* as it is to know where branches of it may be found.

EXAMPLE 4: Find the excluded regions for each of the following equations.

a) $x^2 + y^2 = 1$ b) $x = y^2$
c) $3x^2 + 4y^2 = 1$ d) $x^2 - y^2 = 1$
e) $x = y$ f) $y > |x|$
g) $y = -\sqrt{x - 1}$

Solution:

a) Exclude $x < -1, x > 1, y < -1, y > 1$. See Figure 4.20a.

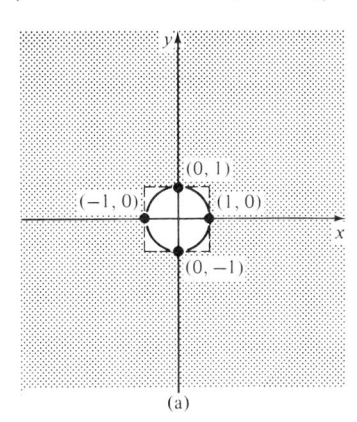

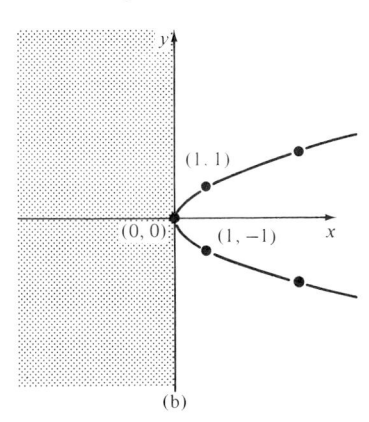

Figure 4.20 (a)(b)

206 *Relations and Functions*

b) Exclude $x < 0$. See Figure 4.20b.
c) Exclude $x < -\sqrt{1/3}$, $x > \sqrt{1/3}$, $y < -1/2$, $y > 1/2$. See Figure 4.20c.

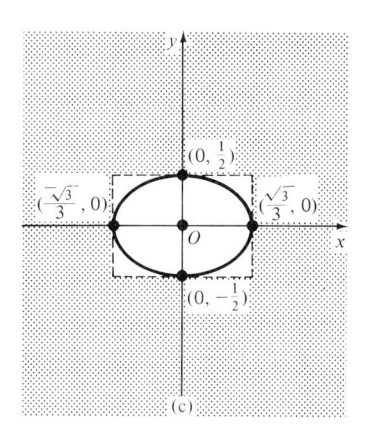

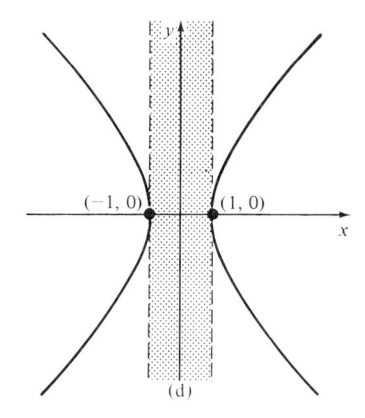

Figure 4.20 (c)(d)

d) Exclude $-1 < x < 1$. See Figure 4.20d.
e) No exclusions $\mathscr{D} = \mathscr{E} = \mathscr{R}$.
f) Exclude $y \leq 0$.

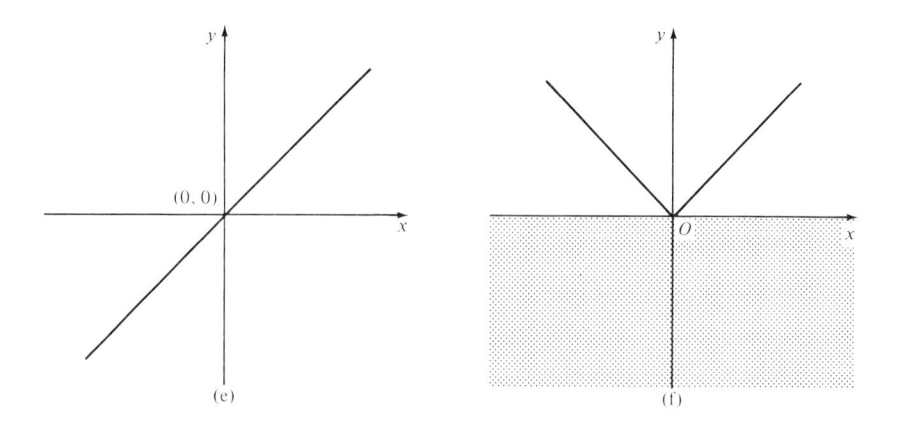

Figure 4.20 (e)(f)

g) Exclude $x \leq 1$, $y > 0$.

§4g   *Curve Sketching*   207

(g)

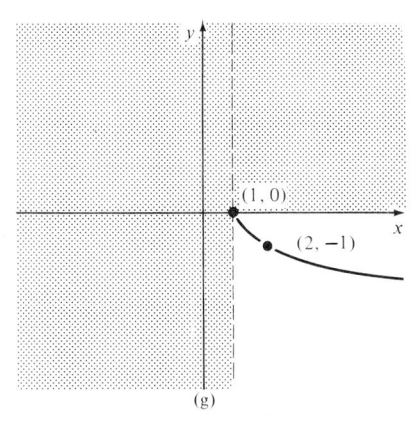

Figure 4.20 (g)

EXERCISE SET

1. Test each open sentence for symmetry of its locus about the axes and origin, and find the intercepts.
   a) $xy - x - 2 = 0$
   b) $4x^2 + 9y^2 - 36 = 0$
   c) $xy = 1$
   d) $x = y^3$
   e) $x^2 + y^2 - 2x + y = 0$
   f) $2x^2 - y^2 - 10 = 0$
   g) $x^2 + y + 7 = 0$
   h) $y^2 + 3x + 4 = 0$

2. Find the excluded regions and intercepts for each locus specified.
   a) $x^2 + y^2 = 4$
   b) $xy = 4$
   c) $y = x^3$
   d) $y = \dfrac{x}{(x-2)}$
   e) $\dfrac{x^2}{4} - \dfrac{y^2}{9} = 1$
   f) $y^2 + 4x + 8 = 0$
   g) $y = |x - 2|$
   h) $xy^2 + x^2 - 4 = 0$
   i) $y = -\sqrt{x+4}$

3. Sketch the graph of each specified locus.
   a) Exercise 1c
   b) Exercise 1d
   c) Exercise 1g
   d) Exercise 2b
   e) Exercise 2c
   f) Exercise 2d
   g) Exercise 2a
   h) Exercise 2e
   i) Exercise 2f
   j) Exercise 2g
   k) Exercise 2i
   l) $y \leq |x + 3|$

5
AN INTRODUCTION TO PROBABILITY

§5a WHAT IS PROBABILITY?

One very useful application of the ideas of sets is to the concept of probability. We often use the word "probability" in such expressions as the "probability of passing this course," or the "probability of good weather," or the "chances of getting a date are slim." In each of these cases we get an intuitive idea of what probability is, but, as in all other mathematical topics, formal definitions must be given to make the concept precise.

Often what actually happens in reality is not what was predicted theoretically. The subject of probability offers no exception. There is a difference between theoretical probability and empirical probability. If an experiment is performed n times and a certain outcome is obtained f times, the empirical probability is the ratio f/n.

Definition (5-1): The **empirical probability** $\hat{p}$ of the occurrence of an outcome is the ratio of the times f the outcome is obtained to the number of times n the experiment was performed. That is, $\hat{p} = f/n$.

It may be possible to use $\hat{p}$ to predict the number of successful future outcomes of the experiment. That is, the empirical probability will be close to the theoretical probability as more experiments are performed.

EXAMPLE 1: In fifteen of the last hundred years it has rained in Salem on June 23rd. Then $\hat{p} = \frac{15}{100} = \frac{3}{20}$. Here we could predict that in the future it would rain in Salem on that day $\frac{3}{20}$ of the time.

Theorem (5a-1): The empirical probability $\hat{p}$ is positive or zero, and less than or equal to 1, i.e., $0 \leq \hat{p} \leq 1$.

Proof: Since f is positive or zero and n is positive, f/n cannot be negative. Thus $\hat{p} \geq 0$. Also $f \leq n$. Thus $f/n \leq 1$, whence $\hat{p} \leq 1$. Q. E. D.

Theorem (5a-2): If $\hat{q}$ is the probability that a result will not occur and $\hat{p}$ is the probability that it will occur, then $\hat{p} + \hat{q} = 1$.

Proof: Let r be the number of unsuccessful outcomes, i.e., $\hat{q} = r/n$. Then $\hat{p} + \hat{q} = f/n + r/n = (f + r)/n$. But $f + r = n$, since the successful outcomes plus the unsuccessful ones equal the number of trials. Hence $\hat{p} + \hat{q} = (f + r)/n = n/n = 1$.   Q. E. D.

In all problems for which $\hat{p}$ can be computed, we must keep in mind that there is some theoretical or true probability p which is unknown to us. We should expect that $\hat{p}$ would be close to p as the number of performances of the experiment are increased.

§5b SAMPLE SPACES AND EVENTS

The way in which set concepts are used in the study of probability begins to be evident with introduction of the concept of sample spaces.

Definition (5–2): The set of all possible outcomes of an experiment is called the **sample space**.

Definition (5–3): The elements of the sample space are called **sample points**.

EXAMPLE 1: Two coins are tossed. A set of possible outcomes is how the two fall: "No tails, one tail, or two tails." This set can be expressed as the sample space

$$S_1 = \{0T, 1T, 2T\}$$

Another set of possible outcomes is how each coin falls, resulting in the following sample space:

$$S_2 = \{(H, H), (H, T), (T, H), (T, T)\}$$

EXAMPLE 2: Two dice are thrown; the possible sums show a range from 2 to 12. Thus one sample space is

$$S_1 = \{2, 3, 4, 5, 6, 7, 8, 9, 10, 11, 12\}$$

Since the sums can occur in more than one way, a more useful sample space would be to use all the ordered pairs. That is, if $S_2 = \{1, 2, 3, 4, 5, 6\}$ is the sample space of a die, $S_2 \times S_2$ would be the sample space of all possible ordered pairs. The 36 possible outcomes are shown in Figure 5.1.

Certain outcomes of an experiment might be considered favorable while others might not. The set of favorable outcomes would be a subset of the sample space.

Definition (5–4): The set of favorable outcomes to an experiment is called an **event**.

Figure 5.1

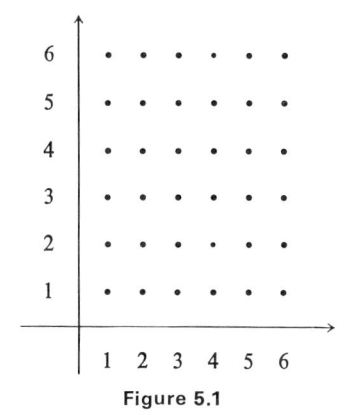

(grid of dots at integer coordinates from 1 to 6 on both axes)

EXAMPLE 3: Using S_2 from Example 1, we are interested in getting at least one tail. The event E would be the following set:

$$E = \{(H, T), (T, H), (T, T)\}$$

EXERCISE SET I

1. A coin and a die are thrown. List the sample space.
2. Two coins are selected from a set consisting of a penny, a nickel, a dime, and a quarter. List a sample space.
3. A bag contains three red marbles and two black marbles. List the sample space if we select
   a) one marble at random
   b) two marbles at random
4. A group of three people sit in five chairs. List the sample space.
5. Using the sample space shown in Figure 5.1, describe the following events in roster form.
   a) The sum of the dice is greater than 12
   b) The sum is 5 or 9
   c) The sum is less than 7
   d) The dice show the same total
6. A card is drawn from a regular deck. Describe the following event in roster form or use set builder form if too large.
   a) A red card is drawn
   b) A heart is drawn
   c) A face card is drawn
   d) An ace is drawn

e) A black ace is drawn
f) The ace of spades is drawn

7. From a set of five different apples v, w, x, y, z a pie using three of them is to be made. Describe the following events in roster form.

   a) v is in the pie (and any two others)
   b) v and w are in the pie
   c) v or w are in the pie
   d) w is not in the pie
   e) v and w are not in the pie

The elements of the sample space that are not members of a specific event E constitute another subset of the sample space called, obviously, the complementary event E'. Thus, if E is an event and E' is its complement relative to the sample space S, we have

$$E \cup E' = S \quad \text{and} \quad E \cap E' = \varnothing$$

Parts of Exercises 5 and 7 describe events that depend on the logical use of *and* and *or*. The event "A and B" is understood to occur if both A, B occur. Hence, we denote the event "A and B" as $A \cap B$. Similarly, the event "A or B" is understood to occur if either one of A, B occurs. Hence we denote the event "A or B" as $A \cup B$.

EXAMPLE 4: Suppose two dice are thrown. Let x denote the number showing on die 1, y denote the number showing on die 2, A the event $x + y = 7$, B the event $y \leq 4$. The compound C is described as "A total of 7 *and* die $2 \leq 4$," i.e., $A \cap B$. The compound D is described as "A total of 7 *or* die $2 \leq 4$," i.e., $A \cup B$. See Figure 5.2.

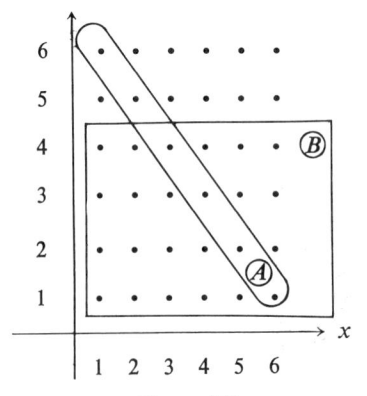

Figure 5.2

EXERCISE SET II

8. Two coins and a die are thrown. Describe the following events.
   a) at least one head and a 3
   b) at most one tail and a number greater than 3
   c) two heads or a 5
   d) two tails or any number

9. A bag contains three marbles; two of them are red and one of them is black. If we draw two at a time, describe the following events.
   a) at least one black
   b) one black and one red
   c) one black or one red
   d) two blacks

§5c PROBABILITY

Empirical probability has been defined and described as the actual or real occurrence. We are now in a position to discuss the theoretical probability of an outcome of an experiment. Two assumptions will be made to facilitate our work:

   i) The sample space is finite;
   ii) Every sample point is equally likely to occur.

Definition (5-5): The **probability** $P(E)$ of an event E occurring is defined as the number of occurrences that will result in the event $n(E)$ divided by the number of outcomes in the sample space $n(S)$. Thus $P(E) = n(E)/n(S)$.

EXAMPLE 1: A coin is tossed. What is the probabity of getting a head?
Solution:
$$S = \{H, T\} \Rightarrow n(S) = 2$$
$$E = \{H\} \Rightarrow n(E) = 1$$

Therefore,
$$P(E) = \frac{n(E)}{n(S)} = \frac{1}{2}$$

EXAMPLE 2: Two coins are tossed. What is the probability of getting at least one head?

Solution:

$$S = \{(H, H), (H, T), (T, H), (T, T)\} \Rightarrow n(S) = 4$$
$$E = \{(H, H), (H, T), (T, H)\} \Rightarrow n(E) = 3$$

Therefore,

$$P(E) = \frac{n(E)}{n(S)} = \frac{3}{4}$$

EXAMPLE 3: A die is thrown. What is the probability of getting a seven?

Solution:

$$S = \{1, 2, 3, 4, 5, 6\} \Rightarrow n(S) = 6$$
$$E = \{\ \} \Rightarrow n(E) = 0$$

Therefore,

$$P(E) = \frac{n(E)}{n(S)} = \frac{0}{6} = 0$$

EXAMPLE 4: A coin is tossed; what is the probability of getting a head or a tail?

Solution:

$$S = \{H, T\} \Rightarrow n(S) = 2$$
$$E = \{H, T\} \Rightarrow n(E) = 2$$

Therefore,

$$P(E) = \frac{n(E)}{n(S)} = \frac{2}{2} = 1$$

Examples 3 and 4 indicate that theoretical probability, like empirical probability, is any number between zero and one including either zero or one.

Theorem (5c-1): $0 \leq P(E) \leq 1$.

Theorem (5c-2): $P(S) = 1;\ P(\emptyset) = 0$.

The proofs of these theorems are left to the exercises. An interpretation of Theorem (5c-2) is that an event with probability one is certain to happen and an event with probability zero will never occur.

Because of the way probability is defined, much use is made of the number of elements in a set. In particular we need to recall the expression

(1) $$n(A \cup B) = n(A) + n(B) - n(A \cap B)$$

See Definition (4-6). Using this principle, we can establish the following theorem.

Theorem (5c-3): If A' is the complement of event A relative to the sample space S, then $P(A) + P(A') = 1$.

Proof:

$$\begin{aligned} P(A) + P(A') &= \frac{n(A)}{n(S)} + \frac{n(A')}{n(S)} \\ &= \frac{n(A) + n(A')}{n(S)} \\ &= \frac{n(A \cup A')}{n(S)} \quad \text{(Note: } A \cap A' = \emptyset \text{)} \\ &= \frac{n(S)}{n(S)} \\ &= 1 \end{aligned}$$

Q. E. D.

Since it is sometimes easier to compute the probability of an event by asking when it will *not* occur, Theorem (5c-3) can be very useful.

EXERCISE SET

1. A card is drawn from a regular deck. Find the probability of the following events.
   a) A red card is drawn
   b) A heart is drawn
   c) A face card is drawn
   d) An ace is drawn
   e) A black ace is drawn
   f) The ace of spades is drawn

2. Three coins are tossed. Find the probability of the following events.
   a) Something besides three tails
   b) Two heads or a tail
   c) At least one head

3. Two dice are thrown. Find the probability of the following events.
   a) The sum is even
   b) The sum is odd
   c) The sum is more than eight
   d) The sum is eight
   e) The sum is less than eight

4. A family has three children of different ages. Find the probability that
   a) the oldest is a boy or a girl
   b) the youngest is a girl

c) the oldest is a boy or the youngest is a girl
d) the oldest is a girl and the youngest is a girl

5. A box contains three white, four blue, and two black balls. Two balls are drawn. What is the probability that
    a) the balls are of the same color
    b) the balls are of different colors

6. Prove:
    a) Theorem (5c-1)
    b) Theorem (5c-2)

§5d MORE PROPERTIES OF PROBABILITY

We now resume consideration of the events that use the logical connectives *or* and *and*.

Theorem (5d-1): $P(A \cup B) = P(A) + P(B) - P(A \cap B)$.

Proof:

$$P(A \cup B) = \frac{n(A \cup B)}{n(S)} \qquad \text{Definition (5-5)}$$

$$= \frac{n(A) + n(B) - n(A \cap B)}{n(S)} \qquad \text{Definition (4-6)}$$

$$= \frac{n(A)}{n(S)} + \frac{n(B)}{n(S)} - \frac{n(A \cap B)}{n(S)}$$

$$= P(A) + P(B) - P(A \cap B) \qquad \text{Definition (5-5)}$$

Q. E. D.

If A and B have no outcomes in common, then $A \cap B = \emptyset$ and

$$P(A \cup B) = P(A) + P(B)$$

In case $A \cap B = \emptyset$, we say that A and B are **mutually exclusive events**: If A occurs, then B cannot occur.

EXAMPLE 1: A card is drawn from a regular deck. Let D be the event "A diamond is drawn." Let K be the event "A king is drawn." What is the probability that a diamond or a king is drawn?

Solution:

$$P(D \cup K) = P(D) + P(K) - P(D \cap K)$$

$$P(D) = \frac{13}{52} = \frac{1}{4}$$

$$P(K) = \frac{4}{52} = \frac{1}{13}$$

Since there is only one king of diamonds,

$$P(D \cap K) = \frac{1}{52}$$

Therefore,

$$P(D \cup K) = \frac{13}{52} + \frac{4}{52} - \frac{1}{52} = \frac{16}{52} = \frac{4}{13}$$

EXAMPLE 2: A card is drawn from a regular deck. Let Q be the event a queen is drawn and T be the event a ten is drawn. What is the probability that a queen or a ten will be drawn?

Solution:

$$P(Q \cup T) = P(Q) + P(T) - P(Q \cap T)$$
$$P(Q) = \frac{4}{52} = \frac{1}{13}$$
$$P(T) = \frac{4}{52} = \frac{1}{13}$$

Since no queen can be a ten, these events are mutually exclusive, and

$$P(Q \cap T) = 0$$

Therefore,

$$P(Q \cup T) = \frac{1}{13} + \frac{1}{13} - 0 = \frac{2}{13}$$

EXERCISE SET I

1. A bag contains two red, three white, four blue, and five black marbles. One marble is drawn. Find the probability that
    a) the marble is red or white
    b) the marble is red or white or blue
    c) the marble is white or blue or black
    d) the marble is red or white or blue or black.

2. A box contains tennis balls and ping-pong balls in assorted colors as follows: three red, one blue, and two green ping-pong balls; two blue and two green tennis balls. A single ball is selected. Find the probability of selecting
    a) a ping-pong ball
    b) a green ball
    c) a blue ball or a ping-pong ball
    d) a red ping-pong ball
    e) a red ball or a tennis ball.

3. A die is rolled. Let D be the event the outcome is odd, B the event the outcome is prime.

   a) What is the probability that the outcome is odd?
   b) What is the probability that the outcome is prime?
   c) What is the probability that the outcome is even and not prime?
   d) What is the probability that the outcome is odd or prime?

It is easier to consider the arrow connective before the "and" connective. Sometimes the probability of one event occurring depends upon whether or not another event has already occurred. Such a dependent probability is called *conditional probability* and the two events are said to be dependent events.

EXAMPLE 3: Two dice are thrown. If the first die turns up a 5, what is the probability that the sum showing is at least 9?

Solution 1: The first die shows a 5. Hence we need a 4, 5, or 6 for the second. Since the probability of obtaining a 4 or a 5 or a 6 for the second die is $\frac{3}{6}$ or $\frac{1}{2}$, the probability that the sum will be at least 9 is $\frac{1}{2}$.

If we indicate the number on die one by x and the number on die two by y, the solution to the problem can be written as follows:

$$P(x + y \geq 9 \mid x = 5) = \frac{1}{2}$$

The symbol | can be read "given" or "under the condition that." A second approach to the solution of the problem given in Example 3 is more useful for our later work.

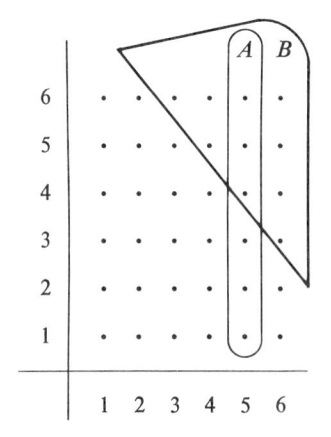

Figure 5.3

Solution 2: Figure 5.3 shows a diagram of the sample space S. A and B are subsets of S with $A = \{(x, y) | x = 5\}$, $B = \{(x, y) | x + y \geq 9\}$. The question is, what is $P(B|A)$? Since we are given that die one shows a 5, we know that only the subset A of S can include possible outcomes of the experiment. That is, we now consider A as a new sample space S'. Thus,

$$S' = \{(5, 1), (5, 2), (5, 3), (5, 4), (5, 5), (5, 6)\}$$

and $N = \{(5, 4), (5, 5), (5, 6)\}$ where N is the event that the die sum is nine or more. Hence

$$P(N) = \frac{n(N)}{n(S')} = \frac{3}{6} = \frac{1}{2}$$

However, $P(N)$ is $P(B \mid A)$ and $N = A \cap B$; hence,

$$P(B \mid A) = (A \cap B) \text{ in } S'$$

Now

$$P(A \cap B) = \frac{n(A \cap B)}{n(S')}$$

where $S' = A$; hence,

$$P(B \mid A) = \frac{n(A \cap B)}{n(A)}$$

EXERCISE SET II

4. Show that $P(A \mid B)$ in Example 3 is $\frac{3}{10}$.
5. What is the probability of tossing a 7
    a) with a pair of dice?
    b) with a pair of dice if the first one shows 5?
    c) with a pair of dice nine times in a row if you were successful for the first eight tries?
6. Three coins are tossed. What is the probability of
    a) at least two heads if the first is a head?
    b) at least two heads if the first is a tail?
    c) three heads?
    d) three heads if the first is a tail?
7. An urn contains five blue and three green marbles. Two marbles are drawn one at a time. What is the probability of
    a) drawing a blue if the first was blue?
    b) drawing a blue if the first was green?
    c) drawing a green if the first was green?
    d) drawing a blue or green if the first was green?

§5e THE FORMULA FOR $P(A \cap B)$

The conditional probability of the event A occurring given that B has occurred is given by

$$P(A \mid B) = \frac{n(A \cap B)}{n(B)}$$

while the conditional probability that B occurs given that A has occurred is given by

$$P(B \mid A) = \frac{n(B \cap A)}{n(A)}$$

It is important to see that

$$\frac{n(A \cap B)}{n(B)} = \frac{n(A \cap B)}{n(S)} \cdot \frac{n(S)}{n(B)} = \frac{n(A \cap B)}{n(S)} \div \frac{n(B)}{n(S)}$$

However,

$$\frac{n(A \cap B)}{n(S)} = P(A \cap B)$$

and

$$\frac{n(B)}{n(S)} = P(B)$$

Hence

$$P(A \mid B) = \frac{n(A \cap B)}{n(B)} = \frac{P(A \cap B)}{P(B)}$$

Thus

$$P(A \cap B) = P(B) \cdot P(A \mid B)$$

Theorem (5e-1): If an experiment results in the event $(A \cap B)$, the probability of $A \cap B$ is given by $P(A \cap B) = P(B) \cdot P(A \mid B)$ or $P(A) \cdot P(B \mid A)$.

The proof follows immediately from the preceding discussion and the fact that $A \cap B = B \cap A$.

EXAMPLE 1: A bag contains three white balls and two red balls. Two balls are drawn simultaneously. Find the probability that both are white.

Solution: Let A be the event that the first ball is white and B the event that the second ball is white.

$$P(A \cap B) = P(A) \cdot P(B \mid A)$$
$$P(A) = \frac{3}{5}$$
$$P(B \mid A) = \frac{2}{4} \text{ or } \frac{1}{2}$$

§5e The Formula for P (A ∩ B)

and
$$P(A \cap B) = \frac{3}{5} \cdot \frac{1}{2} \text{ or } \frac{3}{10}$$

If the occurrence of event A does not depend on event B, we say the two events are independent of each other. Thus $P(A|B)$ would be the same as $P(A)$.

Definition (5-6): If A and B are **independent events**, $P(B) = P(B|A)$ and $P(A) = P(A|B)$.

Theorem (5e-2): If A and B are independent events, $P(A \cap B) = P(A) \cdot P(B)$.

Proof:
$$P(A \cap B) = P(A) \cdot P(B|A)$$
$$= P(A) \cdot P(B)$$
because $P(B|A) = P(B)$.

Q. E. D.

EXAMPLE 2: A card is drawn from a regular deck. It is then replaced and a second card is drawn. Find the probability of drawing two hearts.

Solution: Since the card is drawn and then replaced, the events are independent. Hence

$$P(A) = \frac{13}{52} \text{ or } \frac{1}{4}$$
$$P(B) = \frac{13}{52} \text{ or } \frac{1}{4}$$
$$P(A \cap B) = P(A) \cdot P(B)$$
$$= \frac{1}{4} \cdot \frac{1}{4} \text{ or } \frac{1}{16}$$

EXAMPLE 3: Two cards are drawn from a regular deck without replacement. What is the probability of drawing two hearts?

Solution: Since the card is not replaced, the events are dependent. Hence,

$$P(A) = \frac{1}{4}$$
$$P(B|A) = \frac{12}{51}$$
$$P(A \cap B) = \frac{1}{4} \cdot \frac{12}{51} \text{ or } \frac{1}{17}$$

EXERCISE SET

1. A coin is tossed and a child is born. What is the probability of a head and a boy?
2. Two cards are drawn from a regular deck. What is the probability that both are aces?
3. Two TV sets are being sold in a store that contains two sets that are defective and eight that work. What is the probability that of the two sets bought,
   a) both are defective?
   b) neither is defective?
   c) exactly one is defective?
4. A game has five dice; three are red and two are white. What is the probability of selecting
   a) a red die?
   b) a white die and rolling a three with the die selected?
   c) a red die and a white die without replacement?
   d) two red dice without replacement?
5. If A and B are events such that $P(A) = 0.2$, $P(B) = 0.3$, and $P(A \cap B) = 0.05$, find
   a) $P(A \cup B)$
   b) $P(A \mid B)$
   c) $P((A \cap B)')$
   d) $P(B \mid A)$
   e) $P(A')$
   f) $P(B')$
   g) $P(A' \mid B')$
   h) $P(A' \cup B')$
   i) $P(A' \cap B')$
6. One box contains three red and four white balls; the second box holds two red and five white balls. From each box two balls are drawn without replacement. What is the probability of obtaining
   a) four red balls?
   b) two red and two white balls?
   c) four white balls?

§5f SUMMARY

This chapter has attempted to introduce only the most elementary ideas of probability and statistics. It is the authors' hope that you might find the subject interesting enough to want to pursue it further where such topics as permutations and combinations, areas under curves, and the normal curve distribution would be considered. What we have covered in this chapter is summarized as follows:

Empirical probability:	$\hat{p} = \dfrac{f}{n}$
Probability of event A:	$P(A) = \dfrac{n(A)}{n(S)}$
Probability of event S:	$\dfrac{n(S)}{n(S)} = 1$
Probability of impossible event:	$P(\varnothing) = \dfrac{n(\varnothing)}{n(S)} = 0$
Probability of A or B:	$P(A \cup B) = P(A) + P(B) - P(A \cap B)$
Probability of A if B:	$P(A \mid B) = \dfrac{P(A \cap B)}{P(B)}$
Probability of A and B (dep.):	$P(A \cap B) = P(A \mid B) \cdot P(B)$
Probability of A and B (ind.):	$P(A \cap B) = P(A) \cdot P(B)$

6

THE NUMBER SYSTEM

§6a   COUNTING

Counting of one kind or another is thought to be the oldest mathematical activity humans learned to do. Almost every child attempts to learn to count at such an early age that he simply learns words from others and gradually begins to make the same associations of meaning with the words as do those who speak the language regularly. Because each of us has learned this ancient art in this way, there can be real value in taking a good look at the principles underlying it. However, when terms have been learned by association there is often an extra difficulty in learning to use them in a precise way as specialized terms in a logical discourse.

To begin the process of understanding precise meanings for number terms, let us consider Max Shepherd, preliminary man, who, rising with the sun from his comfortable cave, opens his sheep pen and begins to place one stone in his leather bag for each sheep that leaves. Then at nightfall he takes out a stone for each sheep that enters the pen, gets worried if he has stones left over, is happy if the number of stones matches the number of sheep, and is ecstatic if there are more sheep than stones.

We all recognize Max's method of making use of the one-to-one correspondence of matching sets, but at the same time it is worthwhile for us to take a place beside Max at the sheep pen and discover whether or not our method of counting is any different from Max's method, and, if it is different, whether it is any better. Let us begin these thoughts by saying that Max has thirty sheep. Hence Max puts thirty stones in his bag; what do we put in ours? Hopefully we answer, "one numeral representing thirty." Why can we do this? What is special about our counting method which enables us to carry one symbol for thirty when Max must actually carry thirty? Max uses thirty stones, each indistinguishable from the other. Our symbolic stone is different; it is the one that comes after twenty-nine. Now, twenty-nine comes after twenty-eight which comes after twenty-seven... which comes after two which comes after one. Since each counter follows one and only one other counter in a prescribed way, we always *know* exactly which counters have come before it and thus have no need to remember any but the *last* one.

226   *The Number System*

Max's method of counting is really a process of comparison which uses the operation of pairing to achieve a visual result. It certainly was a long step forward on the road to developing true number concepts, but it had serious disadvantages. The need to carry around one counter corresponding to each element of some tallied set in order to know and show "how many" has been mentioned. Also it is difficult to gain any really accurate idea of the actual tally of a set by the comparison method even though it is possible to tell precisely whether two sets are matched. A third difficulty is that it is not possible to gain an accurate idea of *how much* larger or *how much* smaller one set is than another unless the tally of one of the sets is close to the tally of the other.

The perfectly marvelous thing about our counting, then, is that we have represented Max's set of thirty stones by our one "stone." Careful consideration of the points made in our discussion of the two methods leads one to realize that in order to "count" in our usual sense, three things are essential: something to count with, the concept of "follows," and a definite counter with which to begin to count. More formally, we can say that a set of counters must have the following three properties:

(1)
- (i) The set of counters must be nonempty.
- (ii) There must be a definite starting counter.
- (iii) The counters must be usable in some definite order.

These are the properties which make our method of *ordered counting* better than Max's comparison method.

A set of counters having just these properties would suffice for any tally provided we do not try to count some large set like the stars in the Milky Way. Note that as long as the sets to be counted remain small, counters could be some sets of dishes, varicolored beads, or other sets having distinguishable elements.

Returning to our task of gaining a more precise understanding of the counting process, we must convert the principles of (1) into a mathematical system which will give us a counting set. One of the most successful ways to do this is to introduce the idea of a *successor*. By successor we mean an element of a set which "follows" a given element. In this sense, in the set of names for days of the week, Tuesday is the successor of Monday. Likewise, in the set of names for months of the year, June is the successor of May. In terms of this concept we could call January a nonsuccessor because it is not the successor of any other name in the set of names for months. In terms of these concepts of successor and nonsuccessor, the principles of (1) can be restated as follows:

(2)
- (i) The set of counters must be nonempty.
- (ii) The set must contain a single nonsuccessor.
- (iii) Each counter must have a unique successor.

The requirement that a successor must be unique guarantees that each counter is distinguishable from other counters.

If we examine the set {January, February, ...}, we see that January is the required nonsuccessor. Its successor is February; the successor of February is March, etc. However, there is one difficulty. Certainly the set of names of months satisfies the requirements for a counting set as stated in (1), but stating these requirements in terms of successor has actually changed them in an important way. According to Requirement (iii) of (2), December must also have a successor. The set of names for months would have to be endless to satisfy this requirement.

For mathematical purposes, the requirements of (2) are excellent. If a counting set with these properties can be established, there will always be enough counters in the set so that the tally of any set from nature can be found. However, the fact that we can actually establish such an endless counting set points up one of the major differences between mathematics and "reality." As far as we know, nowhere in nature does one find sets which are infinite. Only in mathematics does this concept have a literal meaning.

It will be remembered that our goal was to establish some ordered counting numbers. When we introduced the word "successor" above, we stated what it means in terms of "follows." This latter word certainly implies that some order is known. Hence we cannot use "successor" to define "order" without being circular. One of these terms must be used as a primitive term.

The following five axioms containing three primitive terms can be used to build our entire number system. They were stated by Giuseppe Peano (1858–1932) in 1899. They are given here in the form used by E. Landau in his *Grundlagen der Analysis* (*Foundations of Analysis*).* We are presupposing logic and sets here.

The set of counting numbers $\mathscr{P}$ is to have the properties:

P-1: α is a counting number.

P-2: With each counting number n there is associated a unique counting number n' called the successor of n.

P-3: For all counting numbers n, $n' \neq \alpha$.

P-4: If $m' = n'$, then $m = n$.

P-5: Whenever a proposition is true

  (i) for the counting number α and
  (ii) is also true for k' if it is true for an arbitrary counting number k,

  then the proposition is true for *all* counting numbers.

---

*Edmund Landau, *Grandlagen der Analysis*. (New York: Chelsea Pub. Co., 1948). Also available in English translation.

Axiom 5 is one of the most important and famous principles in all of mathematics. It is called the *Axiom of Induction*. The name is unfortunate, for the axiom has nothing to do with reasoning inductively as described in Chapter 1. Since it appears as a part of a logical system, its use is actually *de*ductive.

The requirement for a set to be a counting set $\mathscr{P}$ is that it satisfy P-1 through P-5. Let us list a few sets of numbers having these properties, assuming that their arithmetic is known.

(3)
- a) $1, 3, 5, 7, \ldots$
- b) $1, 4, 9, 16, \ldots$
- c) $0, 1, 2, 3, \ldots$
- d) $2, 4, 6, 8, \ldots$
- e) $2, 4, 8, 16, \ldots$
- f) $-1, -2, -3, -4, \ldots$
- g) $\dfrac{1}{2}, \dfrac{1}{4}, \dfrac{1}{8}, \dfrac{1}{16}, \ldots$
- h) $-3, -2, -1, 0, \ldots$

Here each number symbol is to have its usual mathematical meaning for the sake of example.

Now each of these sets is certainly a counting set. We are used to considering each of these sets as different from each other and different from our counting numbers in some essential way. These differences, however, are the result of how each set acts under addition, multiplication, or involution. Nowhere in Peano's axioms is there any mention of these operations. Hence such differences have no meaning when considering whether or not a set might be usable as a counting set.

As far as counting is concerned, each of the sets of (3) is just a different set of symbols for our counting set $\mathscr{P}$ with a particular substitute symbol for the single nonsuccessor α. Thus each set has the required single nonsuccessor, and each element has a unique successor as required. Once order of successors in any of the sets was learned, counting could be done with any one of them.

The true number concept actually depends on the recognition of the fact that matched sets have an abstract property in common besides the possibility of putting them into one-to-one correspondence. There may have once been an unknown genius who finally recognized the abstract property shared by such sets as his hands, a pair of shoes or sandals, a span of oxen, a couple (husband and wife), and other sets having the same tally as each of these. The abstract property shared by these sets is, of course, something we can call "twoness." Strange as it may seem to us, there are said to be peoples who still do not recognize that two trees and two people do have this property in common. Instead they use words like "pair," "span," "couple," "brace," and others and have no single word like "two" for the shared property. The fact that we have this assortment of words in our own language indicates that our ancestors also may have lacked a general concept of "twoness."

The number system with which we are somewhat familiar can be developed from Peano's axioms. Any such development takes a great deal of patience on the part of both students and teachers and will not be attempted here. An excellent account of one such development is given in E. Landau's *Grundlagen der Analysis.**

EXERCISE SET

1. Consider the statement: "*Counting* is the operation of establishing a one-to-one correspondence between the set to be counted and a subset of counters used in order beginning from some specified counter." Explain why this could never be considered a good definition.

2. Could
   a) tally marks alone or   b) pebbles

   be used to do counting in the sense of (1)?

3. The set of symbols: I, II, III, IV, ... is matched with what other sets? Are these sets ordered?

4. If one counts the days of the week, which two sets have been put into one-to-one correspondence?

5. Consider the three listed disadvantages of using comparison counting. Exactly how does a set of counters having the properties given in (1) overcome these difficulties?

6. Explain how some set of dishes could be set up for use as counters of the type described in (1). Explain exactly how one would tally the months of the year using this set of dishes as counters.

7. The languages used by some peoples have names only for the numbers "one," "two," and "many." How could a scout from an illiterate tribe using only such a language report to his chief the exact number of men (18) in an advancing war party?

8. State which three terms are the primitive terms in Peano's axioms.

---

*Ibid.

§6b THE NATURAL NUMBERS—AXIOMS

The real number system can be developed in many ways. The approach chosen by a particular author for his text usually depends on the background and goals of the reader. Hopefully, each reader of the following presentation will come away with a better understanding of his number system.

In the preceding section we indicated that a complete but somewhat cumbersome number development could be based on counting and that a counting set $\mathscr{P}$ does not need to have familiar additive and multiplicative properties. Our approach here is to take a short cut by treating addition and multiplication as two abstract operations and stating rules for their use in such a way that the result will be the familiar set of numbers

$$(1) \qquad \mathscr{N} = \{1, 2, 3, \ldots\}$$

which can be added and multiplied as we learned in the early grades.

The natural number system $\mathscr{N}$ is asserted to be the set of counters of (1), with the binary operations $+$ and $\cdot$ that satisfy the following axioms (Note: The properties of element equality are the same as given in §3c).

N-1: $\forall a, b \in \mathscr{N},\ \exists x \in \mathscr{N} \cdot \ni \cdot a + b = x$

N-2: $\forall a, b \in \mathscr{N},\ \exists x \in \mathscr{N} \cdot \ni \cdot a \cdot b = x$

N-3: $\forall a, b, c \in \mathscr{N},\ a = b \Rightarrow a + c = b + c$

N-4: $\forall a, b, c \in \mathscr{N},\ a = b \Rightarrow a \cdot c = b \cdot c$

N-5: $\forall a, b \in \mathscr{N},\ a + b = b + a$

N-6: $\forall a, b \in \mathscr{N},\ a \cdot b = b \cdot a$

N-7: $\forall a, b, c \in \mathscr{N},\ (a + b) + c = a + (b + c)$

N-8: $\forall a, b, c \in \mathscr{N},\ (a \cdot b) \cdot c = a \cdot (b \cdot c)$

N-9: $\forall a, b, c \in \mathscr{N},\ a \cdot (b + c) = a \cdot b + a \cdot c$

N-10: $\exists 1 \in \mathscr{N} \cdot \ni \cdot \forall a \in \mathscr{N},\ a \cdot 1 = a$

N-11: $\forall a, b \in \mathscr{N}$, exactly one of

    (i) $\exists x \in \mathscr{N} \cdot \ni \cdot a + x = b$,
    (ii) $a = b$,
    (iii) $\exists y \in \mathscr{N} \cdot \ni \cdot a = y + b$ holds.

N-12: (i) If $p(1)$ is true for an open sentence $p(m)$ and
      (ii) when $p(k)$ is true for any $k \in \mathscr{N}$ then $p(k + 1)$ is true,

then $p(m)$ is true for *all* $m \in \mathscr{N}$, i.e.,

$$[p(1) \wedge (p(k) \Rightarrow p(k + 1))] \Rightarrow p(m)$$

We should note carefully that N-1, N-2, and N-11 are existentially quantified and hence would not necessarily be true on an arbitrary subset of $\mathscr{N}$.

§6b The Natural Numbers—Axioms

N-1 and N-2 are called the axioms of *closure* and assert that the two primitive binary operations on natural numbers have a natural number result. N-3 and N-4 have a variety of names; perhaps the best way to remember them is by the phrase "equals added to equals result in equals" and "equals times equals result in equals." They assert that the operations are "well defined" or "uniquely defined." Hence they are sometimes called the axioms for *uniqueness of sum* and *uniqueness of product*, respectively. Axioms N-5 and N-6 are the *commutative* properties, while N-7 and N-8 are the *associative* axioms. Axiom N-9 is the one that brings both operations together and is called the *distributive* axiom ($\cdot$ over $+$). The *identity* or *neutral element* axiom for $\cdot$ is the name usually given to N-10, while *trichotomy* axiom is the name for N-11. N-12 is the *Axiom of Induction* P-5 restated for this number set.

To illustrate a probable use of these axioms requires imagination and humor on the part of both teacher and student. You will know that most examples and theorems are true before you start; the point here is that you must also be able to explain *why* each is true.

EXAMPLE 1: Multiply: $(5 \cdot 21) \cdot 2$ using principles directly stated in the axioms.

Solution: Usually we would look at this expression and decide that it would be easier to rewrite it as $(5 \cdot 2) \cdot 21$, obtaining $10 \cdot 21$ or 210. However, no one principle directly stated in the axioms allows this. In abstract language we have said "if $a, b, c \in \mathcal{N}$, then $(a \cdot b) \cdot c = (a \cdot c) \cdot b$." This, of course, is a conjecture that requires proof.

Theorem (6b-1): If $a, b, c \in \mathcal{N}$, then $(a \cdot b) \cdot c = (a \cdot c) \cdot b$

Proof:

1.	$a, b, c \in \mathcal{N}$	1.	C.P.
2.	$(a \cdot b) \cdot c = a \cdot (b \cdot c)$	2.	N-8
3.	$b, c \in \mathcal{N} \Rightarrow (bc) \in \mathcal{N}$	3.	N-2
4.	$b, c \in \mathcal{N} \Rightarrow bc = cb$	4.	N-6
5.	$a(bc) = a(cb)$	5.	N-4
6.	$a(cb) = (ac)b$	6.	N-8
7.	$(ab)c = (ac)b$	7.	Transitivity of equals (2, 5, 6)
8.	$a, b, c \in \mathcal{N} \Rightarrow (ab)c = (ac)b$	8.	C.I.

Q. E. D.

Using this theorem, we can justify the solution proposed for the problem in Example 1 by writing

$$(5 \cdot 21) \cdot 2 = (5 \cdot 2) \cdot 21 \qquad \text{by (6b-1)}$$
$$= 10 \cdot 21$$

assuming that we know that $5 \cdot 2 = 10$.

EXAMPLE 2: If $2, 3, x, y \in \mathcal{N} \wedge 3x = y$, show that $6x = 2y$ assuming that we know that $2 \cdot 3 = 6$.

Solution: We may use

$$3x = y$$

as C.P. Now by closure for multiplication (N-2),

$$3x \in \mathcal{N}$$

whence

(2) $\qquad\qquad\qquad 2(3x) = 2y$

by uniqueness of product (N-4). Now the associative axiom (N-8) allows us to write

$$2(3x) = (2 \cdot 3)x$$

which becomes $6x$ by using the given number fact. Hence $6x = 2y$ by the symmetric and transitive properties of equality.

Note that we justified step (2) by quoting the uniqueness property (N-4). This principle was also used in Step 5 of the proof of (6b-1) and will be used in Step 4 of the proof of (6b-2). The writing of proofs is often abbreviated by omitting mention of this principle.

In the proof of (6b-1), many texts omit Steps 3 and 4, going from 2 to 5 directly. What is being satisfied by these steps is the wording of N-6. That is, $a(bc) = a(cb)$ is not a direct application of commutativity because three numbers are involved on each side of the equal sign instead of two. A second theorem may help clarify this point.

Theorem (6b-2): If $a, b, c \in \mathcal{N}$, then $(ab)c = c(ba)$.

Proof:

1. $a, b, c \in \mathcal{N}$ 1. C. P.
2. $a, b \in \mathcal{N} \Rightarrow a \cdot b \in \mathcal{N}$ 2. N-2
3. $ab = ba$ 3. N-6
4. $c(ab) = c(ba)$ 4. N-4
5. $c, ab \in \mathcal{N} \Rightarrow c(ab) \in \mathcal{N}$ 5. N-2

§6b The Natural Numbers—Axioms

6. $c(ab) = (ab)c$ 6. N-6
7. $(ab)c = c(ba)$ 7. Transitivity of equals, (4, 6)
8. $a, b, c \in \mathcal{N} \Rightarrow (ab)c = c(ba)$ 8. C. I.

Q. E. D.

The word "substitution" is often used in mathematics, particularly in algebra and geometry. For example,
$$(2 \cdot 3)x = 12$$
is changed to
$$6x = 12$$
by substituting 6 for $2 \cdot 3$. This type of substitution is actually accomplished by using the uniqueness and transitivity properties as follows:
$$(2 \cdot 3)x = 12$$
but
$$2 \cdot 3 = 6$$
and if $x \in \mathcal{N}$, then
$$(2 \cdot 3)x = 6x$$
by uniqueness. Hence by transitivity
$$6x = 12$$

Another type of substitution is stated in the following theorem. The reasons are omitted to help you learn the axioms.

Theorem (6b-3): $\forall a, b, c \in \mathcal{N}$,
$$a = b \wedge c = d \Rightarrow a + c = b + d.$$

Proof:

1. $a, b, c, d \in \mathcal{N}, a = b \wedge c = d$ 1. ?
2. $a + c = b + c$ 2. ?
3. $c + b = d + b$ 3. ?
4. $b + c = c + b$ 4. ?
5. $d + b = b + d$ 5. ?
6. $a + c = b + d$ 6. ?
7. $a, b, c, d \in \mathcal{N} \wedge a = b \wedge c = d \Rightarrow a + c = b + d$ 7. ?

Q. E. D.

The following theorem states the corresponding property for multiplication. The proof will be left to the exercises.

Theorem (6b-4): $\forall a, b, c, d \in \mathcal{N}$,
$$a = b \wedge c = d \Rightarrow ac = bd$$

The next two theorems are necessary for the solution of equations using natural numbers. These theorems are usually called the *cancellation* theorems and are the converses of N-3 and N-4.

Theorem (6b-5): $\forall a, b, c \in \mathcal{N}$,
$$a + c = b + c \Rightarrow a = b$$

Theorem (6b-6): $\forall a, b, c \in \mathcal{N}$,
$$ac = bc \Rightarrow a = b$$

The proofs for these theorems are a little more difficult than the ones just completed and will be postponed until later. For now, let us accept these propositions as true. An easy corollary to (6b-6) serves well as a first illustration of their use.

Corollary (6b-6a): $\forall a, b, c, d \in \mathcal{N}$,
$$ab + ac = ad \Rightarrow b + c = d$$

Proof:

1. $a, b, c, d \in \mathcal{N}$, $ab + ac = ad$ 1. ?
2. $ab + ac = a(b + c)$ 2. ?
3. $a(b + c) = ad$ 3. ?
4. $b, c \in \mathcal{N} \Rightarrow (b + c) \in \mathcal{N}$ 4. ?
5. $b + c = d$ 5. (6b-6)
6. $a, b, c, d \in \mathcal{N}$, $ab + ac = ad \Rightarrow b + c = d$.   6. ?

Q. E. D.

We have stated that the cancellation properties are used in the solution of equations like those found in an Algebra I text. Before showing an example, recall that the only known natural number constant mentioned in the axioms is the neutral element 1 of N-10. We introduce further constants as follows.

Definition (6-1): $2 = 1 + 1, 3 = 2 + 1, 4 = 3 + 1, \ldots$.

By the closure principle N-1 each of these is a natural number. In particular, 2 is the x such that $1 + 1 = x$. Then since $2 \in \mathcal{N}$, 3 is the x such that $2 + 1 = x$, and so forth. Now it would be profitable to do some numerical examples.

EXAMPLE 3: Show that $x \in \mathcal{N} \wedge 2x = 2 \Rightarrow x = 1$.

Solution:

1.	$2x = 2, x \in \mathcal{N}$	1.	Hypothesis (C. P.)
2.	$2 = 2 \cdot 1$	2.	N-10
3.	$x2 = 2x$	3.	N-6
4.	$2 \cdot 1 = 1 \cdot 2$	4.	N-6
5.	$x2 = 1 \cdot 2$	5.	Transitivity (3c-10) (used 3 times)
6.	$x = 1$	6.	(6b-6)

Q. E. D.

EXAMPLE 4: Assuming that we know that $6 = 2 \cdot 3$ and that $8 = 2 \cdot 4$, find the solution set of $2x + 6 = 8$ on $\mathcal{N}$.

Solution:

1.	$x \in \mathcal{N} \wedge 2x + 6 = 8$	1.	?
2.	$2x + 2 \cdot 3 = 2 \cdot 4$	2.	Assumed number fact
3.	$2, 3, 4 \in \mathcal{N}$	3.	N-1
4.	$x + 3 = 4$	4.	?
5.	$x + 3 = 1 + 3$	5.	Definition of 4 (6-1)
6.	$1 \in \mathcal{N}$	6.	N-10
7.	$x = 1$	7.	(6b-5)

Hence $\{1\}$ is the solution set of the open sentence $2x + 6 = 8$ for $x \in \mathcal{N}$.

Q. E. D.

The phrase "number fact" used as a reason for Step 2 of this example and in previous cases is meant to indicate that we could construct both an addition table and a multiplication table in which such facts would be listed. This phrase would then refer to a fact presented in the table. You long ago committed to memory facts from both tables. However, it might be instructive to go through the steps a suspicious Martian could use in *proving* to himself that you were not kidding him when you told him that $3 + 2 = 5$.

EXAMPLE 5: Calculate the values of $3 + 2$ and $2 + 3$ and enter them in the addition table of Figure 6.1.

Solution:

$3 + 2 = 3 + (1 + 1)$	Definition of 2 (6-1)
$ = (3 + 1) + 1$	N-7
$ = 4 + 1$	Definition of 4 (6-1)
$ = 5$	Definition of 5 (6-1)

$$\text{Q. E. D.}$$

$2 + 3 = 3 + 2$	N-5
$ = 5$	Just done.

The entries have been made in the table.

+	1	2	3	4	⋯
1	2				
2	3	5			
3	4	5			
4	5				
⋮					

Figure 6.1

Our Martian should now realize that you were not misleading him.

The two operations $+$ and $\cdot$ are really all that is needed for the system of natural numbers. However, there are three other operations—subtraction, division, and involution—which are learned by the school child and which we, too, should consider.

Definition (6-2): For all $a, b \in S$,

the **difference** $a - b = x$ iff $\exists x \in S \cdot \ni \cdot a = b + x$

The operation of finding a difference is called *subtraction*.

This definition should be studied carefully. The first thing to note is that this new operation is defined on an arbitrary set S. It will be used unchanged for any number set S including our set $\mathcal{N}$. Second, the difference $a - b$ is only meaningful when the required $x \in S$ exists for that number set. Consider, for example, the expression

$$5 - 2$$

for $5, 2 \in \mathcal{N}$. By the definition, $5 - 2 = x$ iff there is a natural number x such that

§6b The Natural Numbers—Axioms

$$5 = 2 + x$$

As an addition fact from Figure 6.1, we find that

$$2 + 3 = 5$$

Thus

$$5 - 2 = 3$$

Now let us consider the expression

$$2 - 5$$

This has meaning iff there is a natural number x such that

$$2 = 5 + x$$

In our whole addition table, however far extended, there is no natural number which can be added to 5 to equal 2. Thus

$$2 - 5$$

has no meaning in the set of natural numbers.

One way to state the fact which we have just observed is to say that the set of natural numbers is *not* closed under subtraction. Our basic operations of $+$ and $\cdot$ have no such restriction. The axioms of closure (N-1, N-2) guarantee that the natural number sum or product of two natural numbers will always exist.

EXAMPLE 6: $A = \{x \mid x \in \mathcal{N} \wedge 5 - 3 = x\}$. Make a roster of the solution set A.

Solution:

1. $5 - 3 = x$ iff $\exists\, x \in \mathcal{N} \cdot \ni \cdot 5 = 3 + x$ 1. Definition (5-2)
2. $3, 5 \in \mathcal{N}$ 2. N-1
3. $2 + 3 = 5$ 3. Number fact
4. $2 + 3 = 3 + x$ 4. ?
5. $3 + x = x + 3$ 5. ?
6. $2 + 3 = x + 3$ 6. ?
7. $x = 2$ 7. ?
8. $5 - 3 = 2$, $\therefore A = \{2\}$ 8. Transitivity of equals (1, 7)

The next theorem states a proposition which is also useful in working with equations.

Theorem (6b-7): $\forall\, a, b \in \mathcal{N}, (a + b) - b = a$. (Subtraction undoes addition.)

Proof:

1.	$\forall a, b \in \mathcal{N}, a + b = b + a$	1.	N-5
2.	$(a + b) \in \mathcal{N}$	2.	N-1
3.	$a + b = b + a \Rightarrow$ $(a + b) - b = a$	3.	Definition of subtraction
4.	$\forall a, b \in \mathcal{N}, (a + b) - b = a$	4.	R. D.

Q. E. D.

EXAMPLE 7: $A = \{x \mid x \in \mathcal{N} \wedge (2 + 5) - 5 = 2x\}$. Make a roster of the solution set.

Solution:

1.	$2, x \in \mathcal{N} \Rightarrow 2x \in \mathcal{N}$	1.	N-2
2.	$(2 + 5) - 5 = 2x$	2.	Hypothesis
3.	$(2 + 5) - 5 = 2$	3.	(6b-7)
4.	$2 = 2x$	4.	Transitivity of equals

Now from Example 3 we know that $x = 1$. Therefore, $A = \{1\}$.

It is now possible to prove Theorem (6b-5). The reasons are left for the exercises.

Theorem (6b-5): $\forall a, b, c \in \mathcal{N}, a + c = b + c \Rightarrow a = b$.

Proof:

1.	$a, b, c \in \mathcal{N}, a + c = b + c$	1.	?
2.	$b + c = c + b$	2.	?
3.	$a + c = c + b$	3.	?
4.	$(a + c) - c = b$	4.	?
5.	$(a + c) - c = a$	5.	?
6.	$a = b$	6.	?
7.	$\forall a, b, c \in \mathcal{N},$ $a + c = b + c \Rightarrow a = b$	7.	?

Q. E. D.

The fourth operation to be introduced has a definition and a problem of existence similar to those in the case of subtraction.

Definition (6–3): $\forall a, b \in S$,

the **quotient** $a \div b = x$ iff $\exists x \in S \ni a = b \cdot x$.

The operation of finding a quotient is called *division*. This operation will only be meaningful when the required element x is found in the number set S in question. Consider, for example, the expression

$$6 \div 3$$

By the definition,

$$6 \div 3 = x$$

iff there is a natural number x such that

$$6 = 3 \cdot x$$

The multiplication table in Exercise 3 will show that $3 \cdot 2 = 6$. Thus

$$6 \div 3 = 2$$

Now let us consider the expression

$$3 \div 6$$

This has meaning iff there is a natural number x such that

$$3 = 6 \cdot x$$

In our whole multiplication table, however far extended, there is no natural number which, when multiplied by 6, equals 3. Thus

$$3 \div 6$$

has no meaning in the set of natural numbers. One way to state this fact is to say that the set of natural numbers is *not* closed under division.

EXAMPLE 8: $A = \{x \mid x \in \mathcal{N} \land 8 \div 2 = x\}$. Make a roster of A.

Solution:

1.	$8 \div 2 = x$ iff $8 = 2 \cdot x$, $\exists\, x \in \mathcal{N}$	1.	Definition of division
2.	$2, x \in \mathcal{N} \Rightarrow 2x \in \mathcal{N}$	2.	N-2
3.	$8 = 4 \cdot 2$	3.	Number fact
4.	$4 \cdot 2 = 2x$	4.	Transitivity (1, 3)
5.	$2x = x \cdot 2$	5.	N-6
6.	$4 \cdot 2 = x \cdot 2$	6.	Transitivity
7.	$4 = x$	7.	(6b-6)
8.	$8 \div 2 = 4$	8.	Definition of division

Therefore, $A = \{4\}$

This new operation is also useful in the solution of equations since it is used to undo multiplication.

Theorem (6b-8): $\forall a, b \in \mathcal{N}, (a \cdot b) \div b = a$.
Proof:

1.	$\forall a, b \in \mathcal{N}, a \cdot b = b \cdot a$	1.	N-6
2.	$(a \cdot b) \in \mathcal{N}$	2.	N-2
3.	$ab = ba \Rightarrow (ab) \div b = a$	3.	Definition of division
4.	$\forall a, b \in \mathcal{N}, (ab) \div b = a$	4.	R. D.

Q. E. D.

It is now very easy to prove Theorem (6b-6). The reasons are again omitted for use in the exercises.

Theorem (6b-6): $\forall a, b, c \in \mathcal{N}, ac = bc \Rightarrow a = b$.
Proof:

1.	$a, b, c \in \mathcal{N} \wedge ac = bc$	1.	?
2.	$bc = cb$	2.	?
3.	$ac = cb$	3.	?
4.	$(ac) \div c = b$	4.	?
5.	$(ac) \div c = a$	5.	?
6.	$a = b$	6.	?
7.	$\forall a, b, c \in \mathcal{N}, ac = bc \Rightarrow a = b$	7.	?

Q. E. D.

In high school algebra a great deal of time was spent factoring numbers either as monomials or as polynomials.

Definition (6-4): For all numbers $a, b, c \in S$, a is a **factor** of c iff
$$\exists b \in S \cdot \ni \cdot a \cdot b = c.$$

When a is a *factor* of c, we write $a \mid c$; also, c is called a *multiple* of a. For example, *factors* of 12 are 1, 2, 3, 4, 6, 12 because $1 \cdot 12 = 12$, $2 \cdot 6 = 12$, $3 \cdot 4 = 12$. The numbers 12, 24, 36, ... are *multiples* of 12 because $12 \cdot 1 = 12$, $12 \cdot 2 = 24$, $12 \cdot 3 = 36, \ldots$. Be careful not to confuse the concepts of "factor" and "multiple."

Factoring a number, then, is a process of finding the number's factors. This process of factoring depends on our knowledge of the tables of multiplication. Factoring of polynomials also depends directly on the distributive axiom N-9 and its results.

EXAMPLE 9: Factor $8a + y^2a$ over $\mathcal{N}$.

Solution:

$$8a + y^2a = a8 + ay^2 \qquad \text{N-6}$$
$$= a(8 + y^2) \qquad \text{N-9}$$

By using the following proposition, this expression could have been factored with somewhat different steps.

Theorem (6b-9): $\forall a, b, c \in \mathcal{N}, (b + c)a = ba + ca$.

Now we could factor the expression $8a + y^2a$ of Example 9 by doing

$$8a + y^2a = (8 + y^2)a$$

by this theorem, whence

$$a(8 + y^2)$$

follows by N-6.

Many other theorems about algebraic factoring can be established. A few are now listed for the sake of completeness. The first one states that, with proper restrictions, multiplication is distributive over subtraction.

Theorem (6b-10): $\forall a, b, c, (b - c) \in \mathcal{N}, a(b - c) = ab - ac$.

EXAMPLE 10: Simplify $3(2x - 4)$.

Solution: So long as $(2x - 4)$ is a natural number, we can write

$$3(2x - 4) = 3(2x) - 3(4) \qquad \text{(6b-10)}$$
$$= (3 \cdot 2)x - 3 \cdot 4 \qquad \text{N-8}$$
$$= 6x - 12 \qquad \text{Arithmetic facts}$$

Theorem (6b-11): $\forall a, b, c, d \in \mathcal{N}, (a + b)(c + d) = (ac + ad) + (bc + bd)$.

As the following example indicates, this principle is basic to polynomial factorization.

EXAMPLE 11: Factor $x^2 + 5x + 6$ over $\mathcal{N}$.

Solution:

$$x^2 + 5x + 6 = x \cdot x + (2 + 3)x + 2 \cdot 3$$
$$= (x \cdot x + 3x) + (2x + 2 \cdot 3)$$
$$= (x + 2)(x + 3) \qquad \text{(6b-11)}$$

EXERCISE SET

1. Which of the following are not meaningful in the system of natural numbers? Explain.
   a) $3 - 5$
   b) $10 \div 9$
   c) $8 - 8$
   d) $5 \div 5$
   e) $17 - 12$
   f) $51 \div 17$

2. Fill in the rest of the addition table of Figure 6.1 and justify your steps.

3. Justify steps needed to fill in the multiplication table of Figure 6.2 for natural numbers.

Figure 6.2

4. Solve the following equations over $\mathcal{N}$.
   a) $3 + x = 8$
   b) $x + 4 = 7$
   c) $3 + 9 = y$
   d) $2 + x = 2$
   e) $2x = 8$
   f) $5x + 3 = 13$
   g) $x \cdot 7 = 24$
   h) $x \cdot 17 = 51$

5. Factor the following expressions over $\mathcal{N}$.
   a) $2x + 4$
   b) $ax + ay^2 + 4az$
   c) $21 + 15 + 3 \cdot 4$
   d) $4a + 96$
   e) $51ab + 119b$
   f) $(x + y)x + (x + y)y$
   g) $(2 + a)x - (a + 2)y$

6. Fill in the missing reasons for Theorem (6b-3).

7. Fill in the missing reasons for Corollary (6b-6a).

8. Fill in the missing reasons for a) Example 4; b) Example 6.

9. Fill in the missing reasons for Theorem (6b-5).

10. Fill in the missing reasons for Theorem (6b-6).

11. Show how a child who has learned the multiplication tables only through 5×5 but understands distributivity can determine for himself that $4 \times 7 = 28$.

12. Show that multiplication is not distributive over division, i.e.,
    $\forall a, b, c \in \mathcal{N} \land (b \div c) \in \mathcal{N}$,
    $$a(b \div c) \neq (a \cdot b) \div (a \cdot c)$$
13. Prove the following theorems about natural numbers. (All small letters used stand for natural numbers.)
    a) $a(bc) \in \mathcal{N}$
    b) $a + (b + c) = c + (a + b)$
    c) $(a + b) + c \in \mathcal{N}$
    d) $a(b + c) \in \mathcal{N}$
    e) $(a + b) + c = c + (b + a)$
    f) $a(bc) = b(ac)$
    g) $(a + b) + (c + d) = (c + d) + (a + b)$
    h) $(ab)c = (ac)b$
    i) $(ab)(cd) = (ac)(bd)$
14. Prove each of the following where all letters and all parenthetical expressions stand for natural numbers.
    a) $(a - b) + b = a$
    b) $(a - c) = (b - c) \Rightarrow a = b$
    c) $(a = b) \Rightarrow (a - c) = (b - c)$
    d) $(a \div b) \cdot b = a$
    e) $a = b \Rightarrow (a \div c) = (b \div c)$
    f) $(a \div c) = (b \div c) \Rightarrow a = b$
    g) $c + a = c + b \Rightarrow a = b$
    h) $ab = a \Rightarrow b = 1$
15. Prove: $\forall x \in \mathcal{N}, x + x = 2x$
16. Prove: Theorem (6b-4).
*17. Prove:
    a) (6b-10)   b) (6b-11)

§6c ORDER PROPERTIES

Not much has been said yet about N-11, the trichotomy axiom, although it is necessary to the natural number system as we know it. For one thing, it allows us to make meaningful statements about cases in which two numbers or expressions are unequal. Sometimes such statements involve the "$\neq$" concept of the axioms; others involve what are called "inequalities."

An *absolute inequality* is a proposition which relates two elements of a set by *is less than* or *is greater than*. Relations determined by such sentences are called *order relations*.

Definition (6-5): $\forall a, b \in \mathcal{N}, a < b$ iff $\exists x \in \mathcal{N} \cdot \ni \cdot a + x = b$.

The symbol $<$ is read "is less than." ($b > a$ is another way of writing $a < b$.) Special note should be taken of the fact that this relationship depends on showing that the natural number x which satisfies the *equation* exists.

The following theorems will be useful in working with inequalities. The first is called the *transitive* property of order.

Theorem (6c-1): $\forall a, b, c \in \mathcal{N}$,
$$a < b \wedge b < c \Rightarrow a < c$$

Proof:

1. $a, b, c \in \mathcal{N}, a < b, b < c$ 1. ?
2. $\exists x \in \mathcal{N} \cdot \ni \cdot a + x = b$
   $\exists y \in \mathcal{N} \cdot \ni \cdot b + y = c$ 2. Definition (6-5)
3. $b = c - y$ 3. ?
4. $a + x = c - y$ 4. ?
5. $(a + x) \in \mathcal{N}$ 5. ?
6. $(a + x) + y = c$ 6. Definition (6-2)
7. $(a + x) + y = a + (x + y)$ 7. ?
8. $a + (x + y) = c$ 8. ?
9. $(x + y) \in \mathcal{N}$ 9. ?
10. $a < c$ 10. Definition of $<$ (6-5)

Q. E. D.

EXAMPLE 1: What can we conclude if we are given that
$$x < y + 4 \wedge y + 5 < 7$$
for $x, y \in \mathcal{N}$?

Solution:

$y + 5 = y + (4 + 1)$ Definition of 5 (6-1)
$ = (y + 4) + 1$ Associativity (N-7)
$\therefore y + 4 < y + 5$ Definition of $<$ (6-5)

Hence
$$y + 4 < 7 \quad \text{(6c-1)}$$
whence we can conclude that
$$x < 7 \quad \text{(6c-1)}$$

Theorem (6c-2): $\forall a, b, c \in \mathcal{N}$,
$$a < b \Rightarrow a + c < b + c$$

Theorem (6c-3): $\forall a, b, c \in \mathcal{N}$,
$$a < b \Rightarrow ac < bc$$

The proofs of these two propositions can be constructed similarly to that for (6c-1) and are left to the exercises.

EXAMPLE 2: Show that $7 < 8$ and $12 < 16$ if we know that $3 < 4$.

Solution:

$3 < 4$	Hypothesis
$3 + 4 < 4 + 4$	(6c-2)
$7 < 8$	Addition facts
	Q. E. D.
$3 < 4$	Hypothesis
$3 \cdot 4 < 4 \cdot 4$	(6c-3)
$12 < 16$	Multiplication fact
	Q. E. D.

Cancellation properties also hold for the order relation.

Theorem (6c-4): $\forall a, b, c \in \mathcal{N}$,
$$a + c < b + c \Rightarrow a < b$$

Theorem (6c-5): $\forall a, b, c \in \mathcal{N}$,
$$a \cdot c < b \cdot c \Rightarrow a < b$$

EXAMPLE 3: Solve $2x + 8 < 12$.

Solution:

$2x + 8 < 12$	Hypothesis
$2x + 8 < 4 + 8$	Addition fact
$2x < 4$	(6c-4)
$2x < 2 \cdot 2$	Multiplication fact
$x < 2$	(6c-5)

Hence, $\{x \mid x < 2\}$ is the solution set in simplest form.

The next two theorems are the first which we have stated which make explicit use of the Trichotomy Axiom. Proofs are left to the exercises.

Theorem (6c-6): $\forall a, b \in \mathcal{N}$, exactly one of the following is true: $a < b, a = b, a > b$.

Theorem (6c-7): $\forall a, b \in \mathcal{N}, a \neq a + b$.

246 The Number System

EXAMPLE 4: Prove: $3 \neq 5$.

Solution:

$$\begin{aligned} 5 &= 4 + 1 & &\text{Definition of 5 (6-1)} \\ &= (3 + 1) + 1 & &\text{Definition of 4 (6-1)} \\ &= 3 + (1 + 1) & &\text{Associativity, N-7} \\ &= 3 + 2 & &\text{Definition of 2 (6-1)} \end{aligned}$$

Now,

$$3 < 5 \qquad \text{Definition of } < \text{ (6-5)}$$

Therefore,

$$3 \neq 5 \qquad \text{Theorem (6c-6)}$$

EXERCISE SET

All letters represent natural numbers.

1. Solve for x over $\mathcal{N}$.
   a) $x + 5 < 8$ b) $x - 4 > 5$
   c) $2x < 12$ d) $27 > 9x$
   e) $x + 2 < 1$ f) $2x - 3 < 3$

2. Prove each proposition.
   a) $2 < 4$
   b) $5 \neq 1$
   c) $2x + 1 < 6 \Rightarrow 4x < 10$

3. Supply the missing reasons in the proof of (6c-1).

4. Prove the following theorems:
   a) $\forall a \in \mathcal{N}, a < a + a$
   b) $\forall a, b, c \in \mathcal{N} \wedge c \mid (a + b) \wedge c \mid a \Rightarrow c \mid b$
   c) $x + 1 \neq x$
   d) $x < x + 1$

5. Prove Theorem (6c-2).

6. Prove Theorem (6c-3).

7. Prove Theorem (6c-4).

8. Prove Theorem (6c-5).

*9. Prove Theorem (6c-6).

*10. Prove Theorem (6c-7).

*11. Prove each of the following without making any use of either subtraction or division.

    a) (6b-5)   b) (6b-6)

§6d ARITHMETIC OF THE NATURAL NUMBERS

In many of the preceding theorems, number facts have been used. For example, $8 = 4 \cdot 2$ was used in Example 8, §6b. The basic number facts underlying our arithmetic were introduced in Definition (6-1). A few further addition and multiplication facts were demonstrated in §6b. This section provides a more extensive development.

EXAMPLE 1: Show that $1 + 2 = 3$.

Solution:

1. $1 + 2 = 2 + 1$ 1. ?
2. $2 + 1 = 3$ 2. ?
3. $1 + 2 = 3$ 3. ?

                                                       Q. E. D.

EXAMPLE 2: Show that $3 \cdot 2 = 6$.

Solution:

1. $2 = 1 + 1$ 1. ?
2. $3 \cdot 2 = 3(1 + 1)$ 2. ?
3. $3(1 + 1) = 3 \cdot 1 + 3 \cdot 1$ 3. ?
4. $3 \cdot 1 + 3 \cdot 1 = 3 + 3$ 4. ?
5. $3 = 2 + 1$ 5. ?
6. $3 + 3 = 3 + (2 + 1)$ 6. ?
7. $3 + (2 + 1) = (3 + 2) + 1$ 7. ?
8. $(3 + 2) = 5$ 8. ?
9. $(3 + 2) + 1 = 5 + 1$ 9. ?
10. $5 + 1 = 6$ 10. ?
11. $3 + 3 = 6$ 11. ?
12. $3 \cdot 2 = 6$ 12. ?

                                                       Q. E. D.

EXAMPLE 3: Prove: $3 \neq 4$.

Solution:

1.	$3 = 4$	1.	I. P.
2.	$3 = 3 + 1$	2.	Definition (6-1)
3.	$3 \neq 3$	3.	N-11
4.	$3 = 3$	4.	E-1
5.	$3 \neq 3 \wedge 3 = 3$	5.	C. A.
6.	$3 \neq 4$	6.	I. I.

This is the *basic* method for proving two natural numbers unequal. An appeal to Theorem (6c-7) can also be effective in such work.

Now it is possible to introduce fixed exponents as well as prime and composite numbers.

Definition (6-6): For all numbers $a \in S$, $\mathbf{a}^2 = a \cdot a$.

The following theorem states familiar factoring properties and can easily be proved.

Theorem (6d-1): $\forall a, b \in \mathcal{N}$,
$$(a + b)^2 = a^2 + 2ab + b^2$$

The proof of this theorem can be shortened considerably by application of the following proposition, which you were asked to prove in Exercise 15, §6b.

Theorem (6d-2): $\forall a \in \mathcal{N}$,
$$a + a = 2a$$

EXAMPLE 4: Rewrite $(x + 7)^2$ where $x \in \mathcal{N}$.

Solution: $(x + 7)^2 = x^2 + 2 \cdot 7 \cdot x + 7^2$

Prime and composite numbers will be defined in terms of factors. Recall that a is a factor of c ($a \mid c$) iff $a \cdot b = c$ for some number b.

Definition (6-7): $\forall p \in \mathcal{N}$, p is a **prime number** iff $1 < p$ and p has no factor other than itself and 1.

For example, 7 is a prime number because $1 < 7$ and 1, 7 are its only factors.

Definition (6-8): $\forall c \in \mathcal{N}$, c is a **composite number** iff $1 < c$ and c is not prime.

§6d   Arithmetic of the Natural Numbers

For example, 15 is a composite number because $1 < 15$ and 15 is not prime since $3 \mid 15$ and $5 \mid 15$.

EXAMPLE 5: Make a roster of $A = \{x \mid x \in \mathcal{N} \wedge x < 25 \wedge x \text{ is prime}\}$.

Solution:

$$A = \{2, 3, 5, 7, 11, 13, 17, 19, 23\}$$

It should be noted that all the natural numbers except 1 are prime or composite. There are many applications in arithmetic involving primes and composites, some of which are considered here. The following principle is stated for convenience.

Theorem (6d-3): If c is a composite, then c can be expressed as a unique product of primes.

This proposition is called the *Fundamental Theorem of Arithmetic*. Proofs can be found in texts on number theory.

EXAMPLE 6: Express 45 as a product of primes.

Solution:

$$5 \mid 45; \quad 45 = 5 \cdot 9 \wedge 5 \text{ is prime}$$
$$3 \mid 9; \quad 9 = 3 \cdot 3 \wedge 3 \text{ is prime}$$

Thus $\quad 45 = 5 \cdot 3^2$

Definition (6-9): $\forall a, b, c \in \mathcal{N}$, a is called a **common divisor** of b, c iff $a \mid b \wedge a \mid c$.

For example, $2 \mid 6$ and $2 \mid 8$, hence 2 is a common divisor of 6, 8. Also $x \mid xy^2 \wedge x \mid 2x$; hence x is a common divisor of xy^2, $2x$.

Definition (6-10): $\forall a, b, c \in \mathcal{N}$, a is called the **greatest common divisor** (gcd) of b, c, denoted $a = \gcd(b, c)$, iff a is a common divisor of b, c and $x < a \vee x = a$ where x is any common divisor of b, c.

EXAMPLE 7: Find the gcd of 75 and 200.

Solution:

$$75 = 3 \cdot 5^2$$
$$200 = 2^3 \cdot 5^2$$

The prime factors common to both numbers will be the gcd. Hence the gcd is 5×5 or 25. Since finding the gcd of two numbers calls for finding the factors they have in common, we are in effect being asked to find the intersection of their sets of factors. Thus if we let

$$A = \{3, 5^2\} \text{ or } \{3, 5_1, 5_2\}$$
$$B = \{2^3, 5^2\} \text{ or } \{2_1, 2_2, 2_3, 5_1, 5_2\}$$

then
$$A \cap B = \{5_1, 5_2\}$$
and the gcd $(75, 200) = 5 \cdot 5$ or 25.

Definition (6-11): $\forall a, b, c \in \mathcal{N}$, c is a **common multiple** of a, b iff $a \mid c \wedge b \mid c$.

For example, $2 \mid 6 \wedge 3 \mid 6$, hence 6 is a common multiple of 2, 3. Also $x \mid xy \wedge y \mid xy$, hence xy is a common multiple of x, y.

Definition (6-12): $\forall a, b, c \in \mathcal{N}$, c is called the **lowest common multiple** (lcm) of a, b, denoted $c = \text{lcm}(a, b)$, iff c is a common multiple of a, b and $c < x \vee c = x$, where x is any common multiple of a, b.

EXAMPLE 8: Find the lcm of 75 and 200.

Solution: A common multiple of two numbers must at least contain all the factors of each number, and the lowest common multiple does not repeat the factors common to both. The lcm of 75, 200, then, is given by the union of sets A, B from Example 7. That is,
$$A \cup B = \{2_1, 2_2, 2_3, 3_1, 5_1, 5_2\}$$
Hence, $\text{lcm}(75, 200) = 2 \cdot 2 \cdot 2 \cdot 3 \cdot 5 \cdot 5 = 600$.

We have defined a particular case of raising to a power in Definition (6-6). The general operation of raising to a power is called *involution*.

Definition (6-13): $\forall a \in S, n \in \mathcal{N}$,
  (i) $\mathbf{a}^1 = a$
  (ii) $\mathbf{a}^{n+1} = a^n \cdot a$

The a of this definition is called the *base*, n is called the *exponent*, and a^n is called the nth *power of a*.

This type of definition is called a *recursive* definition. A recursive definition defines some entity in terms of (i) the nonsuccessor, and (ii) the successor of arbitrary n, instead of defining the entity for n itself.

The following theorems about involution state familiar number facts. Proofs require use of the Axiom of Induction N-12 and will not be given here.

Theorem (6d-4): $\forall a, n \in \mathcal{N}, a^n \in \mathcal{N}$.

Theorem (6d-5): $\forall a, b, n \in \mathcal{N}, a = b \Rightarrow a^n = b^n$.

Theorem (6d-6): $\forall a, m, n \in \mathcal{N}, a^n \cdot a^m = a^{m+n}$.

The result stated in this theorem is the familiar rule of adding exponents. For example, $a^4 \cdot a^3 = a^{4+3} = a^7$.

Theorem (6d-7): $\forall a, m, n \in \mathcal{N}, (a^m)^n = a^{mn}$.

Theorem (6d-8): $\forall a, b, n \in \mathcal{N}, (ab)^n = a^n b^n$.

EXAMPLE 9: Rewrite the following using Theorem (6d-7) or (6d-8): (a) $(x^2)^3$; (b) $(2x)^4$.

Solution:

$$\begin{align} &\text{a)} \quad (x^2)^3 = x^{2 \cdot 3} = x^6 &&(6d\text{-}7)\\ &\text{b)} \quad (2x)^4 = 2^4 x^4 = 16x^4 &&(6d\text{-}8) \end{align}$$

Before trying the exercises or reading further, let us outline what has been accomplished in our development of the number system. We now have a set $\mathcal{N} = \{1, 2, 3, \ldots\}$ and have made progress toward being able to derive the usual tables of addition and multiplication as well as solve certain open sentences. We have introduced an additional operation of involution. The operations of subtraction and division were defined in an existential manner.

We have accomplished a great deal in our systematic examination of the number system. Yet, as the following examples indicate, we have a great deal left to do.

EXAMPLE 10: $A = \{x \mid x \in \mathcal{N} \land x + 5 = 4\}$. Simplify the expression for A.

Solution:

$$x + 5 = 4 \longleftrightarrow 5 < 4 \qquad \text{Definition of } <$$

Recall that $5 = 4 + 1$, whence $4 + 1 < 4$. However, $4 + 1 \not< 4$. Therefore, $A = \varnothing$.

EXAMPLE 11: $A = \{x \mid x \in \mathcal{N} \land x^2 = 2\}$. Simplify the expression for A.
Solution: $A = \varnothing$.

For many reasons we would prefer nonempty solution sets for open sentences of the relatively simple types considered in these two examples. For this reason it is advantageous to develop new sets of numbers.

EXERCISE SET

1. Rewrite the following expressions using the principle of Theorem (6d-1) or Definition (6-6):

a) $(y + 3)^2$
b) $(2z + a)^2$
c) $(2z)^2$
d) $(2^2 + 5x)$

2. Find
   a) gcd (52, 106)
   b) lcm (52, 106)

3. Find
   a) gcd (3570, 4032)
   b) lcm (3570, 4032)

4. Supply the reasons in Example 1.

5. Supply the reasons in Example 2.

6. Prove the following theorems:
   a) $2 + 4 = 6$
   b) $2 \cdot 4 = 8$
   c) $3^2 = 9$
   d) $6 \neq 4$

7. Prove Theorem (6d-1).

*8. Prove:
   a) (6d-4)
   b) (6d-5)
   c) (6d-6)
   d) (6d-7)
   e) (6d-8)

§6e THE INTEGERS

We saw in §6d that it is impossible to find natural number solutions for some problems involving subtraction, division, and involution. What this really means is that our axiom set was good but not good enough for some purposes. To obtain a number set more suitable for these purposes we specify an axiom set that will allow us to solve some or all of these problems. We first consider subtraction. We call the new system in which subtraction is omnipossible the system of integers.

The primitive terms will be $+$, $\cdot$, and $\mathscr{Z}$.

Z-1: $\forall a, b \in \mathscr{Z}, \exists x \in \mathscr{Z} \cdot \ni \cdot a + b = x$
(Closure of $\mathscr{Z}$ under $+$)

Z-2: $\forall a, b \in \mathscr{Z}, \exists x \in \mathscr{Z} \cdot \ni \cdot a \cdot b = x$
(Closure of $\mathscr{Z}$ under $\cdot$)

Z-3: $\forall a, b, c \in \mathscr{Z}, a = b \Rightarrow a + c = b + c$
(Uniqueness of $+$)

Z-4: $\forall a, b, c \in \mathscr{Z}, a = b \Rightarrow a \cdot c = b \cdot c$
(Uniqueness of $\cdot$)

Z-5: $\forall a, b \in \mathscr{Z}, a + b = b + a$
(Commutativity of $+$)

Z-6: $\forall a, b \in \mathscr{Z}, a \cdot b = b \cdot a$
(Commutativity of $\cdot$)

Z-7: $\forall a, b, c \in \mathscr{Z}, (a + b) + c = a + (b + c)$
(Associativity of $+$)

Z-8: $\forall a, b, c \in \mathscr{Z}, (ab)c = a(bc)$
(Associativity of $\cdot$)

Z-9: $\forall a, b, c \in \mathscr{Z}, a(b + c) = ab + ac$
(Distributivity of $\cdot$ over $+$)

Z-10: $\forall a, b \in \mathscr{Z}$ in that order, $\exists x \in \mathscr{Z} \cdot \ni \cdot a = b + x$
(Solvability for $+$)

Z-11: $\exists$ a subset $\mathscr{Z}^+$ of $\mathscr{Z} \cdot \ni \cdot \forall a, b \in \mathscr{Z}$, exactly one of the following holds:

  (i) $a = b$
  (ii) $\exists x \in \mathscr{Z}^+ \cdot \ni \cdot a = b + x$
  (iii) $\exists y \in \mathscr{Z}^+ \cdot \ni \cdot a + y = b$

Z-12: $\exists 1 \in \mathscr{Z}^+ \cdot \ni \cdot \forall a \in \mathscr{Z}, a = a \cdot 1$
(Neutral element for $\cdot$)

Z-13: $\forall a, b \in \mathscr{Z}^+, \exists x \in \mathscr{Z}^+ \cdot \ni \cdot (a + b) = x$
(Closure of $\mathscr{Z}^+$ under $+$)

Z-14: $\forall a, b \in \mathscr{Z}^+, \exists x \in \mathscr{Z}^+ \cdot \ni \cdot a \cdot b = x$
(Closure of $\mathscr{Z}^+$ under $\cdot$)

Z-15:   If for any open sentence $p(n)$
  (i) $p(1)$ is true and
  (ii) when $p(k)$ is true for $k \in \mathscr{Z}^+$, then $p(k + 1)$ is true,
then $p(n)$ is true for *all* $n \in \mathscr{Z}^+$.

These axioms will look very familiar to us, since Z-1 through Z-9 simply restate N-1 through N-9 for the new set $\mathscr{Z}$. What makes this fact important is that any theorem which was proved using only N-1 through N-9 and theorems which resulted from them is provable in $\mathscr{Z}$ by simply changing the set name. In particular, the following theorem can be accepted as proved.

Theorem (6e-1): $\forall a, b, c \in \mathscr{Z}, a + c = b + c \Rightarrow a = b$. See Theorem (6b-5).

The major difference between the two axiom sets is the solvability property Z-10. This principle states that for the specific pair (2, 3), for example, there is an $x \in \mathscr{Z}$ such that

(1) $$2 = 3 + x$$

None of Z-1 through Z-16 states which integer can be used to replace x to make (1) true. In fact, we cannot even be certain at this point that there is only one such solution. Z-10 states that there is at least one such solution, but it does not say that there is only one. In Theorem (6c-7) we learned that $\forall a, b \in \mathcal{N}, a \neq a + b$. Thus for the ordered pair (a, a),

(2) $$a = a + x$$

is false in $\mathcal{N}$. Z-10 states that this proposition is true in $\mathcal{Z}$, and the following proposition names this new number.

Theorem (6e-2): A unique integer 0 exists having the property that
$$\forall a \in \mathcal{Z}, a = a + 0.$$

Proof (*One*):

$$\forall (a, a) \cdot \ni \cdot a \in \mathcal{Z}, \; \exists x \in \mathcal{Z} \cdot \ni \cdot a = a + x \qquad \text{Z-10}$$

We adopt the symbol 0 for this number x, i.e., $a = a + 0$, $\forall a \in \mathcal{Z}$.

Done.

(*Only One*):

Note: We shall assume two different symbols and show that they must be the same.

1. $\exists 0 \in \mathcal{Z} \cdot \ni \cdot a = a + 0$, $\forall a \in \mathcal{Z}$ 1. "One" part
2. $\exists 0' \in \mathcal{Z} \cdot \ni \cdot a = a + 0'$, $\forall a \in \mathcal{Z}$ 2. "One" part
3. $a = 0 \Rightarrow 0 = 0 + 0'$ 3. Substitution in 2
4. $a = 0' \Rightarrow 0' = 0' + 0$ 4. Substitution in 1
5. $0 + 0' = 0' + 0$ 5. Z-5
6. $0 = 0'$ 6. Transitivity (3, 4)

Done
Q. E. D.

This theorem establishes the existence of a unique **neutral** or **identity element for addition** in $\mathcal{Z}$.

We now have two particular elements of $\mathcal{Z}$, the multiplicative neutral element 1 provided by Z-12 and the additive neutral element 0 provided by this theorem. Axiom Z-12 states that $1 \in \mathcal{Z}^+$. A natural conjecture would be whether $0 \in \mathcal{Z}^+$.

Theorem (6e-3): $0 \notin \mathcal{Z}^+$.

Proof:

1. $0 \in \mathscr{L}^+$	1. I. P.
2. $0 = 0 + 0$	2. Theorem (6e-2)
3. $0 = 0$	3. Reflexivity (3c-8)
4. $0 = 0 + 0 \wedge 0 = 0$	4. C. A.
5. $0 = 0 + 0 \wedge 0 = 0$ is false	5. Z-11
6. $0 \notin \mathscr{L}^+$	6. I. I.

Q. E. D.

Corollary (6e-3a): $1 \neq 0$.

We have looked at the meaning of the solvability axiom Z-10 for the ordered pair (a, a) and found that the required x of (2) is the new number 0. It is now reasonable to consider more general ordered pairs (a, b). For example, what if $a = 0$ and b is a variable? That is, we are asking about the x in

(3) $$0 = b + x$$

We know that there is no natural number which satisfies this equation. However, Z-10 states that there is such an $x \in \mathscr{L}$, and the following proposition names this new number.

Theorem (6e-4): $\forall a \in \mathscr{L}$, $\exists$ a unique $\bar{a} \in \mathscr{L} \cdot \ni \cdot 0 = a + \bar{a}$.

Proof (*One*):

$a, 0 \in \mathscr{L}$, hence for the ordered pair $(0, a)$ an integer x exists such that $0 = a + x$. We adopt the symbol $\bar{a}$ for this x and call it the **additive inverse of a**. Thus

$$\forall a \in \mathscr{L}, 0 = a + \bar{a} \qquad \text{Done}$$

(*Only One*):

Note: Assume two different symbols for the additive inverse of a and show that they must be the same.

1. $\exists \bar{a} \in \mathscr{L} \cdot \ni \cdot 0 = a + \bar{a},$ $\forall a \in \mathscr{L}$	1. "One" part
2. $\exists \char"5E a \in \mathscr{L} \cdot \ni \cdot 0 = a + \char"5E a,$ $\forall a \in \mathscr{L}$	2. "One" part
3. $a + \char"5E a = 0$	3. Symmetry (3c-9)

4. $a + {}^\wedge a = a + {}^- a$	4. Transitivity (3c-10)
5. ${}^\wedge a = {}^- a$	5. (6e-1)

Done
Q. E. D.

Sometimes the additive inverse of x is called *the negative of x*. Care must be taken to keep the notions of "negative number" and "negative of a number" separate. The symbol ${}^- a$ stands for the negative of the number a, but ${}^- a$ may be positive or it may be negative.

EXAMPLE 1: If $x \in \mathscr{Z}$, what is the additive inverse of $(x + 3)$?

Solution: The additive inverse of $(x + 3)$ is ${}^-(x + 3)$ according to this theorem because $(x + 3) + {}^-(x + 3) = 0$.

We have examined solutions to the equations of (2) and (3). These correspond to the ordered pairs (a, a), $(0, b)$. The last step in considering the solvability axiom would be to examine the ordered pair (a, b) where a, b are distinct variables. The axiom provides us with a solution x for the equation

(4) $$a = b + x$$

Proof that this x is unique is left to the exercises. Thus the equation of (1) has a solution, and it is unique. In the following example we find this value of x.

EXAMPLE 2: Make a roster of $A = \{x \mid 2 = 3 + x \wedge x \in \mathscr{Z}\}$.

Solution:

1.	$2 = 3 + x$	1.	Hypothesis
2.	$2 = (2 + 1) + x$	2.	Definition of 3 (6-1)
3.	$2 = 2 + (1 + x)$	3.	Z-7
4.	$2 + 0 = 2 + (1 + x)$	4.	Theorem (6e-2)
5.	$0 = 1 + x$	5.	Theorem (6e-1)
6.	$1 + {}^- 1 = 0$	6.	Theorem (6e-4)
7.	$1 + {}^- 1 = 1 + x$	7.	Transitivity (3c-10)
8.	${}^- 1 = x$	8.	Theorem (6e-1)

Therefore, $A = \{{}^- 1\}$.

Recall that our main purpose here is not to solve equations but to study the principles which underlie the solution of all equations. Thus, rather than try for maximum efficiency, we include those steps which stress important principles. For this reason many more steps have been used in carrying out this solution than you might have used.

Now let us investigate what our new number zero does when multiplied. For example, $5 \cdot 0$ is equal to some integer x by closure, but so far that is all we know about it. One of the rules we learn early in arithmetic is that any number times zero is zero. We now show that this rule is justifiable.

Theorem (6e-5): $\forall a \in \mathscr{L}, a \cdot 0 = 0$
Proof:

1.	$a = a + 0$	1.	(6e-2)
2.	$a \cdot a = a(a + 0)$	2.	?
3.	$a(a + 0) = a \cdot a + a \cdot 0$	3.	?
4.	$a \cdot a = a \cdot a + 0$	4.	?
5.	$a \cdot a + 0 = a \cdot a + a \cdot 0$	5.	?
6.	$0 = a \cdot 0$	6.	(6e-1)

Q. E. D.

Thus

$$5 \cdot 0 = 0$$

The theorem states that any integer times zero is zero. This is an extremely useful result, but it also causes some restrictions which were not necessary for $\mathscr{N}$. For example, $a \cdot 0 = 0$ and $b \cdot 0 = 0$, and even if $a \neq b$, we have

$$a \cdot 0 = b \cdot 0$$

Yet

$$a \cdot 0 = b \cdot 0 \Rightarrow a = b \quad \text{is false.}$$

As an example $2 \cdot 0 = 3 \cdot 0 \Rightarrow 2 = 3$ is false. That is, the cancellation property for multiplication in $\mathscr{L}$ has the restriction that we may not cancel zero.

Theorem (6e-6): $\forall a, b, c \in \mathscr{L} \wedge c \neq 0$
$$ac = bc \Rightarrow a = b$$

For convenience, the proof of this theorem will be delayed until after division has been introduced.

EXAMPLE 3: Suppose $a(x - 1) = 2(x - 1)$. Is $a = 2$?
Solution: We can conclude that $a = 2$ provided $x \neq 1$.

The next theorem finds wide application in factoring polynomials and finding the solutions of factorable quadratic equations.

The Number System

Theorem (6e-7): $\forall a, b \in \mathscr{Z}$,
$$a \cdot b = 0 \Rightarrow a = 0 \lor b = 0$$

Proof:

1. $a = 0 \lor a \neq 0$ 1. $p \lor \sim p \subseteq t$

   Case I:
   - a) $a = 0$
   - b) $a = 0 \lor b = 0$

        a) One of $p, \sim p$ must be true.
        b) D. A.
                                      Done

   Case II:
   - a) $a \neq 0$
   - b) $a \cdot b = 0$
   - c) $a \cdot 0 = 0$
   - d) $a \cdot b = a \cdot 0$
   - e) $b = 0$
   - f) $a = 0 \lor b = 0$

        a) One of $p, \sim p$ must be true.
        b) Hypothesis
        c) Theorem (6e-5)
        d) Transitivity
        e) Theorem (6e-6)
        f) D. A.
                                      Done

2. Hence, in either case,
$$a \cdot b = 0 \Rightarrow a = 0 \lor b = 0$$
     2. C. I.

                                                                                          Q. E. D.

EXAMPLE 4: Solve: $2(x + 2) = 0$.

Solution: By the theorem,
$$2 = 0 \lor x + 2 = 0$$

Since, $2 \neq 0$, we obtain $x + 2 = 0$ by D. S. We also know by (6e-4) that $^-2 + 2 = 0$, whence we have $\{^-2\}$ as the solution set.

As we have seen, the set of integers has made it possible for us to solve many types of equations which were not solvable over the natural numbers. To make further comparisons between these number sets, we need to consider the operations of subtraction and division and the relation $<$ as they apply to $\mathscr{Z}$.

We turn our attention to the operation of subtraction first. For the specific number set $\mathscr{Z}$, Definition (6-2) would read as follows:

(5) $\forall a, b \in \mathscr{Z}, a - b = x$ iff $\exists x \in \mathscr{Z} \cdot \ni \cdot a = b + x$

Recall that the important question about subtraction is, "When does the required x exist?" In the set $\mathscr{Z}$, the solvability axiom Z-10 has given us the required x for all possible (a, b). Thus our new axiom set has overcome one of the difficulties with the set $\mathscr{N}$ of natural numbers. One way to restate this fact is to say that the set of integers *is closed under the operation of subtraction*.

EXAMPLE 5: Is $0 - 7$ meaningful in $\mathscr{Z}$? If so, evaluate this difference.
Solution: By the definition,
$$0 - 7 = x \longleftrightarrow 0 = 7 + x$$
As stated, the solvability axiom guarantees that the required integer x exists. By (6e-4) we see that
$$0 = 7 + {}^-7$$
Since the additive inverse of a number is unique,
$$x = {}^-7, \text{ i.e., } 0 - 7 = {}^-7$$
Hence $0 - 7$ is a meaningful expression in $\mathscr{Z}$.

In this example, we see that there is a direct connection between subtraction and the additive inverse of a number.

In number sets such as $\mathscr{Z}$ which contain additive inverses for all elements, it is usual to define order relations differently than was done for the natural numbers.

Definition (6-14): $\forall a, b \in \mathscr{Z}, \mathbf{a} < \mathbf{b} \ (b > a)$ iff $(b - a) \in \mathscr{Z}^+$.

We note that $b > a$ is simply an alternative way of writing $a < b$. We read $b > a$ as *b is greater than a*. By using this definition with Axiom Z-11, the usual statement of trichotomy for any number set containing negatives can be proved.

Theorem (6e-8): $\forall a, b \in \mathscr{Z}$, exactly one of the following holds:
(i) $a = b$; (ii) $a < b$; (iii) $b < a$

Proof:

$$a, b \in \mathscr{Z} \qquad \text{Hypothesis}$$

By Z-11, one possibility is that
$$a = b$$
This proves (i). According to Z-11, another possibility is that
$$\exists x \in \mathscr{Z}^+$$
such that
$$a = b + x$$
By (5) this becomes
$$a - b = x$$

Since $x \in \mathscr{Z}^+$, we obtain $b < a$ by Definition (6-14) and (iii) is proved. Proof of the third possibility (ii) is left to the exercises. We note that by Z-11 these possibilities are mutually exclusive.

Corollary (6e-8a): $\forall a \in \mathscr{Z}$, exactly one of the following holds:

(i) $a = 0$; (ii) $a > 0$; (iii) $a < 0$

EXAMPLE 6: Show whether $0 < 1$ or $1 < 0$.

Solution: We know that

$$1 = 1 + 0 \qquad \text{(6e-2)}$$
$$= 0 + 1 \qquad \text{Z-5}$$

whence

$$1 - 0 = 1 \qquad \text{Definition of } (-), (5)$$
$$(1 - 0) \in \mathscr{Z}^+ \qquad \text{Z-12}$$
$$0 < 1 \qquad \text{Definition of } < (6\text{-}15)$$

Q. E. D.

Definition (6-15):

$\forall a \in \mathscr{Z}$, a is **negative** iff $a < 0$; a is **positive** iff $0 < a$.

By Corollary (6e-8a) and Definition (6-15) we see that an arbitrary integer is positive or zero or negative *exclusively*. We have also seen that $0 < 1$, i.e., 1 is positive, and we know from Axiom Z-12 that $1 \in \mathscr{Z}^+$. The following theorem shows that $\mathscr{Z}^+$ is indeed the set of positive integers.

Theorem (6e-9): $\forall a \in \mathscr{Z}, 0 < a$ iff $a \in \mathscr{Z}^+$.

Proof:

1. $\forall a \in \mathscr{Z}$; $\quad 0 < a \Longleftrightarrow a - 0 \in \mathscr{Z}^+ \qquad$ (6-14)
   What is $a - 0$?
2. $a - 0 = x$ iff $a = 0 + x \qquad$ Definition of $(-)$, (5)
3. $a = x \qquad$ (6e-2)
4. $a - 0 = a \qquad$ Transitivity
5. $0 < a \Longleftrightarrow a \in \mathscr{Z}^+ \qquad$ Substitution

Q. E. D.

Proofs of the following two theorems are left to the exercises. They restate the provisions of Axioms Z-13 and Z-14 in $<$ notation.

Corollary (6e-9a): $a > 0 \wedge b > 0 \Rightarrow a + b > 0$.

Corollary (6e-9b): $a > 0 \land b > 0 \Rightarrow a \cdot b > 0$.

Proofs of some of the theorems which follow will be left to the exercises. These are familiar properties of the "less than" relation which show us how the additive inverse concept is related to it.

Theorem (6e-10): $\forall a \in \mathscr{L}, {}^-a < 0$ iff $a \in \mathscr{L}^+$.
Proof:

1.	${}^-a < 0 \Longleftrightarrow (0 - {}^-a) \in \mathscr{L}^+$	1.	Definition of $<$, (6-14)
2.	$0 - {}^-a = x$ iff $0 = {}^-a + x$	2.	Definition of $(-)$, (5)
3.	$x = a$	3.	(6e-4)
4.	${}^-a < 0 \Longleftrightarrow a \in \mathscr{L}^+$	4.	Substitution

Q. E. D.

Corollary (6e-10a): $\forall a \in \mathscr{L}, {}^-a < 0$ iff $a > 0$.

Corollary (6e-10b): $\forall a \in \mathscr{L}, a < 0$ iff ${}^-a > 0$.

This proposition states formally that the negative (additive inverse) of an integer may be positive. This is the reason that "negative" and "negative of" are not to be confused.

EXAMPLE 7: Is -1 a negative number?

Solution: $1 \in \mathscr{L}^+$ Z-12
$-1 < 0$ (6e-10)
-1 is a negative number. (6-15)

EXAMPLE 8: Are the integers $-(-1), -2$ negative?

Solution: $2 = 1 + 1$ Definition of (2), (6-1)
$1 \in \mathscr{L}^+$ Z-12
$(1 + 1) \in \mathscr{L}^+$ Z-13
$2 \in \mathscr{L}^+$ Substitution
$2 > 0$ (6e-9)
$-2 < 0$ (6e-10a)
Thus -2 is negative. (6-15)
Considering $-(-1)$, we know that
$-1 < 0$. Example 7
$-(-1) > 0$. (6e-10b)
Thus $-(-1)$ is a *positive* number. (6-15)

To introduce the operation of division, let us recall that we are looking for x's which satisfy the following.

(6) $$a \div b = x \iff a = b \cdot x$$

The question is, "When does the required x exist?" It actually exists in very few cases for $\mathcal{N}$. What can we count on for the integers?

If $b = 0$, the following results:

$$\forall a \in \mathcal{L}, a \div 0 = x \text{ iff } \exists x \in \mathcal{L} \cdot \ni \cdot a = 0 \cdot x$$

However, $0 \cdot x = 0 \ \forall x \in \mathcal{L}$, thus we obtain

(7) $$a = 0 \ \forall a \in \mathcal{L}$$

This is a contradiction provided we are certain that integers different from 0 exist. We have shown that $1 \neq 0$. Thus (7) *is* a contradiction. For this reason there is no x for $b = 0$ in (6). Thus $a \div 0 = x$ *is impossible*. In order to remind ourselves that division by 0 is indeed meaningless, it is reasonable to restate Definition (6-3) for $\mathcal{L}$ as:

Definition (6-3a):

$$\forall a, b \in \mathcal{L} \wedge b \neq 0, \ a \div b = x \text{ iff } \exists x \in \mathcal{L} \cdot \ni \cdot b \cdot x = a$$

The proof of the multiplicative cancellation property (6e-6) can now be given.

Proof (of Theorem (6e-6)):

1. $ac = bc \wedge c \neq 0$ 1. C. P.
2. $(ac) \div c = b$ 2. Definition (6-3a)
3. $(ac) \div c = a$ 3. (6b-8) restated for $\mathcal{L}$
4. $a = b$ 4. Transitivity
5. $ac = bc \wedge c \neq 0 \Rightarrow a = b$ 5. C. I.

Q. E. D.

EXAMPLE 9: What can we conclude from $2x(x - 1) = 2y(x - 1)$ if $x, y \in \mathcal{L}$?

Solution: We have

(8) $$2x = 2y \qquad (6e\text{-}6)$$

provided $x - 1 \neq 0$, i.e., provided

$$x \neq 1$$

Then from (8) we have

$$x = y \qquad (6e\text{-}6)$$

since we are certain that

$$2 \neq 0$$

EXERCISE SET

1. Show that
   a) $3 \in \mathcal{Z}^+$
   b) $4 \in \mathcal{Z}^+$

2. Solve for x, where $a, x \in \mathcal{Z}$.
   a) $x - 1 = 3$
   b) $a + x = a + {}^-x$
   c) $x + 2 = 3$
   d) $2 - x = 2$
   e) ${}^-x = 5$
   f) $3 - x = 1$
   g) $2 \div 1 = x$
   h) $2 \div 2 = x$
   i) $a \div a = x \wedge a \neq 0$
   j) ${}^-a \div {}^-a = x \wedge {}^-a \neq 0$
   k) $7 = x + 4$
   l) $3 = x + 8$
   m) $5 - 6 = x$
   n) $4 = 4 + x$

3. Show whether the given number is positive or negative.
   a) ${}^-3$
   b) ${}^-4$
   c) ${}^-({}^-3)$
   d) ${}^-({}^-4)$

4. Show that the converses of the following propositions do not hold.
   a) (6e-9a)
   b) (6e-9b)

5. a) Show that ${}^-({}^-a) = a$.
   b) Is ${}^-a$ positive or negative?

6. Supply the missing reasons in the proof of (6e-5).

7. Prove each proposition for $x \in \mathcal{Z}$
   a) $5 + x = 3 \Rightarrow x = -2$
   b) $4 + 2x = 5 + x \Rightarrow x = 1$

8. Prove that the x of (4) is unique.

9. Prove that $a < b$ is one possibility in (6e-8).

10. Prove each proposition.
    a) (6e-3a)
    b) (6e-9a)
    c) (6e-9b)
    d) (6e-10a)
    e) (6e-10b)

§6f MAJOR THEOREMS FOR $\mathcal{Z}$

In the preceding section the solvability axiom introduced new numbers called additive inverses. First let us investigate what the additive inverse of a might be.

Theorem (6f-1): $\forall a \in \mathcal{Z}, {}^-({}^-a) = a$.

Proof:

1.	$a \in \mathscr{L}$	1.	C. P.
2.	$\exists\,^-a \in \mathscr{L} \cdot \ni\, \cdot 0 = a +\,^-a$	2.	(6e-4)
3.	$\exists\,^-(^-a) \in \mathscr{L} \cdot \ni\, \cdot 0 =\,^-a +\,^-(^-a)$	3.	(6e-4)
4.	$^-a +\,^-(^-a) = a +\,^-a$	4.	Transitivity
5.	$^-(^-a) = a$	5.	(6e-1)
6.	$\forall a \in \mathscr{L},\,^-(^-a) = a$	6.	C. I.

Q. E. D.

EXAMPLE 1: Write the additive inverse for each of the following: 5, 15, -2, x, -8.

Solution: $-5, -15, 2, -x, 8$.

The next theorem is probably the single most useful one for practical work with integers. It tells us how to subtract in terms of addition.

Theorem (6f-2): $\forall a, b \in \mathscr{L},\, a - b = a +\,^-b$.

Proof:

1.	$\forall (a, b) \cdot \ni \cdot a, b \in \mathscr{L},$ $\exists x \in \mathscr{L} \cdot \ni \cdot a = b + x$	1.	Z-10
2.	$a - b = x$	2.	Definition of Subtraction (5)
3.	$a +\,^-b = (b + x) +\,^-b$	3.	Z-3 (from Step 1)
4.	$a +\,^-b = (x + b) +\,^-b$	4.	Z-5
5.	$\qquad = x + (b +\,^-b)$	5.	Z-7
6.	$\qquad = x + 0$	6.	(6e-4)
7.	$\qquad = x$	7.	(6e-2)
8.	$a - b = a +\,^-b$	8.	Transitivity (2, 7)

Q. E. D.

EXAMPLE 2: Make a roster of
$$A = \{x \mid x \in \mathscr{L} \wedge 5 - (^-4) = x\}$$

Solution:

1.	$5 - (^-4) = x$	1.	Hypothesis
2.	$5 - (^-4) = 5 +\,^-(^-4)$	2.	(6f-2)
3.	$^-(^-4) = 4$	3.	(6f-1)
4.	$5 +\,^-(^-4) = 5 + 4$	4.	Substitution

§6f Major Theorems for $\mathscr{Z}$

5.	$5 + 4 = 9$	5.	Addition fact
6.	$x = 9$	6.	Transitivity

Therefore, $A = \{9\}$

The following corollaries are fairly easy to prove.

Corollary (6f-2a): $0 = {}^-0$.

This proposition states that the additive inverse of zero is zero. It does *not* state that zero is a negative number.

Corollary (6f-2b): $\forall a \in \mathscr{Z}, a - 0 = a$.

Corollary (6f-2c): $\forall a \in \mathscr{Z}, 0 - a = {}^-a$.

EXAMPLE 3: Can we find a single value for x if $x = {}^-3 + {}^-4$?

Solution: Yes.

1.	${}^-3 + {}^-4 = x$	1.	Hypothesis
2.	${}^-3 - 4 = x$	2.	(6f-2)
3.	${}^-3 = 4 + x$	3.	Definition of Subtraction
4.	$0 - 3 = 4 + x$	4.	(6f-2c)
5.	$0 = 3 + (4 + x)$	5.	Definition of Subtraction
6.	$= (3 + 4) + x$	6.	Z-7
7.	$= 7 + x$	7.	Addition fact
8.	$x = {}^-7$	8.	(6e-4)

In this example we have seen that the additive inverse of a sum is equal to the sum of the additive inverses. As stated in the following theorem, this principle holds in general. A proof can be constructed similarly to the way the solution to Example 3 was obtained.

Theorem (6f-3): $\forall a, b \in \mathscr{Z}, {}^-a + {}^-b = {}^-(a + b)$.

Corollary (6f-3a): $a < 0 \wedge b < 0 \Rightarrow a + b < 0$.

That is, the sum of two negative integers is a negative number.

EXAMPLE 4: Make a roster of
$$A = \{x \mid x \in \mathscr{Z} \wedge {}^-5 + {}^-7 = x\}$$

Solution:

1.	${}^-5 + {}^-7 = x$	1.	Hypothesis

2. $^-5 + {}^-7 = {}^-(5+7)$ 2. (6f-3)
3. $^-(5+7) = {}^-(12)$ 3. Addition fact
4. $x = {}^-12$ 4. Transitivity
    Therefore, $A = \{-12\}$

Note that this argument is considerably shorter than that of Example 3.

We now turn our attention to the multiplication of additive inverses.

Theorem (6f-4): $\forall a, b \in \mathscr{L}, a(^-b) = {}^-(ab) = (^-a)b$.

Proof:

1. $^-b + b = 0$ 1. ?
2. $a(^-b + b) = a \cdot 0$ 2. ?
3. $a \cdot 0 = 0$ 3. ?
4. $a(^-b + b) = 0$ 4. ?
5. $a(^-b + b) = a(^-b) + ab$ 5. ?
6. $a(^-b) + ab = 0$ 6. ?
7. $ab + {}^-(ab) = 0$ 7. ?
8. $a(^-b) + ab = ab + {}^-(ab)$ 8. ?
9. $a(^-b) = {}^-(ab)$ 9. ?
10. $^-(ab) = {}^-(ba)$ 10. ?
11. $^-(ba) = b(^-a)$ 11. ?
12. $b(^-a) = (^-a)b$ 12. ?
13. $^-(ab) = (^-a)b$ 13. ?

Q. E. D.

Corollary (6f-4a): $\forall a \in \mathscr{L}, {}^-a = (^-1)(a)$.

EXAMPLE 5: Rewrite the products $(^-3)(2), (^-1)(7), (^-3)(^-2)$, using the principle of (6f-4) or of (6f-4a).

Solution:

$(^-3)(2) = {}^-(3 \cdot 2)$ (6f-4)
$\qquad\quad = {}^-6$ Multiplication fact
$(^-1)(7) = {}^-7$ (6f-4a)
$(^-3)(^-2) = {}^-(^-3)(2)$ (6f-4)
$\qquad\quad\; = {}^-[^-(3 \cdot 2)]$ (6f-4)
$\qquad\quad\; = {}^-(^-6)$ Multiplication fact
$\qquad\quad\; = 6$ (6f-1)

This last result suggests the following proposition.

Theorem (6f-5): $\forall a, b \in \mathscr{Z}, (^-a)(^-b) = ab$.

Proof:

1.	$(^-a)(^-b) = ^-[(^-a)b]$	1.	(6f-4)
2.	$^-[(^-a)b] = ^-[^-(ab)]$	2.	(6f-4)
3.	$^-[^-(ab)] = ab$	3.	(6f-1)
4.	$(^-a)(^-b) = ab$	4.	Transitivity

Q. E. D.

This proposition states that the product of two integers is the same as the product of their additive inverses. In particular, the product of two negative integers is a positive integer.

Corollary (6f-5a): $(^-1)(^-1) = 1$.

Now we turn our attention to some important properties of the order relation among the integers. Proofs of these theorems are left to the exercises.

Theorem (6f-6): $\forall a, b \in \mathscr{Z}, a < b$ iff $^-b < {}^-a$.

EXAMPLE 6: Solve $-x < -(-3)$.

Solution: By the theorem $-3 < x$, i.e., $x > -3$.

Theorem (6f-7): $\forall a, b, c \in \mathscr{Z}$,
$$a < b \land b < c \Rightarrow a < c$$

Theorem (6f-8): $\forall a, b, c \in \mathscr{Z}$,
$$a < b \Leftrightarrow a + c < b + c$$

Theorem (6f-9): $\forall a, b, c \in \mathscr{Z}$,
$$a < b \land c > 0 \Rightarrow ac < bc$$

Theorem (6f-10): $\forall a, b, c \in \mathscr{Z}$,
$$a < b \land c < 0 \Rightarrow ac > bc$$

This last principle is one which must be carefully noted. Failure to use it correctly accounts for many mistakes.

Corollary (6f-10a): $a > 0 \land c < 0 \Rightarrow ac < 0$.

Thus, the product of a negative number and a positive number is negative.

Corollary (6f-10b): $a < 0 \land c < 0 \Rightarrow ac > 0$.

Thus, the product of a negative number and a negative number is positive.

EXAMPLE 7: Simplify the solution set

$$\{x \mid 5 - 2x < 13 \land x \in \mathscr{L}\}$$

Solution:
$$5 - 2x < 13 \Longleftrightarrow -2x < 8 \tag{6f-8}$$
$$-2x < 8 \Longleftrightarrow x > -4 \tag{6f-10}$$

Thus $\{x \mid x > -4\}$ is the required simplification.

We have not mentioned involution over $\mathscr{L}$ to this point. The following example should be carefully noted.

EXAMPLE 8: Determine whether the proposition $a = b \Rightarrow a^n = b^n$ is true or false. Is the converse of the proposition true?

Solution: Since $3 = 3 \Rightarrow 9 = 9$ is true and since $-3 = -3 \Rightarrow -27 = -27$ is true, the proposition seems true.
However, $(-3)^2 = 3^2 \Rightarrow -3 = 3$ is false as is $(-3)^4 = 3^4 \Rightarrow -3 = 3$. Thus, the converse does not hold.
This principle is stated as a theorem for the squared case only.

Theorem (6f-11): $\forall a, b \in \mathscr{L}, a = b \Rightarrow a^2 = b^2$

As we have seen, the converse of Theorem (6f-11) is definitely false.

EXERCISE SET

1. Solve the following equations for x over $\mathscr{L}$.
    a) $5 - 8 = x$
    b) $^-5 - 8 = x$
    c) $^-5 - {^-8} = x$
    d) $5 - {^-5} = x$
    e) $x + 7 = {^-7}$
    f) $8 - x = 4$

2. Solve the following inequalities for x over $\mathscr{L}$.
    a) $x + 5 < 4$
    b) $x - 5 < 4$
    c) $-x + 5 < 4$
    d) $-x - 5 < 4$
    e) $2x + 3 < 5$
    f) $5 - 2x < 13$
    g) $2x - 5 > -7$
    h) $(x + 2)(x - 3) > 0$

3. Solve for x, making a roster if possible.
    a) $\{x \mid x \in \mathscr{L} \land 3x + 5 = 8\}$
    b) $\{x \mid x \in \mathscr{L} \land x^2 - 16 = 0\}$
    c) $\{x \mid x \in \mathscr{L} \land x^2 + 5x + 9 = 2\}$

d) $\{x \mid x \in \mathscr{L} \wedge 2x^2 - 4x - 6 = 0\}$
e) $\{x \mid x \in \mathscr{L} \wedge x^2 - 6x + 9 = 0\}$

4. Supply the missing reasons in the proof of (6f-4).
5. Prove the following:
    a) $\forall a \in \mathscr{L}, a + a = a \Rightarrow a = 0$
    b) $^-8 - {}^-x = 7 \Rightarrow x = 15$
    c) $\forall a, b, c \in \mathscr{L}, a(b - c) = ab - ac$
    d) $\forall a, b \in \mathscr{L}, (a + b)(a - b) = a^2 - b^2$
6. Prove each proposition.
    a) (6f-2a)   b) (6f-2b)
    c) (6f-2c)   d) (6f-3a)
    e) (6f-4a)   f) (6f-5a)
    g) (6f-7)   h) (6f-8)
    i) (6f-9)   j) (6f-10a)
    k) (6f-10b)
*7. Prove each proposition.
    a) (6f-3)   b) (6f-6)
    c) (6f-10)   d) (6f-11)

§6g THE NUMBER LINE

In extending the number system from $\mathscr{P}$ to $\mathscr{N}$ to $\mathscr{L}$, we have obtained a valuable new number set. Before we consider a few more comparisons and more properties of $\mathscr{L}$ itself, we shall list the important properties of integers which have been established based upon the axioms of §6e.

List of Properties

(6e- 1): $\forall a, b, c \in \mathscr{L}, a + c = b + c \Rightarrow a = b$
(6e- 2): $\forall a \in \mathscr{L}, \exists$ a unique $0 \in \mathscr{L} \cdot \ni \cdot a = a + 0$
(6e-3a): $1 \neq 0$
(6e- 4): $\forall a \in \mathscr{L}, \exists$ a unique $^-a \in \mathscr{L} \cdot \ni \cdot 0 = a + {}^-a$
(6e- 5): $\forall a \in \mathscr{L}, a \cdot 0 = 0$
(6e- 6): $\forall a, b, c \in \mathscr{L}, ac = bc \wedge c \neq 0 \Rightarrow a = b$
(6e- 7): $\forall a, b \in \mathscr{L}, a \cdot b = 0 \Rightarrow a = 0 \vee b = 0$
(6e- 8): $\forall a, b \in \mathscr{L}$, exactly one of $a = b, a < b, b < a$ holds
(6e- 9): $\forall a \in \mathscr{L}, 0 < a$ iff $a \in \mathscr{L}^+$
(6e-10): $\forall a \in \mathscr{L}, {}^-a < 0$ iff $a \in \mathscr{L}^+$
(6f- 1): $\forall a \in \mathscr{L}, {}^-({}^-a) = a$
(6f- 2): $\forall a, b \in \mathscr{L}, a - b = a + {}^-b$
(6f- 3): $\forall a, b \in \mathscr{L}, {}^-a + {}^-b = {}^-(a + b)$

(6f- 4): $\forall a, b \in \mathscr{L}, a(^-b) = {}^-(ab) = (^-a)b$
(6f- 5): $\forall a, b \in \mathscr{L}, (^-a)(^-b) = ab$
(6f- 6): $\forall a, b \in \mathscr{L}, a < b$ iff $^-b < {}^-a$
(6f- 7): $\forall a, b, c \in \mathscr{L}, a < b \wedge b < c \Rightarrow a < c$
(6f- 8): $\forall a, b, c \in \mathscr{L}, a < b \Leftrightarrow a + c < b + c$
(6f- 9): $\forall a, b, c \in \mathscr{L}, a < b \wedge c > 0 \Rightarrow ac < bc$
(6f-10): $\forall a, b, c \in \mathscr{L}, a < b \wedge c < 0 \Rightarrow ac > bc$
(6f-11): $\forall a, b \in \mathscr{L}, a = b \Rightarrow a^2 = b^2$

Looking over such a list, one obtains visual evidence of the fact that the integers possess several important properties different from those of the natural numbers. We are just as forcibly reminded that the two number sets have many properties in common.

One obvious goal which we strive toward in making a number extension is to retain as many of the valuable properties of the original set as possible while building in the properties desired for the new set. In making the extension

$$\mathscr{P} \to \mathscr{N}$$

for example, we made sure that $\mathscr{N}$ still has a unique nonsuccessor and a unique successor for each element. Thus counting is still possible with the natural numbers even though each successor in $\mathscr{N}$ has a rigid additive relationship to its predecessor.

In beginning our discussion of the extension

$$\mathscr{N} \to \mathscr{L}$$

we listed the first nine axioms for $\mathscr{L}$ in such a way that they were simply restatements of corresponding axioms for $\mathscr{N}$. This fact assured us that many properties of $\mathscr{N}$ will hold for $\mathscr{L}$ without restriction.

However, it is also important to discuss these two sets from the standpoint of pinpointing some of the outstanding differences between them. For this purpose we introduce a device called a number line, which is often used in elementary school.

To talk about a number line intelligently, we must state something about linear distance. Intuitively, linear distance is a nonnegative number, and we shall leave it as a primitive in this context. To assure ourselves that we have a nonnegative number when needed, we make use of the absolute value concept. The definition is restated here for reference.

Definition (4-20): The **absolute value** of a number a, denoted $|a|$, is a when $a > 0 \vee a = 0$, and is ^-a when $a < 0$. That is,

$$|x| = \begin{cases} x \text{ if } x \geq 0 \\ ^-x \text{ if } x < 0 \end{cases}$$

§6g The Number Line

EXAMPLE 1: $|5| = 5$; $|^-7| = ^-(^-7)$ by the definition, and $^-(^-7) = 7$ by (6f-1).

Now we take a set of points that is a line $\mathscr{L}$ *and assign each number to a point of the line in such a way that no two numbers label the same point.* For this purpose the following definition is desirable.

Definition (6-16): If A and B are distinct points of $\mathscr{L}$ with assigned numbers a, b, respectively, then the **distance** *between A and B* is the nonnegative number given by $|b - a|$.

The number line for $\mathscr{X}$ consists of points of $\mathscr{L}$ whose consecutive distance apart is 1. Also if C and D are consecutive points we want

$$\overline{CD} \cong \overline{AB}$$

(line segment CD is congruent to line segment AB) where A is assigned 1 and B has 2 as its assigned number. See Figure 6.3.

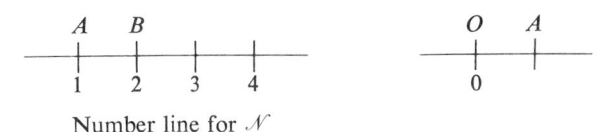

```
    A   B                              O   A
    |   |   |   |                      |   |
    1   2   3   4                      0
    Number line for 𝒩
```

Figure 6.3   Figure 6.4

Now let us consider the problem of producing a number line for the integers. We assign zero to a point, and call the point O. Now, if we assign 1 to a point A to the right of O, where is the point A' to which we assign -1? See Figure 6.4. Since

$$|^-1 - 0| = |1 - 0|$$

we must have $\overline{OA} \cong \overline{OA'}$ according to our distance concept. However, we may not assign two distinct numbers to the one point. Thus, A' must be located to the left of O. Continuing in similar fashion allows us to locate a point for each integer. See Figure 6.5.

```
                    A'  O   A   B
←—+—+—+—+—+—+—+—+—+—+—+—+→
 -5  -4  -3  -2  -1  0   1   2   3   4   5   6
            Number line for 𝒵
```

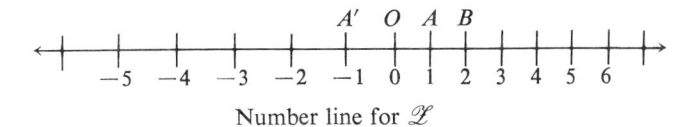

Figure 6.5

By comparing Figures 6.3 and 6.5, we can see an important difference between the sets of natural numbers $\mathscr{N}$ and integers $\mathscr{Z}$. The set $\mathscr{N}$ could be used as a counting set because 1 was a nonsuccessor. There is no nonsuccessor in $\mathscr{Z}$; hence $\mathscr{Z}$ cannot be used as a counting set.

Even though the number line for $\mathscr{L}$ is not the same as the number line for $\mathscr{N}$, let us note that the right side of Figure 6.5 looks the same as Figure 6.3. The right side of Figure 6.5 is the number line for $\mathscr{L}^+$. Now let us compare the mathematical properties of $\mathscr{N}, \mathscr{L}^+$. Recall that statements universally true on a nonempty set will be true existentially. In particular, a statement universally true on $\mathscr{L}$ will be true on $\mathscr{L}^+$. Thus Axioms Z-3 through Z-9 hold for $\mathscr{L}^+$ as they are stated. For $\mathscr{L}^+$ closure for addition is stated in Z-13 and closure for multiplication in Z-14. Z-12 provides $\mathscr{L}^+$ with a neutral element for multiplication, and Z-15 provides it with the property of mathematical induction. Hence $\mathscr{L}^+$ has the same mathematical properties as $\mathscr{N}$. In some developments $\mathscr{L}^+$ and $\mathscr{N}$ are treated as equal sets. For reference, we list representative elements of the three sets.

$\mathscr{N} = \{1, 2, 3, 4, \ldots, n, n+1, \ldots\}.$
$\mathscr{L}^+ = \{1, 2, 3, 4, \ldots, n, n+1, \ldots\}.$
$\mathscr{L} = \{\ldots, ^-(n+1), ^-n, \ldots, ^-3, ^-2, ^-1, 0, 1, 2, 3, \ldots, n, n+1, \ldots\}.$

The set $\mathscr{N}$ has effectively become a subest of the set $\mathscr{L}$. We have stressed the fact that one of the major points of difference between the two axiom sets is the solvability axiom Z-10. Because of this axiom, $\mathscr{L}$ is closed under subtraction whereas $\mathscr{N}$ is not. Thus many equations which have empty solution sets over $\mathscr{N}$ are solvable over $\mathscr{L}$. However, inclusion of the solvability axiom among the properties of $\mathscr{L}$ has forced us to include a restriction on division not found when working with the natural numbers.

Before leaving the number line completely, it would certainly be reasonable to ask whether there are other integers between any two of those on the number line. The answer happens to be "no," but this fact is not too easy to establish mathematically. It requires use of Z-15, the Axiom of Induction, on $\mathscr{L}^+$. Using this axiom, we could establish the principle for $\mathscr{L}^+$ that

(1) $\qquad \forall a \in \mathscr{L}^+$, there is no $b \in \mathscr{L}^+ \cdot \ni \cdot a < b < a+1$

For example, if we let $a = 1$, this principle states that there is no $b \in \mathscr{L}^+$ such that $1 < b < 2$. Thus, there are no numbers to correspond to the points between A, B in Figure 6.5. It is reasonable to assume that this principle must apply to all of $\mathscr{L}$.

We have listed what can be done with the integers. Further consideration of the operation of division will show what cannot be done with them and give an indication of why we find it advantageous to develop a system which for certain purposes has advantages over $\mathscr{L}$.

EXAMPLE 2: Solve $5x + 6 = 18$ over $\mathscr{L}$ if possible.

Solution: From $5x + 6 = 18$, we obtain

$$5x + 6 = 12 + 6$$

whence

(2) $\qquad\qquad\qquad 5x = 12 \qquad\qquad\qquad$ (6e-1)

Consulting our multiplication table, we find that $5 \cdot 2 = 10$ and $5 \cdot 3 = 15$. As we have seen, there are no integers between 2 and 3. Hence this problem cannot be solved in this number system.

The operation of involution points up a further shortcoming of the set of integers. If

(3) $$x^2 = 2$$

what is x? Again consulting our multiplication table, we discover that $\mathscr{Z}$ is lacking any solution for this equation also.

It would be advantageous to have a number set in which the equations of (2), (3) could have solutions. The equation of (2) would have a solution if we had a number set in which an equation of the type

$$a = bx$$

would *always* have a solution. In the following section such a number set is, in fact, considered.

EXERCISE SET

1. Find the solution set of each open sentence where $x \in \mathscr{Z}$.
    a) $|x| = 0$
    b) $|x| = 1$
    c) $|x| = {}^-2$
    d) $|3| = x$
    e) $|x| = 3$
    f) $|3 - 1| = x$

2. Find the distance between the following pairs of points on the number line for $\mathscr{Z}$.
    a) $({}^-2, 5)$
    b) $({}^-3, {}^-7)$
    c) $(0, 4)$
    d) $(7, {}^-5)$
    e) $({}^-\pi, 5)$

3. For each set make a number line which shows the elements of the stated solution set.
    a) $\{x \mid x \in \mathscr{Z} \land x^2 - 4 = 0\}$
    b) $\{x \mid x \in \mathscr{Z} \land x^2 + 4 = 0\}$
    c) $\{x \mid x \in \mathscr{Z} \land x^2 - 5x + 6 = 0\}$
    d) $\{x \mid x \in \mathscr{Z} \land x^3 - 4x = 0\}$

4. List the gains and losses in making the extension $\mathscr{N} \to \mathscr{Z}$.

§6h THE RATIONAL NUMBER SYSTEM

In developing the integers to have additive solvability, a special effort was made to retain as many as possible of the properties of natural numbers.

To all intents and purposes, we could say that $\mathcal{N} \subset \mathcal{Z}$. Now that we have shown a need for a system having multiplicative solvability, we should attempt to keep as many properties of $\mathcal{Z}$ as possible and in addition have $\mathcal{Z}$ as a subset of this new number set.

The new system will be called the rational number system with primitives $+$, $\cdot$, and $\mathcal{Q}$.

Q-1: $\forall a, b \in \mathcal{Q}, \exists x \in \mathcal{Q} \cdot \ni \cdot a + b = x$

Q-2: $\forall a, b \in \mathcal{Q}, \exists x \in \mathcal{Q} \cdot \ni \cdot a \cdot b = x$

Q-3: $\forall a, b, c \in \mathcal{Q}, a = b \Rightarrow a + c = b + c$

Q-4: $\forall a, b, c \in \mathcal{Q}, a = b \Rightarrow a \cdot c = b \cdot c$

Q-5: $\forall a, b \in \mathcal{Q}, a + b = b + a$

Q-6: $\forall a, b \in \mathcal{Q}, a \cdot b = b \cdot a$

Q-7: $\forall a, b, c \in \mathcal{Q}, (a + b) + c = a + (b + c)$

Q-8: $\forall a, b, c \in \mathcal{Q}, (a \cdot b) \cdot c = a \cdot (b \cdot c)$

Q-9: $\forall a, b, c \in \mathcal{Q}, a \cdot (b + c) = a \cdot b + a \cdot c$

Q-10: $\forall a, b \in \mathcal{Q}$ in that order, $\exists x \in \mathcal{Q} \cdot \ni \cdot a = b + x$

Q-11: $\forall a, b \in \mathcal{Q}$ in that order $\wedge a \neq a + b$, $\exists x \in \mathcal{Q} \cdot \ni \cdot a = b \cdot x$

Q-12: $\exists x \in \mathcal{Q} \wedge \exists y \in \mathcal{Q} \cdot \ni \cdot x \neq y$

Q-13: $\exists$ a subset $\mathcal{Q}^+$ of $\mathcal{Q} \cdot \ni \cdot \forall a, b \in \mathcal{Q}$, exactly one of the following holds:

    (i) $a = b$
    (ii) $\exists x \in \mathcal{Q}^+ \cdot \ni \cdot a = b + x$
    (iii) $\exists y \in \mathcal{Q}^+ \cdot \ni \cdot a + y = b$

Q-14: $\forall a, b \in \mathcal{Q}^+ \exists x \in \mathcal{Q}^+ \cdot \ni \cdot a + b = x$

Q-15: $\forall a, b \in \mathcal{Q}^+ \exists x \in \mathcal{Q}^+ \cdot \ni \cdot a \cdot b = x$

This axiom set is the same as the axiom set for the integers except for Q-11, Q-12, the lack of a neutral element for multiplication, and the fact that there is no induction axiom on $\mathcal{Q}^+$. Any theorem for $\mathcal{Z}$ which was not proved using one of these principles is immediately provable for $\mathcal{Q}$ by substituting one set name for the other, and we can accept them as proved.

In trying to determine particular properties of the rational numbers, we thus concentrate on discovering what Q-11 and Q-12 do for us and what restrictions, if any, the lack of an induction axiom places upon us. As for Q-12, it merely assures us that $\mathcal{Q}$ is nonempty and that, in fact, there are at least two numbers in $\mathcal{Q}$. Thus Axiom Q-11 will be our main concern.

Note that Q-11 guarantees that certain equations have solutions. Like Q-10 it is thus a solvability axiom. To distinguish the two, we call Q-11 the **solvability property for multiplication** and refer to Q-10 as the **solvability property for addition**.

Our multiplicative solvability axiom requires that for each ordered pair (a, b) of rational numbers, a rational number x exists such that $a = b \cdot x$ provided $a \neq a + b$. We are not told how to find such an x or whether it is unique; hence, our initial tasks are quite clear. First let us consider the stated restriction

(1) $$a \neq a + b$$

Let us suppose that $a = a + b$. Then $a + 0 = a + b$ or $0 = b$. Hence this restriction is simply another way of saying

(2) $$b \text{ cannot be zero}$$

That is, the solution x of

(3) $$a = b \cdot x$$

is *not* guaranteed to exist when $b = 0$.

The reason for this restriction is clear if we recall that

$$0 \cdot x = 0$$

for all x. Thus, for any *nonzero* a and for $b = 0$, the equation of (3) would become

$$a = 0 \cdot x = 0$$

Since this contradiction cannot be tolerated, we must insist on the restriction of (1) and (2).

Now let us examine the nature of the "solutions" provided us by Q-11 for the ordered pairs (a, b) for which $b \neq 0$.

Theorem (6h-1): The x of Q-11 is unique.

Proof (*One*):

Done, by Q-11.

(*Only One*):

Note: We shall suppose that there are two and show that they must be the same.

1. $\forall (a, b) \cdot \ni \cdot a, b \in \mathcal{Q},$ 1. Q-11
   $\exists x \in \mathcal{Q} \cdot \ni \cdot a = b \cdot x$

2. $\forall (a, b) \cdot \ni \cdot a, b \in \mathcal{Q},$ 2. Q-11
   $\exists \hat{x} \in \mathcal{Q} \cdot \ni \cdot a = b \cdot \hat{x}$

3. $b \cdot x = b \cdot \hat{x}$ 3. E-3

4. $x = \hat{x}$ 4. Cancellation for $\cdot$ holds for $\mathscr{Q}$ as it does in $\mathscr{Z}$ (6e-6)

Done
Q. E. D.

Hence the x is unique. However, we still don't know what kind of thing has been created here. That is, we don't know what kind of number x is. Let us examine some of the particular x's which Q-11 creates. First consider the ordered pair (a, a) for any rational number a. The x we obtain in this case looks very familiar.

Theorem (6h-2): There is a unique rational number 1 such that

$$\forall a \in \mathscr{Q}, a = a \cdot 1$$

Proof:

$$\forall (a, a) \cdot \ni \cdot a \in \mathscr{Q}, a \neq 0, \exists x \in \mathscr{Q} \cdot \ni \cdot a = a \cdot x$$

by Q-11. We adopt the symbol 1 for this $x \in \mathscr{Q}$, i.e., $a = a \cdot 1$, $\forall a \in \mathscr{Q}$. That there is only one such number 1 is guaranteed by (6h-1).

Q. E. D.

As was the case for $\mathscr{N}$ and $\mathscr{Z}$, 1 is called the *neutral element* (identity element) *for multiplication.*

Recall that Axiom Z-12 not only provided us with a neutral element 1 for multiplication in $\mathscr{Z}$, but also stated that it was positive. Theorem (6h-2) makes no such statement about the *rational* number 1. Now if we can also show that $1 \in \mathscr{Q}^+$, this multiplicative neutral element will have all the properties specified for integers, and one more point of correspondence between integers and rational numbers will have been established.

We can establish that $1 \in \mathscr{Q}^+$, that is, $1 > 0$, provided we know that 0 and 1 stand for distinct numbers and not just different symbols for the same thing. The argument that 0 and 1 are distinct rational numbers is based on Q-12.

Theorem (6h-3): $0 \neq 1$.

Proof:

1. $0 = 1$ 1. I. P.
2. $\exists a, b \in \mathscr{Q} \cdot \ni \cdot a \neq b$ 2. Q-12
3. $a = a \wedge b = b$ 3. Reflexivity of equality
4. $a \cdot 0 = a \cdot 1 \wedge b \cdot 0 = b \cdot 1$ 4. Q-4
5. $0 = a \cdot 0 \wedge 0 = b \cdot 0$ 5. (6e-5) (holds for $\mathscr{Q}$)

§6h The Rational Number System

6.	$a \cdot 1 = a \wedge b \cdot 1 = b$	6.	(6h-2)
7.	$a = b$	7.	Transitivity (4, 5, 6)
8.	$a = b \wedge a \neq b$	8.	C. A. (2, 7)
9.	$0 \neq 1$	9.	I. I.

Q. E. D.

If Q-12 were not one of our axioms, the set $\mathscr{Q}$ could consist of 0 only; i.e., we might have $\mathscr{Q} = \{0\}$. This set would satisfy Q-1 through Q-11, Q-13 through Q-15. Its arithmetic would certainly be easy to learn, but it would not be very useful.

Now we are in a position to show that $1 \in \mathscr{Q}^+$.

Theorem (6h-4): $1 > 0$.

Proof: By the trichotomy property (6e-8), which holds for $\mathscr{Q}$, exactly one of the following holds: $1 < 0, 1 = 0, 1 > 0$. By (6h-3) we know that $1 \neq 0$, thus

(4) $\qquad\qquad\qquad 1 < 0 \vee 1 > 0$

holds. Take $1 < 0$ as I. P, then $^-1 \in \mathscr{Q}^+$ by the equivalent of (6e-10) for $\mathscr{Q}$. Hence $(^-1)(^-1) \in \mathscr{Q}^+$ by Q-15. But $(^-1)(^-1) = 1$ by the equivalent of (6e-9) for $\mathscr{Q}$. Thus $1 \in \mathscr{Q}^+$ whence $1 > 0$ by the equivalent of (6e-9) for $\mathscr{Q}$. This contradicts (4). Therefore, $1 \not< 0$ by I. I., and $1 > 0$ is the only possibility by D. S.; that is, $1 \in \mathscr{Q}^+$.   Q. E. D.

In view of (6h-2) and (6h-4), $\mathscr{Q}$ has all properties corresponding to Z-12. That is, only the presence of the multiplicative solvability property Q-11 and the lack of an inductive property corresponding to Z-15 create differences between integers and rational numbers.

Continuing our investigation of Q-11, let us see what kind of $x \in \mathscr{Q}$ results from applying the axiom to the ordered pair $(1, a)$.

Theorem (6h-5): $\forall a \in \mathscr{Q} \wedge a \neq 0 \wedge 1 \in \mathscr{Q}$, there is a unique

$$\frac{1}{a} \in \mathscr{Q} \cdot \ni \cdot 1 = a \cdot \left(\frac{1}{a}\right)$$

Proof: $\forall (1, a) \cdot \ni \cdot 1, a \in \mathscr{Q} \wedge a \neq 0, \exists x \in \mathscr{Q} \cdot \ni \cdot 1 = a \cdot x$ by Q-11. We adopt the symbol $1/a$ for this rational number x. That is,

(5) $\qquad\qquad 1 = a \cdot \left(\frac{1}{a}\right), \forall a \in \mathscr{Q} \wedge a \neq 0$

There is only one $1/a$ for each a by (6h-1).

Q. E. D.

This proposition presents one of the outstanding points of difference between

$\mathscr{I}$ and $\mathscr{Q}$. If we call the number $1/a$ the **multiplicative inverse** of a, it can be shown that 1 is the only *integer* which has such an inverse. As stated in (5), however, *each rational number except* 0 *has a multiplicative inverse*.

If we use the notation of the multiplicative inverse for the "solution" x of any equation

(6) $$a = b \cdot x$$

we can write $x = a/b$. Then this equation can be written as

(7) $$a = b \cdot \frac{a}{b}$$

We call a/b the *ratio* of a to b, which accounts for the name *rational number*.

Recall that the definition of division (6-3) in a number set S stated that $a \div b = x$ iff there is an x in S such that $a = bx$. For the integers we found that there is no such x when $b = 0$. In the case of the rational numbers, the multiplicative solvability property Q-11 *explicitly* fails to provide the x of (6) for the case of $b = 0$. Thus (6-3) can be restated as it was for $\mathscr{I}$.

Definition (6-3a): $\forall (a, b) \cdot \ni \cdot a, b \in \mathscr{Q} \wedge b \neq 0$,

$$\mathbf{a \div b} = x \text{ iff } \exists x \in \mathscr{Q} \cdot \ni \cdot a = bx$$

Since Q-11 guarantees that the x of this definition exists in all cases for which b is not equal to zero, *the set $\mathscr{Q}$ of rational numbers is closed under division for all nonzero divisors*.

Considering the sentence $a \div b = x$, we have

(8) $$x = \frac{a}{b} = a \div b$$

Thus the new rational numbers guaranteed by Q-11 are quotients, which accounts for the designation $\mathscr{Q}$ for the set of rational numbers.

In the development of the rational numbers to this point, the axioms, excluding Q-11, have given us numbers which act like integers. Q-11 has also given us new nonintegral numbers which can be written as in (8) where a, b are integers. Now, if we consider two such rational numbers

$$a = \frac{c}{d}, b = \frac{e}{f}$$

where b, d, f are *not* zero, Q-11 assures us that there is a rational number x such that

$$\frac{c}{d} = \frac{e}{f} x$$

In the following section we shall see that this number can be written as

§6h The Rational Number System

$$x = \frac{cf}{de}$$

Since $\mathscr{L}$ has the property of closure under multiplication, this x is also a ratio of two integers. Thus *all* rational numbers have the form of (8). That is, every rational number can be regarded as a ratio of two integers. When $a \div b$ is written as a/b, we recognize it as what is often called a fraction and see that $\mathscr{Q}$ is a subset of the set of fractions. Also

(9) $$\mathscr{Q} = \left\{ \frac{a}{b} \,\Big|\, a, b \in \mathscr{L} \wedge b \neq 0 \right\}$$

Returning to the operation of involution, we can use the definition and theorems for powers of natural numbers only for $\mathscr{L}^+$ and for that subset of $\mathscr{Q}^+$ which acts like $\mathscr{L}^+$. We must also define what we mean by raising a number to the power zero, to negative powers, and to fractional powers.

Definition (6-17): $\forall a \in \mathscr{Q},\ a \neq 0,\ a^0 = 1$.

EXAMPLE 1: Evaluate $(5^2 \cdot 6x^3)^0$ where $x \in \mathscr{Q}$.
Solution: Let $5^2 \cdot 6x^3 = a$; then we have $a^0 = 1$, i.e. $(5^2 \cdot 6x^3)^2 = 1$.

Definition (6-18): $\forall a \in \mathscr{Q} \wedge \forall n \in \mathscr{L}^+ \wedge a \neq 0,\ a^{-n} = 1/a^n$.

EXAMPLE 2: Rewrite $1^{-1}, (2^0 + 1)^{-5}$ using Definitions (6-19) and (6-20).

Solution: $1^{-1} = \dfrac{1}{1^1} = \dfrac{1}{1} = 1$.

$$(2^0 + 1)^{-5} = (1 + 1)^{-5} = (2)^{-5} = \frac{1}{2^5} = \frac{1}{32}$$

Before defining fractional exponents, it is necessary to define what we mean by something called a root.

Definition (6-19): $\forall x \in \mathscr{Q},\ n \in \mathscr{L},\ x$ is called *an nth root of a* iff $x^n = a$.

Definition (6-20): $\forall x \in \mathscr{Q},\ n \in \mathscr{L}$, the **principal n^{th} root of a**, denoted $\sqrt[n]{a}$ or $a^{1/n}$, is x iff $x^n = a$ and

(i) n even $\Rightarrow a \geq 0 \wedge x \geq 0$;
(ii) n odd $\Rightarrow (a \geq 0 \wedge x \geq 0) \vee (a < 0 \wedge x < 0)$.

Note that $\sqrt[2]{a}$ is most often written as $\sqrt{a}$ without the index 2. The effect of this definition can be seen in the next example. Be sure you know why each result is stated as it is.

EXAMPLE 3: Apply Definition (6-20) to rename the following: $\sqrt{25}$, $\sqrt{36},\ \sqrt[3]{8},\ \sqrt[3]{-27},\ \sqrt[4]{16},\ \sqrt[5]{-32},\ \sqrt[2]{x^2},\ \sqrt[4]{-16}$.

Solution: Respectively: 5, 6, 2, ⁻3, 2, ⁻2, |x|, not possible with numbers discussed to this point.

Note that $\sqrt{a} \geq 0$ in all cases. That is, $\sqrt{a}$ never stands for a negative number. Note also that the article used in Definition (6-19) is *an*, indicating one or more such n^{th} roots, whereas the article of Definition (6-20) is *the*, which indicates that there is at most *one* principal root.

EXAMPLE 4: What numbers are square roots of nine? What is the principal square root?

Solution: Both 3 and ⁻3 are square roots of 9 because

$$3^2 = 9$$
$$(^-3)^2 = 9$$

The principal square root is

$$\sqrt{9} = 3$$

the nonnegative one of the two second roots.

We never write the sentence $\sqrt{9} = {}^-3$. If we want to indicate the negative second root of 9, we write

$$-\sqrt{9} = {}^-3$$

If we have an open sentence in which values of the variable might be negative, we write

$$\sqrt{x^2} = |x|$$

to indicate the principal second root. Compare with Example 3.

Two propositions which provide more information about rational numbers with integral exponents can now be stated.

Theorem (6h-6): $\forall a \in \mathcal{Q} \wedge a \neq 0 \wedge \forall m, n \in \mathcal{Z}, \dfrac{a^m}{a^n} = a^{m-n}$.

EXAMPLE 5: Rewrite $\dfrac{2^5}{2^2}$ according to Theorem (6h-6).

Solution: $\dfrac{2^5}{2^2} = 2^{5-2} = 2^3 = 8$.

EXAMPLE 6: Evaluate $\dfrac{5^2}{5^0}$.

Solution: We can write either

$$\dfrac{5^2}{5^0} = 5^{2-0} = 5^2 = 25, \text{ or } \dfrac{5^2}{5^0} = \dfrac{5^2}{1} = 5^2 = 25.$$

EXAMPLE 7: Evaluate $\dfrac{7^{49}}{7^{51}}$.

Solution: $\dfrac{7^{49}}{7^{51}} = 7^{49-51} = 7^{-2} = \dfrac{1}{7^2} = \dfrac{1}{49}$.

Theorem (6h-7): $\forall \dfrac{a}{b} \in \mathcal{Q},\ m \in \mathcal{L},\ \left(\dfrac{a}{b}\right)^m = \dfrac{a^m}{b^m}$.

EXAMPLE 8: Rewrite $\left(\dfrac{4}{9}\right)^{1/2}$ according to Theorem (6h-7).

Solution: $\left(\dfrac{4}{9}\right)^{1/2} = \dfrac{4^{1/2}}{9^{1/2}} = \dfrac{2}{3}$. Note: $-\dfrac{2}{3}$ is *not* correct.

EXAMPLE 9: Make a roster of $\{x \mid x^2 = 25 \wedge x \in \mathcal{Q}\}$.

Solution:

1. $x^2 = 25$
2. $\sqrt{x^2} = \sqrt{25}$
3. $|x| = 5$
4. $x \geq 0 \Rightarrow x = 5,\ x < 0 \Rightarrow x = {}^-5$
5. $x = 5 \vee x = {}^-5$

Step 2 is the step we were warned about in discussing a previous theorem. What we are saying in Step 2 is that $a^2 = b^2 \Rightarrow a = b$, which does not hold in general. However this principle holds *provided* $a \geq 0 \wedge b \geq 0$.

Theorem (6h-8): For any number set containing negatives,
$$\forall a, b \geq 0 \wedge a^2 = b^2 \Rightarrow a = b$$

Using the concepts of the preceding definitions, we can define what we mean by a fractional exponent.

Definition (6-21): $\forall a \in \mathcal{Q} \wedge \forall m, n \in \mathcal{L} \wedge n \neq 0,\ a^{m/n} = (a^m)^{1/n}$.

EXAMPLE 10: Rewrite $8^{2/3}$ using Definition (6-21).

Solution: $8^{2/3} = (8^2)^{1/3} = 64^{1/3} = 4$.

EXAMPLE 11: Rewrite $({}^-2)^{2/2}$.

Solution: $({}^-2)^{2/2} = (({}^-2)^2)^{1/2} = 4^{1/2} = 2$.

There are two traps you can fall into if you are not careful in doing this example. For one thing, you should resist the temptation to reduce $2/2$ to 1 and thereby obtain ${}^-2$ for the answer. Second, you must be careful to use the definition correctly. We do not write $({}^-2)^{2/2} = (({}^-2)^{1/2})^2$ because there is no rational number x such that $x^2 = {}^-2$. Thus $({}^-2)^{1/2}$ has no meaning in $\mathcal{Q}$.

EXERCISE SET

1. Why is no equivalent for Axiom Q-12 needed for the integers $\mathscr{Z}$? That is, how do we know
   a) that $\mathscr{Z}$ is nonempty?
   b) that there are, in fact, at least two distinct elements of $\mathscr{Z}$?

2. Solve for x.
   a) $3 + x = 7$
   b) $3 - x = 9$
   c) $3x = 16$
   d) $5(^-x) = 2$
   e) $2x + 5 = 13$
   f) $3 - 4x = {}^-17$
   g) $5x + 7 = 3x + 8$

3. Make a roster of each set, if possible.
   a) $\{x \mid x \in \mathscr{Q} \wedge x^2 - 5x = 0\}$
   b) $\{x \mid x \in \mathscr{Q} \wedge x^2 = 1\}$
   c) $\{x \mid x \in \mathscr{Q} \wedge x^2 - 5x + 6 = 0\}$
   d) $\{x \mid x \in \mathscr{Q} \wedge 2x^2 - 7x - 15 = 0\}$
   e) $\{x \mid x \in \mathscr{Q} \wedge 6x^2 - 13x + 6 = 0\}$
   f) $\{x \mid x \in \mathscr{Q} \wedge 2x^3 - 32x = 0\}$
   g) $\{x \mid x \in \mathscr{Q} \wedge (5x + 2)(4x^2 - 1) = 0\}$
   h) $\{x \mid x \in \mathscr{Q} \wedge |2x + 4| = 9\}$

4. Reduce each expression to an integer if possible.
   a) $8 \cdot 2^{-2}$
   b) $(8(2)^{-2})^0$
   c) $4^{1/2}$
   d) $-(4^{1/2})$
   e) $(-4)^{1/2}$
   f) $\sqrt{(-4)^2}$
   g) $(-1)^{2/2}$
   h) $(x^4 y^2)^{1/2}$

5. Simplify as much as possible.
   a) $(3/4)^{0/2}$
   b) $(3/4)^{1/2}$
   c) $(16)^{3/4}$
   d) $4/8^{-2/3}$
   e) $15^2/21^3$
   f) $(4/9)^{-1/2}$
   g) $(a^{1/2} + b^{1/2})^2$
   h) $(9/4)^{-3/2}$
   i) $(a^{-1} + b^{-1})^2$
   j) $(x^2 y^3 / z^0)^n$

6. Prove: $\forall a \in \mathscr{Q}, a/1 = a$.

7. Prove:
   a) (6h-6)
   b) (6h-7)
   c) (6h-8)

§6i DIVISION WITH RATIONALS

In this section many important properties of rational fractions are stated. Most proofs are omitted but may be done as exercises.

Theorem (6i-1): $\forall a, b, c, d \in \mathcal{Q} \wedge b \neq 0 \wedge d \neq 0,$

$$\frac{a}{b} = \frac{c}{d} \text{ iff } ad = bc$$

EXAMPLE 1: Show that $\frac{3}{8} = \frac{9}{24}$.

Solution: $\frac{3}{8} = \frac{9}{24}$ since $3 \cdot 24 = 8 \cdot 9$.

It turns out that multiplication of fractions is somewhat simpler than addition of fractions. The following proposition establishes the most important principle governing products of fractions.

Theorem (6i-2): $\forall \frac{a}{b}, \frac{c}{d} \in \mathcal{Q}, \frac{a}{b} \cdot \frac{c}{d} = \frac{ac}{bd}$.

In words this proposition states that to multiply two rational fractions we multiply numerators and multiply denominators.

EXAMPLE 2: Show that $\frac{3}{8} \cdot \frac{5}{7} = \frac{15}{56}$.

Solution: $\frac{3}{8} \cdot \frac{5}{7} = \frac{3 \cdot 5}{8 \cdot 7} = \frac{15}{56}$.

Perhaps the most useful rule for fraction computation is the one that states: You may change a fraction to an equivalent fraction by multiplying both numerator and denominator by the same nonzero number. The proof of that theorem depends very much on the fact that $x/x = 1$ for all nonzero x.

Theorem (6i-3): $\forall x \in \mathcal{Q} \wedge x \neq 0, \frac{x}{x} = 1$.

Proof:

1. $x \div x = y$ iff $x = xy$ 1. Definition of division, (6-3a)
2. $x = x \cdot 1$ 2. Z-11, applied to $\mathcal{Q}$
3. $x \cdot 1 = xy$ 3. Transitivity
4. $1 = y$ 4. Cancellation, (6f-6) for $\mathcal{Q}$

5. $\dfrac{x}{x} = 1$ 5. Transitivity

Q. E. D.

Now the *fundamental rule of fractions* can easily be proved. The proof will be left to the exercises.

Theorem (6i-4): $\forall a, b, x \in \mathscr{D} \land x \neq 0 \land b \neq 0,$

$$\frac{a}{b} = \frac{ax}{bx}$$

EXAMPLE 3: Change $\dfrac{3}{4}$ to a fraction with denominator 28.

Solution: Since $4 \cdot 7 = 28$, we write $\dfrac{3}{4} = \dfrac{3 \cdot 7}{4 \cdot 7} = \dfrac{21}{28}$, using (6i-4).

Theorem (6i-4) is also used to reduce the value of the numerator and the denominator by removing a common factor.

EXAMPLE 4: Reduce $\dfrac{1071}{84}$ if possible.

Solution:

$$\frac{1071}{84} = \frac{51 \cdot 3 \cdot 7}{4 \cdot 3 \cdot 7} = \frac{51}{4} \qquad \text{By (6i-4)}$$

The next theorem presents the usual rule for adding fractions: "You may add fractions which have a common denominator by adding the numerators and placing the sum over the common denominator."

Theorem (6i-5): $\forall \dfrac{a}{c}, \dfrac{b}{c} \in \mathscr{D} \land c \neq 0, \dfrac{a}{c} + \dfrac{b}{c} = \dfrac{a+b}{c}.$

Proof:

1. $\dfrac{a}{c}, \dfrac{b}{c} \in \mathscr{D}$ 1. ?
2. $\exists x, y \in \mathscr{D} \ni \cdot \dfrac{a}{c} = x, \dfrac{b}{c} = y$ 2. ?
3. $a = cx, b = cy$ 3. ?
4. $a + b = cx + cy$ 4. ?
5. $\qquad = c(x + y)$ 5. ?
6. $(a + b), (x + y) \in \mathscr{D}$ 6. ?

7. $\dfrac{a+b}{c} = x+y$ 7. ?

8. $\dfrac{a+b}{c} = \dfrac{a}{c} + \dfrac{b}{c}$ 8. ?

Q. E. D.

EXAMPLE 5: Add $\dfrac{5}{8}, \dfrac{6}{8}$.

Solution: $\dfrac{5}{8} + \dfrac{6}{8} = \dfrac{5+6}{8} = \dfrac{11}{8}$.

An important corollary to the theorem provides us with justification for our standard technique of changing unlike denominators of fractions to a common denominator before adding.

Corollary (6i-5a):

$$\dfrac{a}{b}, \dfrac{c}{d} \in \mathcal{Q} \wedge b \neq 0 \wedge d \neq 0, \dfrac{a}{b} + \dfrac{c}{d} = \dfrac{ad+bc}{bd}$$

EXAMPLE 6: Add $\dfrac{5}{7}, \dfrac{3}{4}$.

Solution: $\dfrac{5}{7} + \dfrac{3}{4} = \dfrac{5 \cdot 4 + 7 \cdot 3}{28} = \dfrac{41}{28}$.

The next theorem will be used in our discussion of the division of rational numbers. In division we must always exclude the divisor from being zero.

Theorem (6i-6): $\forall a, \dfrac{a}{b} \in \mathcal{Q}, b \neq 0, \dfrac{a}{b} = 0$ iff $a = 0$

Proof:

1. $\dfrac{a}{b} = 0$ iff $a = b \cdot 0$ 1. Definition of Division
2. $b \cdot 0 = 0$ 2. (6f-5) applied to $\mathcal{Q}$
3. $a = 0$ 3. Transitivity
4. $\dfrac{a}{b} = 0$ iff $a = 0$ 4. Substitution (1, 3)

Q. E. D.

EXAMPLE 7: What can we conclude about x if we know that

$$(x^2 - x)/7 = 0?$$

Solution: Since $7 \neq 0$, we have
$$x^2 - x = 0 \qquad (6\text{i-}6)$$
whence
$$x(x - 1) = 0$$
$$x = 0 \lor x = 1 \qquad (6\text{e-}7) \text{ applied to } \mathcal{Q}$$

Theorem (6i-7): $\forall \dfrac{a}{b}, \dfrac{c}{d} \in \mathcal{Q}, c \neq 0,$
$$\dfrac{a}{b} \div \dfrac{c}{d} = \dfrac{a}{b} \cdot \dfrac{d}{c}$$

Proof:

1. $\exists x \in \mathcal{Q} \cdot \ni \cdot \dfrac{a}{b} = \dfrac{c}{d} x$ 1. Q-11
2. $\dfrac{a}{b} \div \dfrac{c}{d} = x$ 2. Definition of $\div$
3. $\dfrac{a}{b} \cdot \dfrac{d}{c} = \dfrac{d}{c}\left(\dfrac{c}{d} x\right)$ 3. Q-4
4. $\phantom{\dfrac{a}{b} \cdot \dfrac{d}{c}} = \left(\dfrac{d}{c} \cdot \dfrac{c}{d}\right)x$ 4. Q-8
5. $\phantom{\dfrac{a}{b} \cdot \dfrac{d}{c}} = \dfrac{dc}{cd} x$ 5. (6i-2)
6. $\phantom{\dfrac{a}{b} \cdot \dfrac{d}{c}} = 1 \cdot x$ 6. (6i-3)
7. $\phantom{\dfrac{a}{b} \cdot \dfrac{d}{c}} = x$ 7. (6h-2)
8. $\dfrac{a}{b} \div \dfrac{c}{d} = \dfrac{a}{b} \cdot \dfrac{d}{c}$ 8. Transitivity (2, 7)

Q. E. D.

Thus, to divide two rationals, simply replace the nonzero divisor by its multiplicative inverse and multiply the resulting rationals.

EXAMPLE 8: Divide $\dfrac{3}{7}$ by $\dfrac{4}{9}$.

Solution: $\dfrac{3}{7} \div \dfrac{4}{9} = \dfrac{3}{7} \cdot \dfrac{9}{4} = \dfrac{3 \cdot 9}{7 \cdot 4} = \dfrac{27}{28}$.

EXAMPLE 9: Make a roster of $A = \left\{x \,|\, x \in \mathcal{Q} \land \dfrac{5x}{4} + 7 = 18\right\}$.

Solution:

$$\frac{5x}{4} + 7 = 18$$

$$\frac{5x}{4} + 7 = 11 + 7$$

$$\frac{5x}{4} = 11$$

$$\frac{4}{5} \cdot \frac{5}{4} x = \frac{4}{5} \cdot 11$$

$$x = \frac{4}{5} \cdot \frac{11}{1}$$

$$x = \frac{44}{5}$$

$$A = \left\{ \frac{44}{5} \right\}$$

The *trichotomy principle* for rational numbers like the trichotomy principle for integers is a direct result of the *less than* definition and Axiom Q-13.

Theorem (6i-8): $\forall x, y \in \mathcal{Q}$, exactly one of $x = y, x < y, x > y$ holds.

By using the following theorem with (6i-1), we have a relatively easy way to tell which of the three relationships allowed by the trichotomy property is the correct one for any arbitrary pair of rational fractions.

Theorem (6i-9): $\forall \frac{a}{b}, \frac{c}{d} \in \mathcal{Q} \wedge b > 0 \wedge d > 0$,

$$\frac{a}{b} < \frac{c}{d} \longleftrightarrow ad < bc$$

Hence, for any two fractions $a/b, c/d$ where $b > c \wedge d > 0$, we have

(1) $\qquad \frac{a}{b} < \frac{c}{d}, \frac{a}{b} = \frac{c}{d},$ or $\frac{a}{b} > \frac{c}{d}$

depending on whether

(2) $\qquad ad < bc, \quad ad = bc, \quad \text{or} \quad ad > bc$

respectively.

EXAMPLE 10: Relate $\frac{9}{11}, \frac{17}{21}$ according to the possibilities allowed by the trichotomy property.

Solution:

$$9 \cdot 21 = 189$$

and

$$11 \cdot 17 = 187$$

Thus

$$9 \cdot 21 > 11 \cdot 17$$

whence

$$\frac{9}{11} > \frac{17}{21} \qquad (6\text{i-}9)$$

A very interesting property called *denseness* can now be demonstrated. A number set is said to be dense when between any two elements of the set there is another number.

Theorem (6i-10): $\forall a, b \in \mathcal{Q} \land a < b,$

$$\exists x = (a+b)/2 \in \mathcal{Q} \cdot \ni \cdot a < (a+b)/2 < b.$$

Proof:

1.	$a, b \in \mathcal{Q}, a < b$	1.	C. P.
2.	$a + a < a + b \land a + b < b + b$	2.	(6f-8) applied to $\mathcal{Q}$
3.	$2a < a + b \land a + b < 2b$	3.	$x + x = 2x$ applied to $\mathcal{Q}$
4.	$\frac{1}{2}(2a) < \frac{a+b}{2} \land \frac{a+b}{2} < \frac{1}{2}(2b)$	4.	(6f-9) applied to $\mathcal{Q}$
5.	$1 \cdot a < \frac{a+b}{2} \land \frac{a+b}{2} < 1 \cdot b$	5.	(6h-5)
6.	$a < \frac{a+b}{2} \land \frac{a+b}{2} < b$	6.	(6h-2)
7.	$a < \frac{a+b}{2} < b$	7.	Meaning of "between" (1), §6g

Q. E. D.

This proposition states that the number $(a + b)/2$ is between the numbers a, b. Thus the set of rational numbers is dense.

$\mathcal{Q}$ is, in fact, the first number set that we have considered which is dense. For $a \in \mathcal{N}$ or for $a \in \mathcal{L}$, there is no number between a and $a + 1$, for example, which shows that $\mathcal{N}$ and $\mathcal{L}$ are *not* dense.

§6i   Division with Rationals

We shall see that the following proposition is also important in showing how the rational numbers differ from other number sets.

Theorem (6i-11): There is no $x \in \mathcal{Q} \cdot \ni \cdot x^2 = 2$.

Proof:

1.	$x \in \mathcal{Q}$	1.	I. P.
2.	$x = \dfrac{a}{b}$, gcd $(a, b) = 1$	2.	(6i-4)
3.	$x^2 = 2$	3.	Hypothesis
4.	$\left(\dfrac{a}{b}\right)^2 = 2$	4.	Transitivity
5.	$\dfrac{a^2}{b^2} = 2$	5.	(6h-7)
6.	$a^2 = 2b^2$	6.	Definition of $\div$
7.	a^2 is even	7.	Definition of even number
8.	a^2 is even $\Rightarrow a$ is even	8.	Exercise 13, §2i
9.	Let $a = 2r$	9.	Definition of even number
10.	$a^2 = 4r^2$	10.	(6g-11) applied to $\mathcal{Q}$
11.	$4r^2 = 2b^2$	11.	Transitivity (6, 10)
12.	$2r^2 = b^2$	12.	Cancellation
13.	b^2 is even	13.	Definition of even number
14.	b^2 is even $\Rightarrow b$ is even	14.	See Step 8
15.	$b = 2s$	15.	Definition of even number
16.	$x = \dfrac{a}{b} = \dfrac{2r}{2s}$	16.	Substitution
17.	gcd $(a, b) = 2$	17.	Definition of gcd
18.	gcd $(a, b) = 1 \wedge$ gcd $(a, b) \neq 1$	18.	C. A.
19.	$x \notin \mathcal{Q}$	19.	I. I.

                                                                                  Q. E. D.

EXERCISE SET

1. Perform the indicated operations.

    a) $\dfrac{2}{3} + \dfrac{3}{4}$ b) $2 + \dfrac{4}{9}$

c) $\dfrac{3}{7} - \dfrac{4}{7}$ d) $\dfrac{5}{17} + \dfrac{2}{3}$

e) $\dfrac{2}{5} \cdot \dfrac{7}{4}$ f) $\left\{\left[\left(1 + \dfrac{1}{2}\right) + \dfrac{1}{3}\right] + \dfrac{1}{4}\right\} + \dfrac{1}{5}$

g) $\dfrac{3}{8} \cdot \left(2 + \dfrac{3}{8}\right)$ h) $\dfrac{5}{7} \div \left(\dfrac{3}{4} \div \dfrac{5}{6}\right)$

i) $\left(\dfrac{5}{7} \div \dfrac{3}{4}\right) \div \dfrac{5}{6}$ j) $\dfrac{2}{3}\left[\left(\dfrac{3}{4} + \dfrac{7}{8}\right) \div \dfrac{5}{4}\right]$

2. Place one of the three symbols allowed by the trichotomy principle between each pair of numbers.

a) $\dfrac{5}{6}, \dfrac{6}{7}$ b) $\dfrac{6}{7}, \dfrac{7}{8}$

c) $\dfrac{4}{5}, \dfrac{22}{25}$ d) $\dfrac{17}{19}, \dfrac{13}{16}$

e) $\dfrac{17}{51}, \dfrac{99}{297}$ f) $\dfrac{8}{-11}, \dfrac{-40}{43}$

g) $\dfrac{3}{2} + \dfrac{4}{9}, \dfrac{2}{9} + \dfrac{31}{18}$ h) $\dfrac{1}{3} + \dfrac{1}{4}, \dfrac{3}{4} - \dfrac{1}{3}$

i) $\dfrac{2}{5} + \dfrac{1}{4}, \dfrac{13}{20}$

3. Arrange the numbers in ascending order.

a) $\dfrac{-5}{-8}, -2, \dfrac{5}{8}, \dfrac{4}{-7}, 0$

b) $\dfrac{17}{23}, \dfrac{105}{79}, 0, -\dfrac{2}{5}, \dfrac{7}{-9}$

4. Reduce each fraction to lowest terms (i.e., so that numerator and denominator have no common factor other than one).

a) $\dfrac{6}{9}$ b) $\dfrac{-8}{12}$

c) $\dfrac{28}{32}$ d) $\dfrac{231}{385}$

e) $\dfrac{65}{85}$ f) $\dfrac{238}{245}$

5. When first learning fractions, school children sometimes try to add them according to the rule

$$\dfrac{a}{b} + \dfrac{c}{d} = \dfrac{a+c}{b+d}$$

Show by numerical example that this proposition is false.

6. Supply the missing "reasons" in the proof of (6i-5).
7. Prove the following:
    a) $\forall a \in \mathcal{Q} \cdot \ni \cdot 0 < a < 1, \frac{1}{a} > 1$
    b) There is no smallest positive rational number.
    c) $\forall a, b, c \in \mathcal{Q} \land c \neq 0, \frac{a}{c} = \frac{b}{c} \Rightarrow a = b$.

*8. Prove:
    a) (6i-1)   b) (6i-2)
    c) (6i-4)   d) (6i-5a)
    e) (6i-8)   f) (6i-9)

§6j DECIMALS

Before taking up a way of writing rational numbers different from that presented in the preceding sections, we list some of the important results which we have obtained.

(6h-2): $\forall a \in \mathcal{Q}, \exists 1 \in \mathcal{Q} \cdot \ni \cdot a = a \cdot 1$

(6h-3): $0 \neq 1$

(6h-4): $1 > 0$

(6h-5): $\forall a \in \mathcal{Q}, a \neq 0 \land$ for $1 \in \mathcal{Q}, \exists \frac{1}{a} \in \mathcal{Q} \cdot \ni \cdot 1 = a \cdot \frac{1}{a}$

(6h-6): $\forall a, b \in \mathcal{Q} \cdot \ni \cdot a \geq 0 \land b \geq 0, a^2 = b^2 \Rightarrow a = b$

(6i-1): $\forall a, b, c, d \in \mathcal{Q}, b \neq 0, d \neq 0, \frac{a}{b} = \frac{c}{d}$ iff $ad = bc$

(6i-2): $\vee \frac{a}{b}, \frac{c}{d} \in \mathcal{Q}, \frac{a}{b} \cdot \frac{c}{d} = \frac{ac}{bd}$

(6i-3): $\forall x \in \mathcal{Q} \land x \neq 0, \frac{x}{x} = 1$

(6i-4): $\forall a, b, x \in \mathcal{Q}, \frac{a}{b} = \frac{ax}{bx}, x \neq 0, b \neq 0$

(6i-5): $\vee \frac{a}{c}, \frac{b}{c} \in \mathcal{Q}, \frac{a}{c} + \frac{b}{c} = \frac{a+b}{c}$

(6i-5a): $\vee \frac{a}{b}, \frac{c}{d} \in \mathcal{Q} \land b \neq 0 \land d \neq 0, \frac{a}{b} + \frac{c}{d} = \frac{ad + bc}{bd}$

(6i-6): $\forall a, \frac{a}{b} \in \mathcal{Q}, \frac{a}{b} = 0$ iff $a = 0$

(6i-7): $\vee \frac{a}{b}, \frac{c}{d} \in \mathcal{Q} \land c \neq 0, \frac{a}{b} \div \frac{c}{d} = \frac{a}{b} \cdot \frac{d}{c}$

(6i-8): $\forall x, y \in \mathcal{Q}$, exactly one of $x = y, x < y, x > y$ holds

(6i-9): $\forall \frac{a}{b}, \frac{c}{d} \in \mathscr{Q} \wedge b > 0, d > 0, \frac{a}{b} < \frac{c}{d}$ iff $ad < bc$

(6i-10): $\forall a, b \in \mathscr{Q} \wedge a < b, \exists x = \frac{a+b}{2} \in \mathscr{Q} \cdot \ni \cdot a < \frac{a+b}{2} < b$.

(6i-11): There is no $x \in \mathscr{Q} \cdot \ni \cdot x^2 = 2$.

In many applications of mathematics, we need some way of writing fractions to make it possible to do calculations with relative ease. Consider, for example, the problem of writing 13/11. We can write this fraction as an integer plus a fractional remainder. That is,

$$\frac{13}{11} = 1 + \frac{2}{11}$$

Now for most calculation purposes the right side of this equation is worse than the left. However, any method which would allow us to write a fraction less than 1, such as 2/11, in a more manageable fashion would also make it possible to write

$$1 + \frac{2}{11}$$

more efficiently.

Since the main difficulty in doing addition with fractions is caused by the necessity of having a common denominator before performing the addition, it would be a powerful advantage to have a system of writing which would always leave a fraction in common denominator form. Being clever at such things, we write the fractional part of any rational number in terms of powers of a particular number called the *base*. If a_1, a_2, a_3, etc., are the digits we use to represent the integral part of a rational fraction and if c_1, c_2, c_3, etc., are the digits we use for the fractional part, then any rational number can be written as

(1) $\qquad \cdots + a_3 b^2 + a_2 b^1 + a_1 b^0 + \frac{c_1}{b^1} + \frac{c_2}{b^2} + \frac{c_3}{b^3} + \cdots$

The powers of the base b provide us with at least a partial common denominator with which to write the fractional part. Since the base b is common to every term of (1), we usually omit the powers of b and use the digits themselves so that the number of (1) becomes

(2) $\qquad \cdots a_3 a_2 a_1 . c_1 c_2 c_3 \cdots$

where the bold face plus sign is replaced by the fraction point. When the base b of (1) is 10, we say that we are writing *decimal fractions* or simply decimals. In contrast, we then call a fraction written in traditional ratio form a *common fraction*.

§6j Decimals

Definition (6-22): A **decimal fraction** is a fraction whose denominator is a power of ten (10^n) and is represented by a number with a decimal point and n digits to the right of that decimal point according to the following scheme:

$$.1 = \frac{1}{10}, .01 = \frac{1}{100}, \text{ etc.}$$

and

$$a \cdot (.1) = .a, \, a \cdot (.01) = .0a, \text{ etc.}$$

EXAMPLE 1: Represent the following common fractions as decimals: a) $\frac{2}{5}$, b) $\frac{1}{8}$.

Solution:

a) $\frac{2}{5} = \frac{2 \cdot 2}{5 \cdot 2} = \frac{4}{10} = 4 \cdot \frac{1}{10}$ or $4 \cdot (.1) = .4$.

b) $\frac{1}{8} = \frac{1 \cdot 125}{8 \cdot 125} = \frac{125}{1000} = 125 \cdot \frac{1}{1000} = .125$.

EXAMPLE 2: Change .532 to a common fraction in lowest terms.

Solution: $.532 = \frac{532}{1000} = \frac{133}{250}$

EXAMPLE 3: Add .03, .27.

Solution: $.03 = \frac{3}{100}$

$.27 = \frac{27}{100}$

$\frac{3}{100} + \frac{27}{100} = \frac{30}{100} = \frac{3}{10} = .3$

EXAMPLE 4: Multiply .07, .36.

Solution: $.07 = \frac{7}{100}$

$.36 = \frac{36}{100}$

$\frac{7}{100} \cdot \frac{36}{100} = \frac{7 \cdot 36}{100 \cdot 100} = \frac{252}{10,000} = .0252$

The algorithms for multiplicative computation with decimals are partly explained by Examples 1 through 4. Every time we multiply by ten, we move the decimal point one place to the right, and every time we divide by

294 *The Number System*

ten (multiply by 1/10), we move the decimal point one place to the left, adding zeros if necessary to indicate the correct position. Returning to Example 1, we showed that $\frac{1}{8} = .125$. Our long division algorithm makes this an easy task, as follows.

$$\begin{array}{r} .125 \\ 8\overline{)1.000} \\ \underline{8} \\ 20 \\ \underline{16} \\ 40 \end{array}$$

We added zeros to the dividend until we obtained a remainder of zero. Thus,

$$\frac{1000}{8} = 125 \quad \text{or} \quad 1000 = 8 \cdot 125$$

so

$$1000 \cdot \frac{1}{10^3} = (8 \cdot 125) \cdot \frac{1}{10^3}$$

or

$$1 = 8 \cdot (.125)$$

or

$$\frac{1}{8} = .125$$

EXAMPLE 5: Change $\frac{5}{8}$ to decimal form.
Solution:

$$\begin{array}{r} .625 \\ 8\overline{)5.000} \\ \underline{4\,8} \\ 20 \\ \underline{16} \\ 40 \\ \underline{40} \end{array}$$

Thus $\frac{5}{8} = .625$.

The success of this algorithm depends on having, after adding sufficient zeros, a zero remainder. Unfortunately, it is not always possible to obtain a zero remainder. We can solve this problem by making certain allowances or assumptions.

EXAMPLE 6: Show that $.637 = \frac{6}{10} + \frac{3}{100} + \frac{7}{1000}$.

Solution:

$$.637 = \frac{637}{1000} = \frac{600 + 30 + 7}{1000}$$
$$= \frac{600}{1000} + \frac{30}{1000} + \frac{7}{1000}$$
$$= \frac{6}{10} + \frac{3}{100} + \frac{7}{1000}$$

Thus, a decimal fraction can be represented as the sum of its successive digits divided by successive powers of ten in the manner of (1).

EXAMPLE 7: Change $\frac{1}{3}$ to a decimal.

Solution:

$$\begin{array}{r} .33 \\ 3\overline{)1.00} \\ \underline{9} \\ 10 \\ \underline{9} \\ 1 \end{array}$$

At this point we can see the futility of continuing. We shall never get a remainder of zero but will continue to get 3's in the quotient and ones as remainders.

To indicate the result found in this example, we write

(3) $$\frac{1}{3} = .33\overline{3}$$

where the bar above one or more digits indicates that those digits will continue indefinitely. Here the 3 will repeat in this way. Now by (1) we should be able to represent this as

(4) $$\frac{1}{3} = \frac{3}{10} + \frac{3}{100} + \frac{3}{1000} + \cdots + \frac{3}{10^n} + \cdots$$

In order to prove (4) we need the notions of limit and convergence. However, at this point it is better for us to accept (4) as provable. Decimals in the form of (3) are called *infinite repeating decimals*, and those which have a remainder of zero like .125 are called *terminating decimals*. Since .125 can be written as .12500... we can consider terminating decimals to be infinite repeating decimals also.

EXAMPLE 8: Write $\frac{2}{5}, \frac{1}{3}$, and 4/11 as infinite repeating decimals.

Solution: $\frac{2}{5} = 0.400\bar{0}$ \qquad See Example 1

$\frac{1}{3} = 0.33\bar{3}$ \qquad See (3)

$\frac{4}{11} = 0.36\overline{36}$ \qquad By method of Example 7

The next argument which we cannot prove here but shall simply accept is that the set of rational numbers $\mathscr{Q}$ equals the set of infinite repeating decimals $\mathscr{D}$. On applying the division algorithm, any fraction in the form a/b will terminate or repeat. If a/b does not terminate there can be at most $b-1$ different remainders. That is, if $\frac{3}{7}$ does not terminate, then the only other possible remainders are 1, 2, 3, 4, 5, and 6 which may occur in some sequence, say 26451. Now, if this division does not terminate and there are at most six possible remainders, then one of the remainders will repeat, which means that the quotient will repeat. Thus, any rational number can be represented by an infinite repeating decimal.

In Example 9 we illustrate an algorithm for changing any infinite repeating decimals to a common fraction. The algorithm depends on the ideas that multiplication by 10^n moves the decimal n places to the right and that addition with infinite decimals is the same as regular addition.

EXAMPLE 9: Change $.18\overline{18}$ to a common fraction.

Solution: Let $n = .18\overline{18}$.

Then
$$10^2 n = 18.18\overline{18}$$
and
$$10^2 n - n = 18.00\bar{0}$$
$$99n = 18$$
$$n = \frac{18}{99} = \frac{2}{11}$$

Now we can write 13/11 from the beginning of this section as
$$\frac{13}{11} = 1 + \frac{2}{11} = 1.18\overline{18}$$

The algorithm of the preceding example seems straight forward, but it does have a peculiarity that must be accounted for.

EXAMPLE 10: Change $.499\bar{9}$ to a rational number.

Solution: Let $n = .499\bar{9}$.

$$10^2 \cdot n = 49.99\overline{9}$$
$$10^2 \cdot n - n = 49.500\overline{0}$$
$$99n = 49.5$$
$$n = \frac{495}{990} = \frac{1}{2}$$

In this example we obtained $.499\overline{9} = \frac{1}{2}$, but $\frac{1}{2} = .5000\ldots$. Hence if transitivity is to hold we must have

$$.4999\ldots = .5000\ldots$$

In a branch of mathematics called the theory of limits, we find that it is perfectly possible to call such numbers equal. Hopefully, the preceding evidence has convinced everyone that $\mathscr{Q}$ the set of rational numbers and $\mathscr{D}$ the set of repeating infinite decimals are the same sets.

What about the infinite decimals which do not repeat? As an example of such a decimal consider

(5) $\qquad\qquad .101001000100001\ldots$

There is an obvious pattern which enables us to keep on writing this number, and the same pattern shows that no block of digits will repeat. From our previous argument, such a number is not rational, and we shall call such numbers *irrational*.

As we have seen, the set $\mathscr{Q}$ of rational numbers is closed under $+, \cdot, -,$ and $\div$ (with a single exception). This is generally sufficient for elementary application of mathematics but unfortunately is not good enough for some purposes. Let us consider equations of the form $x^2 - a = 0$ which can be written as

(6) $\qquad\qquad x^2 = a$

Certainly, $x^2 = 4$ can be solved. We can find an $x \in \mathscr{Q} \cdot \ni \cdot x \cdot x = 4$. However, if $a = 2$ we have some difficulty, as the next example indicates.

EXAMPLE 11: Solve $x^2 = 2$ for x.

Solution: $\sqrt{2}$ is not known. If we square $\frac{3}{2}$, we obtain $(1.5)^2 = (\frac{3}{2})^2 = \frac{9}{4} > 2$. Hence $1.5 > \sqrt{2}$. Since $\sqrt{2} < 1.5$, we try 1.4. Squaring 1.4 yields 1.96. Thus

$$1.96 < 2 < 2.25$$

or

$$(1.4)^2 < 2 < (1.5)^2$$

Since

$$1.4 < \sqrt{2} < 1.5$$

we do not yet have an exact decimal whose square is 2. We can continue our efforts to find a decimal expansion of $\sqrt{2}$ by using the process we have started as follows.

$$(1.41)^2 < 2 < (1.42)^2$$
$$(1.414)^2 < 2 < (1.415)^2$$
$$(1.4142)^2 < 2 < (1.4143)^2$$

From this last step we can see that

$$\sqrt{2} \neq 1.414\ldots$$

However, this process seems to indicate that the x of $x^2 = 2$ has a decimal expansion that does not terminate. More important, it does not seem to repeat. That is, this x does not seem to be rational. We have, in fact, *proved* that this x is not a rational number. See Theorem (6i-11).

Let us try to associate rational numbers with some of the points of the number line of Figure 6.5. It is not too hard to associate $\frac{1}{2}$ with the midpoint of OA, $\frac{3}{2}$ with the midpoint of AB, and so forth. See Figure 6.6. Other rationals such as $\frac{7}{8}$ could be associated in similar fashion.

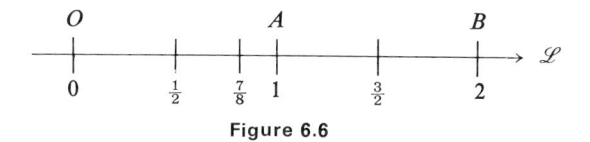

Figure 6.6

Does the process of associating fractions with points between integral points such as A, B ever end? From (6i-10) we see that the answer is no. Between any two rational numbers there is another, and another, and so on ad infinitum. This idea is illustrated in Figure 6.7.

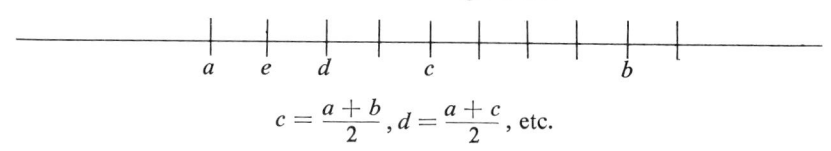

$$c = \frac{a+b}{2}, d = \frac{a+c}{2}, \text{etc.}$$

Figure 6.7

One would think that this property of denseness would allow us to associate a number with each point of the line, but it does not. Recall that $\mathscr{Q}$ lacks solutions to certain equations. There was no solution for

$$x^2 = 2$$

for example. That is, the number $x = \sqrt{2}$ of this equation is not rational. It is not too difficult to show that a point *does* exist on the number line which correspnds to $\sqrt{2}$.

§6j *Decimals*

Thus the number extensions

$$\mathscr{P} \to \mathscr{N}$$
$$\mathscr{N} \to \mathscr{Z}$$
$$\mathscr{Z} \to \mathscr{Q}$$

have failed to label all the points on our number line. We are naturally led to still another extension in order to have numbers to label each of the points we can find on our line and to obtain solutions for equations which result from using involution.

In Figure 6.7 we can easily see one way in which the number set $\mathscr{Q}$ differs from $\mathscr{Z}, \mathscr{N}$. In our continuous effort to extend our various number sets, *we have completely lost the property of induction* (counting). Not only does $\mathscr{Q}$ have no nonsuccessor, there is also no unique successor for any rational number whatever. Let us suppose that b is the successor of a. We have found that c lies between a, b. Hence b is not the successor of a. For the same reason c is not the successor of a. The number a has no unique successor.

EXERCISE SET

1. Rename the following numbers.

    a) $(.01 + .1) + .03$
    b) $(.02 + .07) - .08$
    c) $(.01) \cdot (.01)$
    d) $(.03 + .005) \cdot (.7)$
    e) $(.03) - (.025)$
    f) $[(.7 + 3.5) \cdot (4.2)] - 7$

2. Change to decimals.

    a) $\frac{1}{7}$
    b) $\frac{3}{5}$
    c) $\frac{1}{6}$
    d) $\frac{7}{9}$
    e) $\frac{1}{16}$
    f) $\frac{4}{9}$

3. Change to common fractions.

    a) $.33$
    b) $.67$
    c) $.333\ldots$
    d) $.666\ldots$
    e) $.7373\overline{73}$
    f) $.0101\overline{01}$

4. Label the following as rational or irrational.

    a) $.3131\overline{31}$
    b) $.1234567\ldots n, n+1, \ldots$
    c) $x \cdot \ni \cdot x^2 = 9$
    d) $.101001000\ldots$
    e) $x \cdot \ni \cdot x^2 = 5$
    f) $x \cdot \ni \cdot x^3 = 3$

§6k THE REAL NUMBER SYSTEM

Before we develop a number system which allows us to associate a number with each point of the number line, certain concepts and definitions are needed. In particular, we need to develop the concept of least upper bound.

Definition (6-23): A set S of numbers is **bounded** iff there is a positive number $b \cdot \ni \cdot \forall s \in S, |s| \leq b$. b is called the **upper bound** of S.

Definition (6-24): b is called the **least upper bound** (lub) of S iff b is an upper bound of S and $b \leq y$ where y is any other upper bound of S.

EXAMPLE 1: What are the bounds of $A = \{x \mid x \in \mathscr{L} \wedge x < 5\}$?

Solution: All integers greater than or equal to 4 are bounds of A, and 4 is the lub.

EXAMPLE 2: What are the bounds of $B = \{x \mid x \in \mathscr{Q}^+ \wedge x^2 < 2\}$?

Solution: In Example 11, §6j, we saw that
$$x^2 = 2 \Rightarrow x = \sqrt{2}$$
The upper bounds of B are numbers greater than or equal to $\sqrt{2}$, and the lub is $\sqrt{2}$.

Note that we have proved that $\sqrt{2}$ is not a rational number; so in this case the least upper bound of the set B is not an element of the set. The stage is now set for our new number set.

The *real number system* has as primitive terms $\mathscr{R}, +, \cdot$, and the following axioms.

R-1: $\forall a, b \in \mathscr{R}, \exists x \in \mathscr{R} \cdot \ni \cdot a + b = x$

R-2: $\forall a, b \in \mathscr{R}, \exists x \in \mathscr{R} \cdot \ni \cdot a \cdot b = x$

R-3: $\forall a, b, c \in \mathscr{R}, a = b \Rightarrow a + c = b + c$

R-4: $\forall a, b, c \in \mathscr{R}, a = b \Rightarrow a \cdot c = b \cdot c$

R-5: $\forall a, b \in \mathscr{R}, a + b = b + a$

R-6: $\forall a, b \in \mathscr{R}, a \cdot b = b \cdot a$

R-7: $\forall a, b, c \in \mathscr{R}, (a + b) + c = a + (b + c)$

R-8: $\forall a, b, c \in \mathscr{R}, (a \cdot b) \cdot c = a \cdot (b \cdot c)$

R-9: $\forall a, b, c \in \mathscr{R}, a \cdot (b + c) = a \cdot b + a \cdot c$

R-10: $\forall (a, b) \in \mathscr{R}, \exists x \in \mathscr{R} \cdot \ni \cdot a = b + x$

R-11: $\forall (a, b) \in \mathscr{R} \wedge a \neq b + a, \exists x \in \mathscr{R} \cdot \ni \cdot a = b \cdot x$

R-12: $\exists x \in \mathscr{R} \land \exists y \in \mathscr{R} \cdot \ni \cdot x \neq y$

R-13: $\exists$ a subset $\mathscr{R}^+$ of $\mathscr{R} \cdot \ni \cdot \forall a, b \in \mathscr{R}$, exactly one of the following holds:
$a = b$
$\exists x \in \mathscr{R}^+ \cdot \ni \cdot a = b + x$
$\exists y \in \mathscr{R}^+ \cdot \ni \cdot a + y = b$

R-14: $\forall a, b \in \mathscr{R}^+, \exists x \in \mathscr{R}^+ \cdot \ni \cdot a + b = x$

R-15: $\forall a, b \in \mathscr{R}^+, \exists x \in \mathscr{R}^+ \cdot \ni \cdot a \cdot b = x$

R-16: Any bounded set of real numbers has a least upper bound b where b is also a real number.

This axiom set is the same as that for the rational number system $\mathscr{Q}$ except for R-16. For this reason we know that the rational number set $\mathscr{Q}$ is a subset of the real number set $\mathscr{R}$. Any new numbers can come into the set $\mathscr{R}$ only as a result of R-16. As Example 2 should indicate, these new numbers are the infinite nonrepeating decimals, the irrational numbers among which are the numbers like $\sqrt{2}$. That is,

$$R = \{x \mid x \text{ is an infinite decimal}\}$$
$$= \{x \mid x \text{ is repeating}\} \cup \{x \mid x \text{ is nonrepeating}\}$$

The various parts of the following theorem (stated without proof) provide some of the principles used in coping with the computational peculiarities of these new irrational numbers.

Theorem (6k-1): $\forall \sqrt{a}, \sqrt{b} \in \mathscr{R}$,
(i) $\sqrt{a} + \sqrt{a} = 1 \cdot \sqrt{a} + 1 \cdot \sqrt{a} = (1 + 1)\sqrt{a} = 2\sqrt{a}$
(ii) $\sqrt{a} \cdot \sqrt{b} = \sqrt{a \cdot b}$

EXAMPLE 3: Rename the following real numbers, if possible, according to Theorem (6k-1).

a) $\sqrt{5} + \sqrt{5}$ b) $\sqrt{5} + \sqrt{6}$ c) $5\sqrt{7}$ d) $\sqrt{8} \cdot \sqrt{6}$

Solution:

a) $2\sqrt{5}$ b) $\sqrt{5} + \sqrt{6}$ c) $5\sqrt{7}$ d) $\sqrt{48}$

All the theorems, except those which depend on mathematical induction, that were proved for the integers and for the rationals now hold for the real numbers. Thus the set $\mathscr{R}$ of real numbers is closed for $+, \cdot, -, \div$. In addition it has numbers x such that $x^2 = a$ for any real *nonnegative a*.

All the problems which were posed in the beginning of this chapter have been solved. We now have a system that is extremely useful and is sufficient for a vast number of problems which occur in the physical world. Any sighs

of relief are premature, however. Someone is bound to want numbers x such that
$$x^2 = -a \wedge a > 0$$
for some obscure problem somewhere. We must then note that
$$x^2 = -2$$
would have to have a solution
$$|x| = \sqrt{-2}$$
However, this would require that $\sqrt{-2}$ be a real number having the property that $(\sqrt{-2})^2 < 0$. No such real number exists. That is, there is no real number whose square is negative. Thus the real number system does *not* possess the property of *quadratic* solvability. We must make still another extension to a new number set in order to obtain a number set which has this property.

We shall not state the axioms for this next set $\mathscr{C}$ of complex numbers which does possess quadratic solvability, but shall give an indication of what kinds of numbers it has and what axioms it would need. The *complex number system* has primitives $\mathscr{C}, +, \cdot$, and properties which correspond to R-1 through R-12 and R-16. The set of complex numbers can be described as follows:

(1) $\qquad \exists i \in \mathscr{C} \cdot \ni \cdot \mathscr{C} = \{a + bi \mid a, b \in \mathscr{R} \wedge i^2 = -1\}$

With this concept $x^2 = -a$ can be solved when $a > 0$.

EXAMPLE 4: Make a roster of $A = \{x \mid x \in \mathscr{C} \wedge x^2 = -5\}$.

Solution:
$$x^2 = -5$$
$$x^2 = 5i^2$$
$$x^2 - 5i^2 = 0$$
$$(x + i\sqrt{5})(x - i\sqrt{5}) = 0$$
$$x = i\sqrt{5} \vee x = -i\sqrt{5}$$

Therefore,
$$A = \{i\sqrt{5}, -i\sqrt{5}\}$$

With the introduction of complex numbers we have, in fact, the ability to solve any equation of the form

(2) $\qquad ax^2 + bx + c = 0 \wedge a \neq 0$

This is called a quadratic equation, and by using the method of completion of squares, a formula called the quadratic formula can be deduced which renames the variable x in terms of the coefficients a, b, c.

$$\text{(3)} \qquad x = \frac{-b \pm \sqrt{b^2 - 4ac}}{2a}$$

In the set $\mathscr{C}$ both values of x will always exist, whereas in $\mathscr{R}$ they do *not* exist if $b^2 - 4ac < 0$.

Each time we have extended the number system to gain some desired new property, we lost at least one of the original properties. Hence, we have every reason to ask what we have lost in making the extension

$$\mathscr{R} \to \mathscr{C}$$

You may remember that we stated that $\mathscr{C}$ was only to obey R-1 through R-12 and R-16 for $\mathscr{R}$ in addition to having the property of (1). What about the order axioms R-13 through R-15?

These order axioms unfortunately must be sacrificed in order to obtain the property of (1). That is, $\mathscr{C}$ is not an ordered set. See Exercise 4. In spite of this loss of the order property, it is valuable to have solutions to all the quadratic equations of (2).

Can we now relax and say, "The job of extending the number system is complete; this set does everything"? Tomorrow, someone may find a need for an entirely new number system. We can meet the challenge.

The following table summarizes and compares some of the important basic properties for each number set we have studied in this chapter except a general counting set $\mathscr{P}$.

Properties	$\mathscr{N}$	$\mathscr{L}$	$\mathscr{Q}$	$\mathscr{R}$	$\mathscr{C}$
Closure $+$, $\times$; commutatity $+$, $\times$; associativity $+$, $\times$; uniqueness of sum and product; distributivity of $\times$ over $+$.	yes	yes	yes	yes	yes
Inductive property (counting)	yes	no	no	no	no
Order ($<$)	yes	yes	yes	yes	no
Solvability $+$ (linear)	no	yes	yes	yes	yes
Solvability $\times$ (linear)	no	no	yes	yes	yes
Completeness	no	no	no	yes	yes
Quadratic solvability	no	no	no	no	yes

Figure 6.8

EXERCISE SET

1. Rename each number.

    a) $\sqrt{-18}$ 	 b) $\sqrt{y^2}$

c) $\sqrt[3]{-54}$
e) $2\sqrt{7} + \sqrt{63}$
g) i^3
d) $\sqrt{3} \cdot \sqrt{6}$
f) $(a + bi)(a - bi)$
h) i^4

2. Name the bounds of the following sets.
   a) $\{x \mid x \in \mathscr{I} \wedge x^2 < 0\}$
   b) $\{x \mid x \in \mathscr{I} \wedge x + 5 < 8\}$
   c) $\{x \mid x \in \mathscr{Q} \wedge x^2 + 2x + 1 < 5\}$
   d) $\{x \mid x \in \mathscr{R} \wedge x + 5 < 8\}$

3. Make a roster of each solution set.
   a) $\{x \mid x \in \mathscr{C} \wedge x^2 + x + 1 = 0\}$
   b) $\{x \mid x \in \mathscr{N} \wedge (x + 5)(x - 2) = 0\}$
   c) $\{x \mid x \in \mathscr{I} \wedge (x - 7)(x + 4) = 0\}$
   d) $\{x \mid x \in \mathscr{Q} \wedge (4x + 7)(3x - 4) = 0\}$
   e) $\{x \mid x \in \mathscr{R} \wedge x^2 + 5x + 2 = 0\}$

*4. We have stated that $\mathscr{C}$ is not ordered. Recall that for an ordered set the trichotomy property, $a < b$ or $a = b$ or $a > b$ exclusively, must hold. Show that $\mathscr{C}$ is not ordered by setting $a = i$, $b = 0$.

7
NUMERATION

In the preceding chapter we studied properties of our number system. That is, we investigated how basic axioms govern our numbers and the way we use them. However, it would be impossible to make much use of numbers without having a way of writing them which makes rapid and easy calculation possible. A written symbol for a number name is called a *numeral*. A systematic method for representing each number of an entire set by a particular numeral is called a *numeration system*. Sometimes people have a tendency to confuse properties of numbers with principles which are actually due to the numeration system we use. Hence, a study of our system can be very helpful in gaining a real understanding of the uses we make of the number concept.

§7a   EARLY NUMERATION SYSTEMS

One of the best aids in gaining a clear understanding of principles we use in writing numbers as opposed to the properties of numbers themselves is to investigate some of the numeration devices which have been used by other peoples. As far as we know, early man counted long before recording the results of his counting. When number recording first began, it was done by collecting pebbles, notching a stick, knotting a rope, or making tally marks on stone or wood. Though this may seem a crude method of recording the results of counting, early man needed nothing beyond this method for the simple kind of bartering used in the trade of the times.

As society became more sophisticated and possessions became more bountiful, it behooved man to develop a more systematic, simple, complete, and standardized method for recording "how much" and "how many" things were available. Early Egyptians, for example, were plagued by the floodings of the Nile river; since markers were not sufficient to withstand the tides, written records had to be kept. A calendar became useful in order to keep track of such things as birthdays, planting and harvest times. Keeping track of the number of cattle in the herd, the number of wives to which each man was entitled, etc., were important matters for which accurate records were needed.

Archaeological evidence indicates that the recording of numbers first began over five thousand years ago. The oldest known systems are those of the Egyptians and of the Sumerians, the early inhabitants of Mesopotamia. Let us examine these systems of numeration in the hope that by analyzing them we may learn to appreciate the long, painstaking efforts of mankind to record number concepts. Also, we should look for characteristics of the systems with an eye to seeing how they differ from our present system.

The Egyptians used hieroglyphs for their number symbols. We find these on their tombs and other structures. Early manuscripts such as the "Rhind Papyrus" and the "Moscow Papyrus" provide us with further knowledge of Egyptian numerals. The former was written by Ahmes about 1700 B.C., and contains solutions to practical problems of the times. It is now on display in the British Museum. In Figure 7.1 the number symbols used are shown with their meanings.

Egyptian numeration

Our numeral	Egyptian numeral	Historical meaning
1	│	Stroke (tally)
10	∩	Heel bone
100	ᛞ	Scroll
1000	⌇	Lotus flower
10,000	⌒	Bent finger
100,000	ᗡ	Burbot fish
1,000,000	ᛤ	Astonished man

Figure 7.1

The following are characteristics of the Egyptian numeration system:

(i) Seven symbols were used; they are shown in Figure 7.1.
(ii) Each unique symbol represented a power of ten.
(iii) The system was additive; that is, they *added* one symbol to another to represent numbers between powers of ten (between 1 and 10, 10 and 100, etc.).
(iv) No use was made of place value; that is, the position (place) of the symbols did not affect the value of the numeral.
(v) This system had no zero.

EXAMPLE 1: Determine the value of │ │ ∩ ⌒

Solution: $1 + 1 + 10 + 10{,}000 = 10{,}012$.

EXAMPLE 2: Translate 5,707 into Egyptian numerals.

Solution: 𓏲𓏲𓏲𓏲𓏲 𓐁𓐁𓐁𓐁𓐁𓐁𓐁 |||||||

Addition and multiplication in this system was not too difficult provided one used only natural numbers. Computation with fractions required a truly monumental effort due to the manner in which the Egyptians handled such problems. All fractions except $\frac{2}{3}$ were expressed as sums of unit fractions. For example, $\frac{7}{8}$ was written as $\frac{1}{8} + \frac{1}{4} + \frac{1}{2}$. That is, all unit fractions were required to have different denominators. The special fractions $\frac{2}{3}$ and $\frac{1}{8}$ were expressed as

$$\underset{/ \, / \, \backslash}{\ominus} \quad \text{and} \quad \underset{/ \, / \, / \, / \, / \, \backslash}{\ominus}$$

The Sumerian system required only two symbols called cuneiforms (wedges). With an instrument called a stylus they were able to scratch upright and sidewise wedges into clay. The upright wedge represented one, while the sidewise wedge represented ten of whatever value was indicated by the place it occupied.

EXAMPLE 3: What is the value of ◁▷ ▷▷▷/▷▷▷

Solution: The place on the right contains six symbols representing ones. The next place contains one symbol for one and one symbol for ten. Thus this system makes use of the place value concept. The place on the right is the unit's place while each successive place to the left represents increasing powers of sixty. This numeral represents six ones and eleven sixties, resulting in a value of

$$660 + 6 = 666$$

The concept of base should not be foreign to you since our own system makes use of the same principle. For example, 356 stands for

$$3(100) + 5(10) + 6(1)$$

We use a base of ten whereas the Sumerians used a base of sixty. One difficulty with this system of numeration involves spacing. Spacing indicates the correct place value. For example, let us write the numbers 62 and 3,602.

(1)
$$62 \cdot \cdot \cdot \cdot \cdot \cdot \cdot \cdot \cdot \triangleright \quad\quad\quad \triangleright\triangleright$$
$$3,602 \cdot \cdot \cdot \cdot \cdot \cdot \triangleright \quad\quad\quad\quad \triangleright\triangleright$$

If the writer were careful and the reader knew how much space he intended for each missing place, the symbols would be interpreted correctly. As you might suspect, one could never tell whether a numeral like ◁ might represent 10(1), 10×60, 10×60^2, ... One would have to tell the value intended from the context, i.e., whether, for example, the symbol was intended to represent the population of a small village, a town, or a city.

EXAMPLE 4: Translate ⟨⟨ ⟨⟨⟨⟨ ⟨⟨⟨⟨⟨⟨ into our numerals.

Solution:
```
   20   (3,600)  · · · · · · · ·  72,000
   31     (60)   · · · · · · · ·   1,860
   42      (1)   · · · · · · · ·      42
                                   ──────
                                   73,902
```

EXAMPLE 5: Translate 72,005 into cuneiforms.
Solution:

$$20 \times (3,600) = 72,000 \quad \cdots \quad ⟨⟨$$

$$1 \times \quad (5) = \quad 5 \quad \cdots \quad \text{⟨⟨⟨⟨⟨}$$

In summary, we can say that the Egyptian system was strictly additive whereas the Sumerians used place value on a sexagesimal (base 60) system. Why is the concept of place value important? Simply because it minimizes the number of symbols needed to represent a number. The Sumerians needed only two individual symbols to write a number of any size. On the other hand, the Egyptians would have needed a very large number of different symbols to write a number of great size. As we have seen, however, the Sumerian place value system was sometimes awkward to use and read. Both systems were excessively cumbersome because of having to use the principle of *iteration* (the repetition of individual symbols to tell how many tens, sixties, or other units were to be added to obtain the value represented).

EXERCISE SET I

1. Translate the following Egyptian symbols into our numerals.

    a) 𓋹 𓏺𓏺 ∩ ||

    b) ∩∩∩ ||

    c) 𓆑𓆑𓆑 𓋹 𓏺𓏺 ∩ ||

    d) 𓋹 ∩∩ |||

2. Change to Egyptian numerals.

§7a   Early Numeration Systems   309

    a) 46   b) 5,062
    c) 407   d) 56,742

3. Translate the cuneiforms into our numerals.
   a)
   b)
   c)
   d)

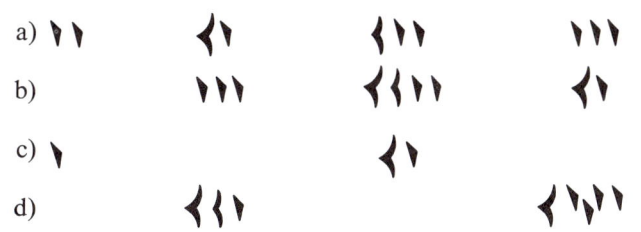

4. Change to Sumerian numerals.
   a) 724   b) 2,465
   c) 5,601   d) 1,174,728

*5. Find the sum of

6. What is the *minimum* number of symbols required for a positional numeration system?

    Some of the later numeration systems managed to overcome one or more of the disadvantages of the two earliest systems. Since some of the resulting improvements are now features of our own system, we shall take a look at three other systems, each of which exhibits at least one important refinement.

    As an aid to understanding how the various systems were used, let us recall that each numeration system other than the simple tally system requires a base b which can be any natural number considered to be convenient. Then symbols for powers of the base are used to help represent larger numbers. Figure 7.2 shows symbols used for the Egyptian and Sumerian systems.

Egyptian	Sumerian	
b = ten	b = sixty	
b^0:	b^0:	in first position (place)
b:	b:	in second position (place)
b^2:	b^2:	in third position (place)
b^3:	b^3:	in fourth position (place)
...	...	
	Special symbol for ten:	

Figure 7.2

We have seen that both of these systems used the additive principle. Thus fifty would be written respectively as

∩∩∩∩∩ , ❬❬❬❬❬

Such iterative numerals are, of course, cumbersome because of their length and the need to count the number of symbols in order to determine the value of each symbol group. Writing of numerals can be simplified and shortened by using a multiplicative principle. If, for example, π were a special multiplicative symbol for five and Δ the symbol for ten, then the numeral

$$\pi\Delta$$

could be used to stand for fifty instead of writing $\Delta\Delta\Delta\Delta\Delta$. We need to know only that a product is intended in order to construct a multiplicative numeration system. For a particular base b, $b - 1$ symbols are selected. Then another set of symbols is chosen to represent powers of the base. The symbols are then combined multiplicatively to represent any number.

This multiplicative principle is, in fact, used in the traditional Chinese-Japanese system. A partial list of the numerals used is found in Figure 7.3.

1	一	4	四	7	七	10	十
2	二	5	五	8	八	10^2	百
3	三	6	六	9	九	10^3	千

Figure 7.3

Using this system, 300 and 40 are written respectively as

$$\frac{三}{百} , \frac{四}{十}$$

Since these symbols are written vertically, one thinks "three times one hundred equals three hundred" and "four times ten equals forty."

EXAMPLE 6: Translate 2043 into Chinese-Japanese numerals.

Solution: We think "2043 equals two times one thousand plus four times ten plus three." Writing this from right to left with the sum to the left of the vertical bar, we obtain:

二
千
四
十
三
2043

三	四	二
	十	千
3	40	2000

§7a Early Numeration Systems 311

EXAMPLE 7: Translate these Chinese-Japanese numerals into our symbols.

Solution: 7000 + 300 + 70 + 6 = 7376

Writing an Egyptian or Sumerian equivalent of the Chinese-Japanese numerals in these examples would obviously be more laborious due to the iteration. Thus the multiplicative principle has allowed some shortening of the required work.

Recall that one of the difficulties in writing a Sumerian numeral was the requirement of an understanding between the reader and the writer regarding spacing. Recall (1) where we saw that 602 might easily be read as 62, or vice versa, in this place value (positional) system. A positional scheme can avoid this difficulty if a symbol is adopted for the number of elements in the empty set.

Our symbol for n ({ }) is 0 (zero). In using zero we must always be mindful of the fact that it represents a *number:* It is on our thermometers; it is at the beginning of our rulers; it tells us how many pencils we have when we have none, i.e., when we have an empty set of pencils. Just because we have no pencils does not mean that we have nothing. The symbol for zero also has an important function in any place value system: In the numeral 706 the 0 indicates that we have an empty set of tens whereas there are 6 ones and 7 hundreds. Without this feature, the 7 and the 6 might telescope into each other leading to ambiguity in meaning. That is

7 6

might be read as seventy-six, seven-hundred-six, or even seven-thousand-sixty. Thus the zero is a symbol which holds a place for n ({ }) just as other digits hold a place for n (S) where S is nonempty.

We are now ready to examine a numeration system developed by the Mayas of Central America which possesses both place value and a symbol for n({ }). It is interesting to note that this Mayan system is a rather sophisticated one in comparison to those of other ancient civilizations. It was an iterative system which used repeated dots (·) and bars (—). It was also positional. Each dot represented one of any given place value while each bar represented five of any place value. When a set of five dots was collected, they were equal to and replaced by a bar. Hence four bars represented twenty. See Figure 7.4. The Mayan system used a base of twenty in its place value with one exception. In the third position, eighteen 20's were used instead of twenty 20's. Researchers attribute this exception to the fact that the Mayan calendar consisted of eighteen months of twenty days each.

312 Numeration

Mayan numerals: ⊘ • •• ••• •••• — —̇ —̈ = =̇ =̈

Their value: 0   1   2   3   4   5   6   7 15 17

Figure 7.4

Mayan numerals were actually written vertically, but we shall write them horizontally in this text. For example, twenty-seven would be written as

• —̈

The unit position is on the extreme right. The value of the numeral is determined as "one twenty plus seven ones."

EXAMPLE 8: Translate 5543 into Mayan numerals.

Solution: Recall that the value of the third position is not 20^2, but 18×20, i.e., 360. Thus we think

$$5543 = 15 \times 360 + 7 \times 20 + 3$$

$$5543 = \equiv \quad —̈ \quad •••$$

EXAMPLE 9: Translate  —̇ —̈ = into our numerals.

Solution:

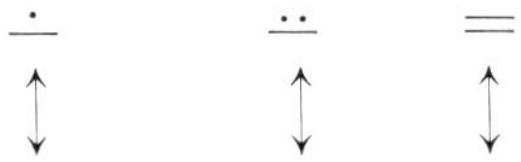

$6 \times 360 + 7 \times 20 + 10 = 2{,}310$

EXAMPLE 10: Translate = ⊘ •• into our numerals.

Solution:

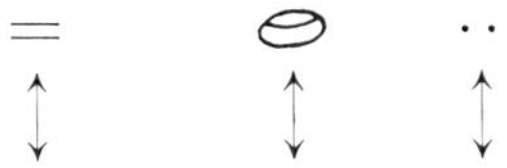

$10 \times 360 + 0 \times 20 + 2 = 3{,}602$

As you can see, the Mayan positional numeral

= ⊘ ••

cannot be misread as a numeral for 362 or 3620 because of the use of ⊘ as a numeral for $n(\{\ \})$. We still have the advantage of a minimum number of symbols as with any other positional system. However, the disadvantage of iteration is still with us.

Very few ancient systems of numeration were able to eliminate the disadvantages of iteration. The Ionic Greek system which came into use after

§7a   **Early Numeration Systems**   313

the classical period of Greek history was one of these few. In this scheme letters were used to represent numbers according to the pattern of Figure 7.5. The symbols for six, ninety, and nine-hundred were the obsolete letters

one	α	ten	ι	one hundred	ρ	one thousand	$,\alpha$
two	β	twenty	κ	two hundred	σ	two thousand	$,\beta$
three	γ	thirty	λ	three hundred	τ	three thousand	$,\gamma$
four	δ	forty	μ	four hundred	υ	four thousand	$,\delta$
five	ϵ	fifty	ν	five hundred	ϕ	five thousand	$,\epsilon$
six	Ϛ	sixty	ξ	six hundred	χ	six thousand	,Ϛ
seven	ζ	seventy	o	seven hundred	ψ	seven thousand	$,\zeta$
eight	η	eighty	π	eight hundred	ω	eight thousand	$,\eta$
nine	θ	ninety	ϙ	nine hundred	ϡ	nine thousand	$,\theta$

Figure 7.5

digamma, koppa, and sampi respectively. With an extended system of diacritical markings, numbers of any size could be represented. You will note, however, that lack of place value makes this system cumbersome because of the extensive number symbols which had to be committed to memory. Even though no special symbol for zero was needed to prevent ambiguity, the order of writing digits was unimportant. That is, the system was not positional so that

(2) $\beta\phi, \phi\beta$

would both be read as five-hundred-two. Iteration is never necessary as seen in Figure 7.6, where various symbols for fifty are represented.

Egyptian _ _ _ _ _ _ _ _ _ _ ∩∩∩∩∩

Sumerian _ _ _ _ _ _ _ _ _ _ ⟨⟨⟨⟨⟨

Chinese-Japanese _ _ _ _ _ _ 五 / 十

Mayan _ _ _ _ _ _ _ _ _ _ _ •• ═

Ionic Greek _ _ _ _ _ _ _ _ _ _ ν

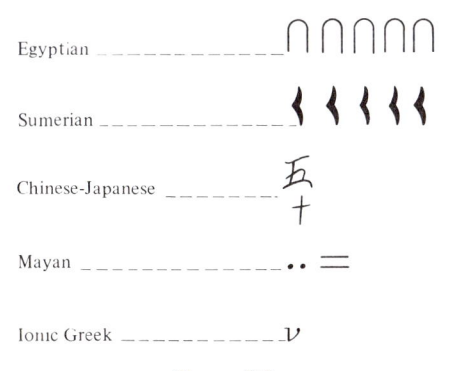

Figure 7.6

The principle which eliminates the use of iteration is called *cipherization*, which is a pattern of writing a numeral as a polynomial in the base *b*. To appreciate what this means, recall that an algebraic polynomial can be written as

(3) $\qquad \ldots a_6 b^5 + a_5 b^4 + a_4 b^3 + a_3 b^2 + a_2 b^1 + a_1 b^0$

Although we have written this polynomial in traditional order, we know that natural numbers are commutative under addition. Thus the order of writing such numerals is of no mathematical concern. This fact accounts for the phenomenon observed in (2).

The base *b* of (3) can be any number larger than 1. In the Ionic system $b = 10$. The cipherization is complete when each coefficient a_i is a *single digit*. With this restriction, iteration is never needed.

EXAMPLE 11: Translate 82 into Ionic Greek numerals.

Solution: We think

$$80 + 2 = 82$$
$$\pi + \beta = \pi\beta \quad \text{or} \quad \beta\pi$$

Note that although $\pi\beta = \beta\pi$, $82 \neq 28$. That is, the *digits* in our numeral system are not commutative.

EXAMPLE 12: Translate, $\epsilon\mu\alpha$ into our numerals.

Solution: We think

$$\begin{array}{r} \epsilon \ldots\ldots 5{,}000 \\ \mu \ldots\ldots\phantom{5{,}0}40 \\ \alpha \ldots\ldots\phantom{5{,}00}1 \\ \hline 5{,}041 \end{array}$$

To illustrate that computation in this system is relatively simple once the many symbols have been memorized, consider the following example.

EXAMPLE 13: Multiply $\upsilon\lambda\zeta$ by $\beta\alpha$

Solution:

$$\begin{array}{rr} \upsilon\lambda\zeta & 437 \\ \beta\alpha & 21 \\ \hline \upsilon\lambda\zeta & 437 \\ \eta\psi\mu & 874 \\ \hline \theta\rho o\zeta & 9177 \end{array}$$

The cipherization principle allows us to arrange the work of this example

§7a Early Numeration Systems 315

in our own familiar way according to powers of the base and to "carry" easily. In order to appreciate the advantages offered by this scheme, one needs only to attempt to do this example using either Sumerian or Egyptian numerals.

To summarize, in studying these five numeration systems, we have attempted to accomplish two things:

(i) discover the important principles which govern numeration;
(ii) acquire an appreciation for the structure of our own system of numeration through consideration of the age-long efforts that have gone into producing it.

EXERCISE SET II

7. Translate into Chinese-Japanese numerals:
   a) 24 b) 152 c) 246
   d) 1984 e) 36 f) 728

8. Translate into our numerals:
   a) 三
      十
      六
   b) 四
      千
      三
      百
      九
   c) 二
      百

9. Translate into Mayan numerals:
   a) 728 b) 54 c) 6742 d) 507

10. Translate into our numerals:
    a) ∵ ∴ ∷
    b) · ≡ ∴
    c) · ∵ ∴ ∵
    d) = ≡ = ⊖

11. Translate into Ionic Greek numerals:
    a) 562 b) 300
    c) 706 d) 8642

12. Translate into our numerals:
    a) κα b) ρμβ c) γϕπ d) ˏβο

13. In Example 13 we see that the digit 7 occurred five times in performing the multiplication our way. Yet these 7's were represented variously by ζ, ο, ψ in Ionic Greek symbolization. Explain.

14. Translate into our numerals:
    a) LVII b) DCXVII
    c) MDDIV d) CCXXXXVII

316   *Numeration*

15. a) Is the Roman numeration system iterative? Explain.
    b) Is the Roman numeration system additive? Subtractive? Explain.
    c) Does the Roman numeration system have place value? Explain.

16. Compare the advantages and disadvantages of our numeration system with that of the Romans.

§7b   THE INDO-ARABIC SYSTEM

From the preceding section we can see that a numeration system having all of the following properties would be very advantageous. It should

(i) be positional, i.e., make use of place value;
(ii) have a zero symbol;
(iii) make use of the cipherization principle.

Recall that (ii) is needed when we insist on (i). A system which makes use of both place value and cipherization will of necessity have other features we have investigated. For example, cipherization in particular demands that the system make use of a fixed base b. The number of symbols can be kept to a minimum by using the positional scheme. Use of both (i) and (iii) requires at least $b - 1$ symbols. Thus with the symbol for $n(\{\ \})$, the total number of individual symbols needed is b.

Using a, d, f as digits, the numeral

$$a\ d\ f$$

is interpreted as

$$a(b^2) + d(b^1) + f(b^0)$$

Note that the system still uses the additive principle but that iteration is *not* necessary.

You are familiar with a system which has all of these features. We call it the *Indo-Arabic* system because the digits

$$0, 1, 2, 3, 4, 5, 6, 7, 8, 9$$

came to us from India by way of the Arabs. You know that we use ten as a base probably because we have ten fingers. Hence we say that our numerals are decimals from the Latin word *decem* which means ten.

Any Indo-Arabic decimal numeral can thus be expanded as

(1) $$\ldots + a_4(10)^3 + a_3(10)^2 + a_2(10)^1 + a_1(10)^0$$

where $a_1, a_2, a_3, a_4, \ldots$ are particular digits. When we actually write a numeral, of course, we only write

(2) $$\ldots a_4 a_3 a_2 a_1$$

EXAMPLE 1: Write 2754 in exponential form (1).
Solution: $2(10)^3 + 7(10)^2 + 5(10)^1 + 4(10)^0$

When we are not stressing principles, we would of course, write 1 for 10^0 and 10 for 10^1.

EXAMPLE 2: State the meaning of each 2 in the numeral 222.
Solution: In exponential notation we have
$$2(10)^2 + 2(10)^1 + 2(10)^0 = 200 + 20 + 2$$
Therefore, the left 2 stands for 200, the middle 2 for 20, the right one for 2.

EXERCISE SET

1. Write in exponential notation
   a) 5274   b) 98795
2. Using each of the digits 2, 5, 7, 3 exactly once,
   a) write a numeral representing the largest possible number;
   b) write a numeral representing the smallest possible number.
3. Write the numeral for the largest possible number which can be represented with three digits in base b.
4. If $384k$ is a numeral in a system of base b, write it in exponential form.
5. Consider 237:
   a) Can this be a numeral in base $b = 5$?
   b) For what values of the base b would this numeral have meaning?
6. Compare the Roman Numeral system feature for feature with the Indo-Arabic Decimal system.
7. Why should we study the Roman Numeral system as part of our arithmetic program since "only dead Romans and the makers of late-late T.V. movies are concerned," e.g., film serial MMCXVIII?
8. Compare the cipherization and place value principles.

§7c   NUMERATION BASES OTHER THAN TEN

Let us consider an arbitrary base b. In the Indo-Arabic system each numeral

(1) $\qquad \ldots a_4 a_3 a_2 a_1$

stands for

(2) $\qquad \ldots + a_4 b^3 + a_3 b^2 + a_2 b^1 + a_1 b^0$

The latter is called expanded exponential form.

318 *Numeration*

If we were to use five as a base, we would need the five digits

(3) $$0, 1, 2, 3, 4$$

in order to write any number. Thus, (2) becomes

$$\ldots + a_4(5)^3 + a_3(5)^2 + a_2(5)^1 + a_1(5)^0$$

where the a's are any of the digits of (3).

EXAMPLE 1: Translate 243_{five} into decimal numerals.

Solution: The subscript *five* is meant to indicate that 243 is written in base 5 notation. The numeral itself is read "two-four-three" rather than "two hundred forty-three" for the very simple reason that 243_{five} does not stand for two hundred forty-three! We think,

$$\begin{aligned}
243_{\text{five}} &= 2(10)^2 + 4(10)^1 + 3(10)^0 \quad \text{in base five} \\
&= 2(5)^2 + 4(5)^1 + 3(5)^0 \quad \text{in base ten} \\
&= 2(25) + 4(5) + 3(1) \quad \text{in base ten} \\
&= 50 + 20 + 3 \quad \text{in base ten} \\
&= 73_{\text{ten}}
\end{aligned}$$

Hence 243_{five} is a name for seventy-three.

EXAMPLE 2: Convert 37_{ten} to a base five numeral.

Solution: Recall that

$$\begin{aligned}
5^0 &= 1 \\
5^1 &= 5 \\
5^2 &= 25 \\
5^3 &= 125 \\
&\ldots
\end{aligned}$$

and thus our base five numeral will be

$$\ldots + a_4(125) + a_3(25) + a_2(5) + a_1(1) = x$$

in expanded form. Our problem is to identify the a_1. Immediately you can see that if $a_4 \neq 0$, x will be too large since $125 > 37$. We next consider a_3. Since

$$(1)(25) < 37$$

and

$$(2)(25) > 37$$

we must have $a_3 = 1$. Thus

$$37 = 1(25) + 12$$

Now

$$(1)(5) < 12$$
$$(2)(5) < 12$$
$$(3)(5) > 12$$

Thus $a_2 = 2$ and

$$37 = 1(25) + 2(5) + 2$$

Since $5^0 < 2 < 5^1$, $a_1 = 2$. Therefore,

(4)
$$\begin{aligned} 37_{\text{ten}} &= 1(25) + 2(5) + 2(1) \text{ base ten} \\ &= 1(5)^2 + 2(5)^1 + 2(5)^0 \text{ base ten} \\ &= 1(10)^2 + 2(10)^1 + 2(10)^0 \text{ base five} \\ &= 122_{\text{five}} \end{aligned}$$

An algorithm which simplifies the process used in this example can be developed. Consider (4) in factored form, that is, $1(5)^2 + 2(5) + 2 = [1(5) + 2]5 + 2$. Successive division of this number and its quotients by 5 will result in remainders 2, 2, 1. These remainder-digits are the reverse of 122, the answer to example 2. Thus we can state the algorithm as follows: To change from base ten numeral to base b numeral

(i) divide the base ten numeral by b
(ii) record the remainder
(iii) divide the quotient of step (i) by b, record its remainder, and
(iv) continue in this manner until a quotient of zero is obtained
(v) write the recorded remainders in reverse order.

Note that each successive remainder obtained in this algorithm is less than b.

EXAMPLE 3: Change 2793_{ten} to a base seven numeral.

Solution:

$$\begin{array}{r} 399 \\ 7\overline{)2793} \\ 21 \\ \hline 69 \\ 63 \\ \hline 63 \\ 63 \\ \hline 0 \end{array} \quad R_1 = 0$$

$$\begin{array}{r} 57 \\ 7\overline{)399} \\ 35 \\ \hline 49 \\ 49 \\ \hline 0 \end{array} \quad R_2 = 0$$

320   *Numeration*

$$\begin{array}{r} 8 \\ 7\overline{)57} \\ \underline{56} \\ 1 \end{array} \qquad R_3 = 1$$

$$\begin{array}{r} 1 \\ 7\overline{)8} \\ \underline{7} \\ 1 \end{array} \qquad R_4 = 1$$

$$\begin{array}{r} 0 \\ 7\overline{)1} \\ \underline{0} \\ 1 \end{array} \qquad R_5 = 1$$

Writing the remainders in reverse order we have 11100. Thus

(5) $\qquad 2793_{\text{ten}} = 11100_{\text{seven}}$

The work involved in using this algorithm may be shortened considerably by doing the division mentally, and writing only the quotient and remainders.

$$\begin{array}{r|l} 7 & 2793 \\ 7 & 399 \quad 0 \\ 7 & 57 \quad\; 0 \\ 7 & 8 \quad\;\; 1 \\ 7 & 1 \quad\;\; 1 \\ 7 & 0 \quad\;\; 1 \end{array}$$

again (following the arrow) we obtain (5).

What is the minimum number of separate digits needed to write numerals in the Indo-Arabic system? Let us try base one numeration. Since we can have only one digit and since we must have a zero, this system can only represent zero. How about base two numeration? Here we would have two digits, namely 0, 1. This will work, and, surprisingly, has wide application in computer technology.

EXAMPLE 4: Change 11011_{two} to its decimal equivalent.

Solution:

$$\begin{aligned} 11011_{\text{two}} &= 1(2)^4 + 1(2)^3 + 0(2)^2 + 1(2)^1 + 1(2)^0 \\ &= 1(16) + 1(8) + 0 + 1(2) + 1 \text{ base ten} \\ &= 16 + 8 + 0 + 2 + 1 \text{ base ten} \\ 11011_{\text{two}} &= 27_{\text{ten}} \end{aligned}$$

EXAMPLE 5: Change 57_{ten} to its base two equivalent.

Solution: Using the algorithm of Example 3, we obtain

$$
\begin{array}{r|l l}
2 & 57 & \\
2 & 28 & 1 \\
2 & 14 & 0 \\
2 & 7 & 0 \\
2 & 3 & 1 \\
2 & 1 & 1 \\
  & 0 & 1 \\
\end{array}
$$

Thus $57_{ten} = 111001_{two}$.

What if we tried to use a base larger than ten? Since we are required to have b digits, a base larger than ten would require that we invent additional symbols. For example, if we wanted to use twelve as a base we would be required to provide two symbols in addition to those we normally use in the decimal system. What number does 10_{twelve} represent? If you said *ten* you were wrong.

$$
\begin{aligned}
10_{twelve} &= 1(12)^1 + 0(12)^0 \quad \text{base ten} \\
&= 12 + 0 \quad \text{base ten} \\
&= 12_{ten}
\end{aligned}
$$

Since 10 represents twelve in base twelve, the two new digits we need in a base twelve system must represent ten and eleven. We shall use

$\bigcap$, for ten;
$\bigcap\!\!\!|$, for eleven.

EXAMPLE 6: Change $2 \bigcap \bigcap\!\!\!| \, 4_{twelve}$ to its decimal equivalent.

Solution:

$$
\begin{aligned}
2 \bigcap \bigcap\!\!\!| \, 4_{twelve} &= 2(12)^3 = 10(12)^2 + 11(12)^1 + 4(12)^0 \text{ base ten} \\
&= 2(1728) + 10(144) + 11(12) + 4(1) \text{ base ten} \\
&= 3456 + 1440 + 132 + 4 \quad \text{base ten} \\
&= 5032_{ten}
\end{aligned}
$$

EXAMPLE 7: Change 2735_{ten} to its base twelve equivalent.

Solution:

```
12 | 2735
12 |  227   ↑
12 |   18   ↑
12 |    1   6
        0   1
```

Thus $2735_{ten} = 16 ⇑⇑_{twelve}$.

Computation in non-decimal bases often helps give us a better understanding of our own decimal computation. In order to do any efficient computation it is necessary to memorize addition and multiplication facts for that particular base. These facts are usually given in tabular form. Tables for addition and multiplication in base two are found in Figure 7.7.

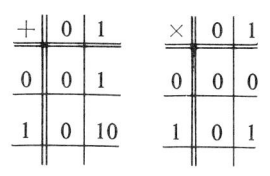

+	0	1		×	0	1
0	0	1		0	0	0
1	1	10		1	0	1

Figure 7.7

EXAMPLE 8: Add 101_{two} to 11_{two}.

Solution:

```
   ¹¹¹
    101
 +   11
   1000
```

→ $1 + 1 = 10$, write 0 carry 1
→ $1 + 0 + 1 = 10$, write 0 carry 1
→ $1 + 1 = 10$, write 0 carry 1
→ $1 = 1$, write 1

EXAMPLE 9: Multiply 101_{two} by 11_{two}.

Solution:

```
   101
    11
   ---
   101    → 1 × 1 = 1, 1 × 0 = 0, 1 × 1 = 1
  101
  ----
  1111
```

Addition and multiplication tables for base twelve.

Figure 7.8

EXAMPLE 10: Add 9⑪⑩ to 435 (base twelve).

Solution:

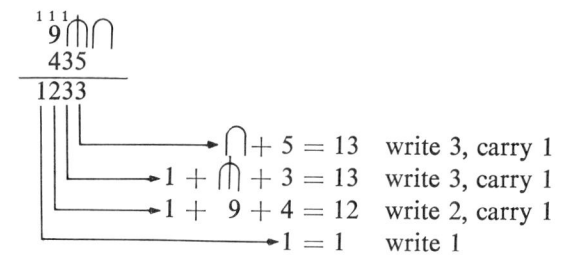

$$\begin{array}{r} \overset{1\,1\,1}{9⑪⑩} \\ 435 \\ \hline 1233 \end{array}$$

⑩ + 5 = 13   write 3, carry 1
1 + ⑪ + 3 = 13   write 3, carry 1
1 + 9 + 4 = 12   write 2, carry 1
1 = 1   write 1

EXAMPLE 11: Multiply 435 by 26 (base twelve).

Solution:

$$\begin{array}{r} 435 \\ 26 \\ \hline 2186 \\ 86⑩ \\ \hline ⑩866 \end{array}$$

6 × 5 = 26, write 6 carry 2
6 × 3 = 16, 16 + 2 = 18, write 8 carry 1
6 × 4 = 20, 20 + 1 = 21, write 1 carry 2

EXERCISE SET

1. Change the values given in various bases to decimal values.
    a) 14_{five}
    b) 34_{twelve}
    c) 1111_{two}
    d) 34_{seven}

2. Change these decimal values to the base numeration indicated.
    a) 56 to base twelve
    b) 87 to base nine
    c) 76 to base two

3. Fill in the missing entries in the base twelve tables of Figure 7.8.

4. Add:
    a) binary addends (base two)   110110
       1110
       10111
       11
    b) duodecimal addends (base twelve)   ⑪⑩9
       87⑩
       95⑪

5. Multiply:
    a) base twelve   17⑩⑪
       4⑩
    b) base two   11011
       11

§7d Common Algorithms for Arithmetic

6. Subtract:
   a) base eight   5742
                          777
   b) base five   1421
                          234

7. Divide:
   a) base twelve   ∩↑↑ ⟌ 47↑↑6∩
   b) base seven   65 ⟌ 4462

8. Construct addition and multiplication tables for base four.

9. Translate the numerals 123, 32 into base four numerals, then
   a) add them
   b) multiply them
   c) subtract them
   d) divide them.

10. Write the number 214 in base eight and then in base two notation. Can you see any relationship between the two bases?

11. What is an obvious way of recognizing an odd number written in base two? An even number?

12. Is it possible to use 0 (zero) as a base for Indo-Arabic numeration. Explain.

13. If b is any Indo-Arabic numeration base, what does 10_b represent?

§7d   COMMON ALGORITHMS FOR ARITHMETIC

An algorithm is a set of rules for finding a particular mathematical result. We have standard algorithms for performing any but the simplest addition, multiplication, subtraction, and division problems. Our purpose here is to examine some of these.

Consider the addition problem

(1)
$$\begin{array}{r} \overset{1}{47} \\ +35 \\ \hline 82 \end{array}$$

The algorithm is:

(i) Write the addends in a column so that the units places are aligned.
(ii) Add the digits in the units column.
(iii) Write the units digits of the sum under the units column.
(iv) If the sum is more than a single digit numeral, use the remaining portion of the sum as an entry in the next column. (This is called *carrying*.)

Numeration

(v) Continue this process for each column in order.

We see that the 35 and 47 have been aligned as required in (i). According to (ii) we have mentally added the units digits 7, 5 to obtain 12. According to (iii) we have written the 2 from the number 12 under the units column. The sum 12 has two digits. Thus, according to (iv) we carry the digit 1 into the second column as an addend. We then continue the process for the digits 3, 4, 1 to obtain 8. Since this is a single digit sum and there are no more columns to the left, the process is complete.

Hidden in this algorithm are several basic principles of arithmetic as stated in the axioms for natural numbers. See §6b. To make this fact clear we shall redo the problem by using the axioms.

1. $47 + 35 =$
    $(4(10) + 7(1)) + (3(10) + 5(1))$
2. $= [(4(10) + 7(1)) + 3(10)] + 5(1)$
3. $= [(4(10) + (7(1) + 3(10))] + 5(1)$
4. $= [4(10) + (3(10) + 7(1))] + 5(1)$
5. $= [(4(10) + 3(10)) + 7(1)] + 5(1)$
6. $= [((10)4 + (10)3) + 7(1)] + 5(1)$
7. $= [10(4 + 3) + 7(1)] + 5(1)$
8. $= 10(4 + 3) + [7(1) + 5(1)]$
9. $= 10(4 + 3) + (7 + 5)$
10. $= 10(7) + 12$
11. $= 10(7) + (10(1) + 2)$
12. $= (10(7) + 10(1)) + 2$
13. $= 10(7 + 1) + 2$
14. $= 10(8) + 2$
15. $= 82$

1. Numeration system
2. Associativity N-7
3. N-7
4. Commutativity N-5
5. N-7
6. Commutativity N-6
7. Distributivity N-9
8. N-7
9. Neutral element N-10
10. Number facts
11. Step 1
12. N-7
13. N-9
14. Step 10
15. Step 1

We are now able to justify the steps of our algorithm by examining in particular steps 9 through 12 in this proof. The carrying has been done in going from step 11 to step 12.

We next consider our usual algorithm for multiplication. For example,

(2)
$$\begin{array}{r} 23 \\ \times 47 \\ \hline 161 \\ 92 \\ \hline 1081 \end{array}$$

The algorithm is:

(i) Align the units digits vertically.
(ii) Multiply the units digits.
(iii) Copy units digit of product under units column.

(iv) Multiply units digit of the lower number by the second digit of the upper number.
(v) Add the remaining digit of (iii) (circled above) to the product of (iv).
(vi) Copy units digit of the sum under the second column.
(vii) Repeat these steps until all digits of the upper number have been used.
(viii) Multiply the second digit of lower number by the units digit of the upper number.
(ix) Place the units digit of this product below the second column.
(x) Multiply the second digit of the lower number by the second digit of the upper number.
(xi) Add the remaining digits of step (ix) to the product of (x).
(xii) Copy units digit of this sum under the third column.
(xiii) Repeat these steps until all digits of the upper number have been used.
(xiv) Repeat the process indicated until all digits of the lower number have been used.
(xv) Add the results obtained in steps (vii) and (xiv).

Returning to (2) we have aligned the units digits 3, 7. According to (ii) we multiplied 7 times 3 to obtain 21 and wrote the digit 1 below the units column as required by (iii). We next multiplied 7 times 2 to obtain 14 according to (iv). We added the remaining digit 2 of the first product to 14 to obtain 16 and wrote the digit 6 under the second column by (v) and (vi). All digits of the upper number have been used as required by (vii). The result of completing (vii) requires that the digit 1 of 16 be written under the third column. The 92 is obtained similarly using (viii) through (xiv). Finally, (xv) is used to obtain the answer 1081.

This algorithm also hides the fact that each step is justified by the axioms of the natural numbers and numeration principles. Let us redo the problem using theorem (6b-11) which states $\forall\, a, b, c, d \in \mathcal{N}, (a + b)(c + d) = (ac + ad) + (bc + bd)$. The proof of this theorem depends heavily on the distributive axiom N-9. Hence the distributive axiom is the fundamental principle underlying the multiplication algorithm.

1. $23 \times 47 =$
   $(2 \times 10 + 3)(4 \times 10 + 7)$
2. $= [(2 \times 10)(4 \times 10) + (2 \times 10)(7)] +$
   $[3(4 \times 10) + (3 \times 7)]$
3. $= [(2 \times 10)(10 \times 4) +$
   $(2 \times 10)(7)] + [(3(4 \times 10) + (3 \times 7)]$
4. $= [(2 \times 10)10] \times 4 + \ldots$
5. $= [2 \times (10 \times 10)] \times 4 + \ldots$
6. $= [(2 \times 10^2) \times 4 + \ldots]$

1. Numeration
2. (6b-11)
3. N-6
4. N-8
5. N-8
6. (6-6)

7. $= [4 \times (2 \times 10^2) + \ldots]$ 7. N-6
8. $= [(4 \times 2) \times 10^2 + \ldots]$ 8. N-8
9. $= 8 \times 10^2 + 26 \times 10 + 21$ 9. Number facts
10. $= 8 \times 10^2 + 2 \times 10^2 + 6 \times 10 + \ldots$ 10. Numeration
11. $= 10 \times 10^2 + 8 \times 10 + 1$ 11. Addition algorithm
12. $= 1 \times 10^3 + 8 \times 10 + 1$ 12. (6-13), N-10
13. $23 \times 47 = 1081$ 13. Numeration

The steps in using the algorithm can be seen better if we do the problem of (2) in the following way.

$$\begin{array}{r} 23 \\ \times\ 47 \\ \hline 21 \longleftarrow\ 7 \times 3 \\ +\ 140 \longleftarrow\ 7 \times 20 \\ +\ 120 \longleftarrow\ 40 \times 3 \\ +\ 800 \longleftarrow\ 40 \times 20 \\ \hline 1081 \end{array}$$

Turning to the subtraction and division algorithms, we make no attempt to justify each step in the process. However, doing each in somewhat expanded form can be valuable in understanding the processes.

Consider the difference $(143 - 74)$. We can arrange the work as follows:

$$\begin{array}{r} 143 = 130 + 13 \\ 74 = \ \ 70 + \ \ 4 \\ \hline 69 = \ \ 60 + \ \ 9 \end{array}$$

To illustrate division, let us divide 887 by 17.

$$\begin{array}{r} 52 \\ +\ \ 2 \\ \hline 50 \\ 17 \overline{)887} \\ 17 \times 50 \longrightarrow 850 \\ 887 - 850 \longrightarrow 37 \\ 17 \times 2 \longrightarrow 34 \\ \hline 3 \end{array}$$

Thus $\dfrac{887}{17} = 52 + \dfrac{3}{17}$.

EXERCISE SET

1. Using the proof of the problem of (1) as a pattern, prove that $96 + 25 = 121$.

2. Using the proof of the problem of (2) as a pattern, prove that $34 \times 32 = 1088$.

§7e Alternative Algorithms

3. Rewrite $\begin{array}{r} 1008 \\ -\ 995 \\ \hline 13 \end{array}$ in such a way that an elementary school child might understand the steps.

4. Carry out the steps of dividing 6475 by 37 so as to help an elementary school child understand how each partial result is obtained.

§7e ALTERNATIVE ALGORITHMS

The standard algorithms which all of us have learned are not the only ones it is possible to use in doing arithmetic processes. It can be both pleasurable and instructive to practice with some others.

For example, the commutative axiom allows us to add up as well as down. This principle is, in fact, used as a check on the accuracy of doing addition in the usual way.

(1)
$$\begin{array}{r} 290 \\ 47 \\ 152 \\ 91 \\ \hline 290 \end{array}$$

Another method of checking addition is sometimes called "lightning addition." Follow the arrow in the following example. Each digit is considered as an addend, and then each arrow shows which addend to use next.

(2)

We are thinking of each digit as having its place value so that we are actually adding as follows:

$$40 + 7 = 47$$
$$47 + 100 = 147;\ 147 + 20 = 167;\ 167 + 6 = 173$$
$$173 + 20 = 193;\ 193 + 9 = 202$$
$$202 + 60 = 262;\ 262 + 4 = 266$$

The *Equal Additions* method for subtracting is used in some European countries. Instead of borrowing, as we do in our standard algorithm, we do the following:

330   *Numeration*

$$\begin{array}{r} 95 \to 9 (10+5) \\ -68 \to (6+1)8 \\ \hline 27 2 7 \end{array}$$

We have added ten ones to the units place of 95 and ten itself to the tens place of 68 (equal additions) using the principle

$$x - y = (x + 10) - (y + 10)$$

Thus the difference, 27, is the same by both methods.

EXAMPLE 1: Subtract 5678 from 7432 by the equal addition method.

Solution:

(3)
$$\begin{array}{r} 7\,\overset{1}{4}\,\overset{1}{3}\,\overset{1}{2} \\ -\,\overset{6}{\cancel{5}}\,\overset{7}{\cancel{6}}\,\overset{8}{\cancel{7}}\,8 \\ \hline 1\,7\,5\,4 \end{array}$$

The advantage of using this algorithm is that it minimizes the 'pains' of borrowing. A somewhat different procedure for achieving this result can be called the *complementation algorithm*. It was once in common use and has returned to some favor recently because computers find it easier to manage than our standard subtraction algorithm. The *complement* $C(n)$ of a natural number n is the difference between n and the next highest power of ten. Thus

(4)
$$\begin{aligned} C(9) &= 10 - 9 = 1 \\ C(50) &= 100 - 50 = 50 \\ C(189) &= 1000 - 189 = 811 \end{aligned}$$

Note that this complement can be found without borrowing by simply subtracting the units digit of n from 10 and subtracting each of the other digits from 9. Thus to find the complement of 189, we think

$$\begin{array}{r} 9\,9\,(10) \\ -1\,8\,9 \\ \hline 8\,1\,1 \end{array} = C(189)$$

to obtain the result in (4) instead of doing the actual subtraction required by the definition.

Now to find the difference $x - y$ by the complementation process, we add the complement of y to x and then subtract 1 from that digit of the sum which is located one position to the left of the digit on the extreme left of the complement. Using this algorithm to do the subtraction of (3) we first find

$$C(5678) = 4322$$

Then we add

§7e   **Alternative Algorithms**   331

$$\begin{array}{r}7432\\+4322\\\hline \cancel{1}1754\\0\end{array}$$

whence 1754 is the required difference.

Our standard algorithm for multiplication is efficient but very sophisticated in its requirements of carrying and correct positioning of partial products. In various times and places less demanding algorithms have been popular. One such process is the so-called *lattice* method of multiplication which was once much used. It is sometimes called the *grating* or *jalousie* method. The lattice is a rectangle ruled into small squares (cells). A set of diagonals is drawn dividing the cells into triangles. The number to be multiplied is written across the top of this rectangular frame, the multiplier down the side of the frame.

EXAMPLE 2: Multiply 2614 by 473 using the lattice method.

Solution: We arrange the problem as in Figure 7.9.

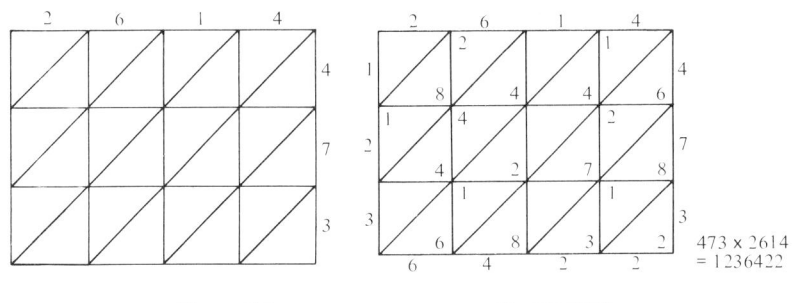

Figure 7.9   Figure 7.10

Each product is written in its cell, the units digit of the product in the lower section and the tens digit of the product in the upper section. The addition is done along the diagonal strips from upper right to lower left. At the end of each diagonal the units digit of the sum is recorded outside the lattice, and the remaining digits of the sum are carried to the next diagonal strip. Now the answer is read from upper left to lower right along the outside of the lattice. See Figure 7.10.

As you can see, the lattice method is quick and easy once the lattice is drawn. Except for that handicap this method might have become the acceptable algorithm for multiplication.

An algorithm called *Peasant Multiplication* places even fewer demands on the user. One must only be able to "halve" and "double" to use it successfully. In order to illustrate the process, we shall place the larger of the two numbers to be multiplied on the left. Then in each step the left factor is halved and the right factor is doubled producing two columns. Any remainder of 1 is discarded and the work continues until 1 is reached on the left. Then each row which has an even entry in the left column is crossed off. The sum of the remaining entries in the right column is the desired product.

EXAMPLE 3: Multiply 47 by 11 using peasant multiplication.

Solution:

```
47        11
23        22
11        44
 5        88
 2       176
 1      +352
         ───
         517
```

EXAMPLE 4: Multiply 17 by 36 using peasant multiplication.

Solution:

```
36        17
18        34
 9        68
 4       136
 2       272
 1      +544
         ───
         612
```

Checking the result of any standard algorithm is usually done by using the inverse. For example, division is checked by multiplication and subtraction is checked by addition. It can be interesting to consider some less common ways of checking.

The method of *casting nines* can be effectively used to check any of the basic operations. The *nines excess* of a number is the remainder obtained when the number is divided by nine. For example, the nines excess of 5 is 5, of 9 is 0, of 12 is 3, of 43 is 7, of 100 is 1. Carrying out an actual division is unnecessary because the *nines excess of a number is equal to the nines excess of the sum of the digits of the number*. The number 9736 has a digital sum of 25. The number 25 has a digital sum of 7. Since this number 7 is less than 9, the nines excess of 9736 is 7. The reason that the sum of the digits can be used to find the nines excess of a number is illustrated in Figure 7.11. Each term in the first brackets has a factor of nine and each term of the second brackets is a digit of the original number. A check of any numerical computation involving addition, subtraction, multiplication, and division can be carried out with the nines excesses in place of the original numbers.

§7e Alternative Algorithms 333

$$9736 = 9(1000) + 7(100) + 3(10) + 6$$
$$= 9(999 + 1) + 7(99 + 1) + 3(9 + 1) + 6$$
$$= [9(999) + 7(99) + 3(9)] + [9 + 7 + 3 + 6]$$

Figure 7.11

EXAMPLE 5: Add 274, 53, 15, 171 and check by casting nines.

Solution:

$$\begin{array}{rll}
274 & 2+7+4=13 & 1+3=4 \\
53 & 5+3=8 & 8 \\
15 & 1+5=6 & 6 \\
+171 & 1+7+1=9 & +0 \\ \hline
513 \longrightarrow & 5+1+3=9 & 18 \quad 1+8=9 \\
& =0 \longleftarrow \quad =0
\end{array}$$

Since the excesses are the same, we are reasonably confident that our original computation was correct.

EXAMPLE 6: Subtract 31 from 245 and check by casting nines.

Solution:

$$\begin{array}{rlll}
245 & 2+4+5=11 & 1+1=2 & \text{or} \quad 11 \\
-\;31 & 3+1=4 & 4 & -\;4 \\ \hline
214 & 2+1+4=7 & \longleftarrow & 7
\end{array}$$

Since 4 cannot be subtracted from 2 in the natural number system, we increased the smaller excess by 9 and then performed the subtraction.

EXERCISE SET

1. Add and check by adding upward.

    a) 315
        2784
        365
        22
        + 8

    b) 263
        56
        1784
        + 95

2. Check the additions in Exercise 1 by lightning addition.

3. Check the additions in Exercise 1 by casting nines.

4. Subtract by the equal additions method.

    a)   4321
        −1234

    b)   5003
        −1274

5. Do Exercise 4 using complementation.

6. Check the subtraction in Exercise 4 by casting nines.

7. Multiply by the lattice method.

   a) 354
       × 26

   b) 632
       ×145

8. Do Exercise 7 using peasant multiplication.
9. Check the multiplication in Exercise 7 by casting nines.
10. Divide 764 by 23 and check by casting nines.
11. Check the commutative property of peasant multiplication by halving the smaller factor instead of the larger in Example 3 and Example 4.
12. List the advantages and disadvantages of each algorithm.
    a) peasant multiplication
    b) lattice multiplication
    c) subtraction by complementation
    d) subtraction by equal additions

8
GEOMETRY

Geometry is one of the oldest organized subjects. There are innumerable ways to use its principles. Claims about its value as a study are many and conflicting. There was a time when geometry was studied no earlier than the tenth grade. Newer classroom materials have children learning considerable geometric information in elementary school. To gain a better appreciation for this remarkable subject, it may be valuable to ask ourselves some questions about it.

§8a WHAT AND WHY

In earliest times geometry was developed simply as a practical tool. The Sumerians and Egyptians developed a number of fairly accurate rules-of-thumb for making a variety of computations and constructions. In their hands geometry was a body of empirical knowledge obtained inductively from consideration of special cases and completely unsupported by anything resembling logical proof. Yet their achievements were considerable. The great pyramid of Gizeh gives some insight into the remarkable ability of the Egyptians to use their really primitive science to get very accurate results.

> Erected about 2900 B.C., it covers 13 acres and contains over 2,000,000 stone blocks. The sides of the square base involve a relative error of less than 1/14,000, and the relative error in the right angles at the corners does not exceed 1/2700.[1]

When the Greeks inherited the knowledge of the more ancient peoples, they made substantial changes. As outlined in §1a, they not only extended geometric principles, but also began to study and organize the subject deductively. Some of the most famous contributors to this transformation were Thales of Miletus (640–546 B.C.), Hippocrates of Chios (circa 440 B.C.), Pythagoras (about 580–500 B.C.), Plato (429–348 B.C.), Eudoxus (circa 370 B.C.), Archimedes (287–212 B.C.), and Euclid (about 300 B.C.). Euclid's text,

---

[1] H. Eves, *An Introduction to the History of Mathematics* (Rinehart and Company, Inc., 1953), p. 37.

The Elements, presented all the geometry then known, not as a disjointed collection of empirical results, but as a well-organized chain of theorems very similar in form to the pattern of formal axiomatics.

Next to the Bible, *The Elements* is probably the most widely distributed book ever written. More than a thousand editions of the work have appeared since the invention of printing. Before the age of printing, manuscript copies of Euclid's work were the major source for learning geometry. To this day most of our traditional school geometry texts are a rewriting (often with little change) of the first six books (i.e., chapters) of *The Elements.*

In spite of the great effort which the Greeks put into working out the logical bases for geometry, Euclid's work does have a few flaws. Some of the great modern mathematicians such as Pasch, Peano, Pieri, and Hilbert have put a lot of time and energy into making the development of Euclidean geometry more logically valid. These men have done their work since 1850; essentially, they rewrote the axioms. The most famous and influential work is by Hilbert, *Grundlagen der Geometrie,* which has had seven revisions, the latest in 1930. All of them put much effort into an attempt to follow the pattern of formal axiomatics. The following quote shows that the strictly logical aspects of the subject have often been advanced as a reason for making geometry a part of the school curriculum.

> One of the great needs of the world today is the need for clear thinking. Too often people accept statements without thinking them through carefully. They are prone to let a few people do their thinking for them. This is probably because they have never learned to think. One of the chief purposes of the study of geometry is to teach the student to think carefully. Theorems in geometry were proved thousands of years ago by mathematicians. The chief reason for proving theorems is that it will enable the student to examine and study the process of clear thinking.[2]

This viewpoint has been seriously challenged recently. For one thing, the vaunted principles of reasoning used in geometric arguments were not overtly used in the traditional geometry course. Instead of learning deductive principles, students have tended to learn by osmosis how to handle geometric theorems. Thus there has been little transfer of clear thinking to other fields. If logic skills are to be learned for themselves, they must be encountered explicitly in more than one context. Today we do not wait until the tenth grade geometry course to practice logic skills. They are used in number arguments in some junior high school classroom materials. Also some attempts are being made to make some explicit use of deductive principles in nonmathematical fields.

---

[2] E. Hemmerling, *Fundamentals of College Geometry* (John Wiley and Sons Inc., 1964), Chapter I.

Even though school geometry has lost its place as *the* course in clear thinking, a review of current classroom materials indicates that we may now be teaching more geometry in the schools rather than less. For one thing, all of the engineering sciences rest ultimately on classical mechanics which presupposes Euclidean geometry. Thus, one reason to teach geometry in the schools is to provide essential background for other studies. However, both the traditional sequence of topics and the style of presentation have been altered.

Contemporary elementary school materials introduce the pupils to such things as straight lines, triangles, circles, radii, and to such relations as perpendicularity, parallelism, and congruence. Most of the topics are nonmetric in nature. That is, measurement with such related concepts as length and distance are not stressed; rather, the accent is on shapes, relationships, and at times, construction. An attempt is made to show the child how such concepts apply to the world in which he lives. Elementary pupils find such geometry very interesting and often display more learning readiness for it than for the more usual number material. If such new topics in the elementary mathematics curriculum do nothing else, they at least provide variety and help to do the all-important job of keeping the pupils' interest and enthusiasm alive.

Granted that geometry has and deserves an important place in the scheme of things, what is it actually about? The word itself was derived from the Greek words *geos*, meaning earth, and *metrein*, meaning measure. This derivation stresses the fact that geometry in the beginning was the art of earth measurement, that is, of surveying. Does this idea of the subject agree with the one you have obtained from your previous work? To get a better picture, let us look at the basis upon which Euclidean geometry rests. The following list presents the axiom set used for *The Elements*.

The postulates:
1. A straight line can be drawn from any point to any point.
2. A finite straight line can be extended continuously in a straight line.
3. A circle can be drawn with any center and distance.
4. All right angles are equal to one another.
5. If a straight line falling on two straight lines makes the interior angles on the same side less than two right angles, the two straight lines, if produced indefinitely, meet on that side on which the angles are less than two right angles.

The common notions:
1. Things which are equal to the same thing are equal to one another.
2. If equals are added to equals, the wholes are equal.
3. If equals are subtracted from equals, the remainders are equal.

4. Things which coincide with one another are equal to one another.
5. The whole is greater than the part.

Some of the language may seem strange. This is in part due to difficulties in translation. In part it is due to different meanings for specialized terms. For example, Euclid meant "line segment" when he used the term "straight line."

Putting these axioms into more familiar and more precise modern terminology, what kind of geometry is produced? Is it really a geometry of "earth measurement"? That is, does it provide us with a mathematical model of the surface of the earth?

Postulate 1 provides a partial answer. If we use our intuitive idea of "straight line," this cannot be a model of a spherical surface such as that of the earth. If an airplane were to fly from Boston to San Francisco, for example, it could not fly in a "straight" path without going *through* the earth.

If we were to interpret "straight line" as "arc" in the axioms, then Postulate 1 might hold for a spherical surface provided "a" means "one or more than one." On any spherical surface there are infinitely many arcs between two points. It is probable that Euclid meant that exactly one straight line can be drawn between two points in Postulate 1. Then he really did not mean his axioms to produce a model of a spherical surface.

As another example, consider the case of parallel lines. We know that parallel lines can be defined in Euclidean geometry. What would they be like on a spherical surface? We call the latitude lines on a globe "parallels" because they are everywhere equidistant as parallel lines must be in the Euclidean plane. Note, however, that each earth latitude line doubles back on itself as it travels around the globe. This could conceivably be a kind of "line" which might satisfy Postulate 2, but it is certainly not what was meant by Euclid if we read how he interpreted this axiom.

Thus Euclidean geometry is not actually a geometric model of the earth's surface. Its properties produce a model more like the top of a table upon which one can set such solid figures as cubes and pyramids. Why is the table top system called "earth measurement"? Remember that most people once thought that the earth was flat. Thus the name harks back to the time when the earth seemed like an extensive table top, and linear measurement was not accurate enough for most people to tell the difference. Of course, any geometric model would do in an application provided that the discrepancies could not be detected by measuring devices available.

Spherical and Euclidean geometries are not the only reasonable models. There are other kinds which would be accurate enough to represent small sections of the earth's surface very well. (See Exercise 1.) When interplanetary space travel becomes so common that distances become great, we may find that a new geometry provides a more accurate model than does our familiar geometry.

EXERCISE SET

1. Write a brief report (about one page) based on supplementary reading on the contribution to geometry of one of the following: Thales, Pythagoras, Eudoxus, Archimedes, Saccheri, Bolyai, Lobachevsky, Gauss, Riemann, or Newton.

2. List the terms which must be considered as primitive in Euclid's axioms.

3. Give a more modern translation of the phrase "can be drawn" as used in Postulates 1 and 3.

4. Can two distinct angles be equal in our sense as is implied by Postulate 4? See Definition (3-4).

5. What distinction did Euclid seem to be trying to make when he put his axioms into two categories?

§8b MODERNIZING EUCLIDEAN GEOMETRY

As mentioned previously, there are flaws in Euclid's development of geometry. However, for the conditions which pertained in the age in which *The Elements* was written, this work can only be considered as a truly monumental achievement. The important thing in any modern development of geometry is to correct the deficiencies according to contemporary standards, not to blame Euclid because he did not adhere to such standards.

For example, the answer to Exercise 4, §8a should have been *No*. If we are to think of angles as sets of points, then, by definition, distinct angles cannot be equal. If we teach what we mean by equal sets, insist that geometric objects are sets of points, and then refer to particular distinct objects as equal upon occasion, we have no right to be amazed when some pupils become confused. Modern terminology as introduced by Hilbert and others would require us to call objects having the same size and shape congruent instead of equal. As far as we have noted, contemporary elementary school classroom materials are consistently careful about such matters as this. Some of the newer high school texts are just as careful.

In order to make a reasonably careful modern presentation of Euclidean table-top geometry, let us try to sort out the concepts which we regard as essentially geometric and see how they should be related in order to produce a mathematical model like an infinite table top. One thing which we all take for granted is that geometry deals with points. Intuitively we think of a point as a dimensionless indicator of position. Since this property is very difficult to specify by other means, we accept *point* as a primitive term and simply make sure that anything our axioms say about points fits this intuitive notion.

A set of points is called a *locus* (plural:loci) or a *figure*. The latter term is also used for a drawing of a locus. We do not regard a drawing as a mathematical device. It is an inductive device used to arrive at conjectures which can then be treated deductively. As our statement of the meaning of locus implies, the notions of set, subset, and equality of sets are freely used in a modern development of geometry.

Now what relationships among points on a table top might be considered to be geometric? Certainly we consider the *size* and *shape* of a figure to be important geometrically. Thus we must have some way of determining mathematically when two *different* loci have the same size and shape and when they do not. When two loci do have the same shape we say they are *similar*. When they have both the same size and the same shape we say they are *congruent*. The latter term is used as a primitive since size and shape are not previously specified explicitly.

In talking about points on a table top, we certainly want to be able to locate one with respect to another. That is, we must have some way to specify *direction*. One way to specify such direction is to use the system employed for maps. There a system of lines is set up, and such terms as *above*, *below*, *to the left of*, and *to the right of* are used relative to the lines to specify location. In mathematical context all these terms would, of course, have to be defined explicitly, and we would have to make sure that lines so used are straight.

We can adequately specify what we mean by all of these concepts if we manage to specify what we mean by saying that one point is *between* two others. To see why this should be so, it is useful to examine some figures. In making such drawings, we use the age-old convention of designating points by capital letters. Loci will be designated by script capitals or by special combinations of point letters standing for points of the locus.

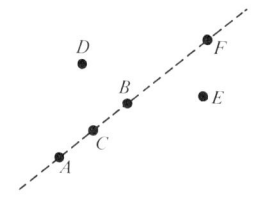

Figure 8.1

Now consider Figure 8.1. If the points A, B, C, D, E, F were on a table top, we would probably agree that B and F are in the same direction from A as C, while C, D, E lie in different directions from A. Also we say that C is between A and B and that C is between A and F. Thus the notion of betweenness specifies the same relationship among points as "same direction" or "opposite direction." We accept "between" as a primitive term and use the notation $\overline{ACB}$ to stand for "Point C is between Point A and Point B." Using the betweenness concept we can also specify what we mean by *straight*. (See Figure 8.1.) Where $\overline{ACB}$, $\overline{ABF}$, but $\sim \overline{AEF}$.

Some of the axioms needed to develop implicitly the concept of betweenness may seem strange at first glance. The following discussion given prior to the beginning of the axiomatic development may be helpful in understanding this notion and why the various axioms about it are needed.

Referring to Figure 8.1 again, our intuitive concept of betweenness should include the idea that since $\overline{ACB}$, the points A, B, C must be distinct. Surely we would not want to say that $\overline{ACB}$ if C were another name for either A or B, or worse, if A, B were two names for the same point. Hence we shall need an axiom to specify that the three points are distinct.

Also, if we have two points specified, we want to be certain that there really are points between the two given points. It will be sufficient to specify that there is *one such point*. Then if $\overline{ACB}$, we must also have $\overline{ADC}$ and $\overline{AED}$,....

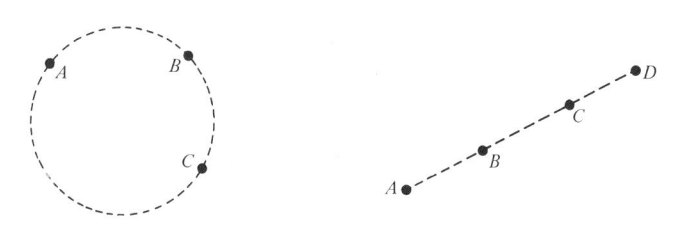

Figure 8.2   Figure 8.3

Now recall that if one looks directly down on a parallel of latitude on a globe, the parallel will look straight. From the side, however, we might have the view in Figure 8.2. This certainly does not represent a situation which should hold in table-top geometry. Note that if we consider the points in clockwise order, we have $\overline{ABC}$ or $\overline{BCA}$ or $\overline{CAB}$. In counterclockwise order we have $\overline{ACB}$ or $\overline{CBA}$ or $\overline{BAC}$. We can eliminate these possible circular interpretations of betweenness by requiring that $\overline{ABC} \Rightarrow \overline{CBA}$ and that at most one of any three points is between the other two. We can accomplish the latter by requiring that $\overline{ABC} \wedge \overline{BCD} \Rightarrow \overline{ABD} \wedge \overline{ACD}$, and $\overline{ABC} \wedge \overline{ACD} \Rightarrow \overline{ABC} \wedge \overline{BCD}$. See Figure 8.3.

Returning to Figure 8.1, we see that Point F has the same direction from A as B. We want to make sure that such points beyond a given two points exist. Hence we must also assert this in an axiom.

To make our mathematical model fit our everyday concept of table-top geometry, there must be points which are *not* between two given points. We shall make all these requirements for table-top geometry about points and the concept of betweenness in the axioms which follow.

Axioms about congruence will be asserted later. Only a partial development of geometry will be made here, for our aim is primarily to reinvestigate a substantial block of material from a more modern viewpoint than Euclid's.

Primitive Terms: *point, between, congruent*
Axioms[3] (first of three groups):

G-1: At least two points exist.

G-2: For any points A, B, C, if $\overline{ABC}$, then A, B, C are distinct.

G-3: For any distinct points A, B, there is at least one point $P \cdot \ni \cdot \overline{APB}$.

G-4: For any two distinct points A, B, there is at least one point $P \cdot \ni \cdot \overline{ABP}$.

G-5: For any points A, B, C, if $\overline{ABC}$, then $\overline{CBA}$.

G-6: For any points A, B, C, D, if $\overline{ABC} \wedge \overline{BCD}$ then $\overline{ABD} \wedge \overline{ACD}$.

G-7: For any points A, B, C, D, if $\overline{ABC} \wedge \overline{ACD}$, then $\overline{ABD} \wedge \overline{BCD}$.

G-8: For any points A, B, C, D, if $\overline{ABC} \wedge \overline{ABD}$ then $\overline{BDC} \vee \overline{BCD}$.

G-9: For any points A, B, there is a point P different from A, $B \cdot \ni \cdot \sim\overline{APB} \wedge \sim\overline{ABP} \wedge \sim\overline{BAP}$.

EXERCISE SET

1. Comment on the meaning of *equal* in Common Notion 1 of Euclid as given in §8a.

2. Rewrite Euclid's Postulates 1, 3, 4 using modern terminology.

3. If $\overline{ACB}$, what direction does point B have from point C compared with the direction of A from C?

4. In what way does the model produced by G-1 through G-9 *not* represent an actual table-top?

5. Draw a figure to represent G-9. Is there a point $Q \cdot \ni \cdot \overline{AQB}$? Why? Is there a point $R \cdot \ni \cdot \overline{ARP}$? Why?

---

[3] M. Keedy and C. Nelson, *Geometry—A Modern Introduction* (Addison-Wesley, 1965), Chapters II, III.

§8c STRAIGHT LINES AND THEIR SUBSETS

Most of us have an excellent idea of what is meant by the phrase *straight line*. Almost all texts on geometry use that phrase as a primitive term, but we should be able to specify or define it with our primitive, *between*.

Definition (8-1): For any distinct points A, B, the set consisting of A, B, together with all the points between them is called a **line segment**, denoted $\overline{AB}$ or simply AB. That is, $\overline{AB} = \{A, B\} \cup \{X \mid \overline{AXB}\}$ or
$$\overline{AB} = \{X \mid X = A \lor X = B \lor \overline{AXB}\}.$$

Each of the points A, B of this definition is called an *end point* of the line segment $\overline{AB}$. Regarding Figure 8.1, for example, we see that $C \in \overline{AB}$ by the definition because $\overline{ACB}$. In the same figure, however, $F \notin \overline{AB}$ because $F \neq A$, $F \neq B$, and $\sim \overline{AFB}$.

You should note carefully that Definition (8-1) specifies that a line segment is a *set* of points, that is, a locus. Accordingly, set sentences such as

(1) $\qquad X \in \overline{AB} \qquad \overline{AB} \subset \overline{CD} \qquad \overline{AB} = \overline{ST}$

are all meaningful ways of stating relationships among points and line segments. In particular we use Figure 8.4 to illustrate the notion that X is any of the points which belong to $\overline{AB}$. The relationship $\overline{AB} \subset \overline{CD}$ is illustrated in Figure 8.5. As for the third sentence of (1), a word of caution may be in order. Since line segments are sets, Definition (3-4) of set equality must apply. Thus the sentence $\overline{AB} = \overline{ST}$ must mean that each element (point) of $\overline{AB}$ also belongs to $\overline{ST}$ and vice versa. Hence, the situation illustrated in Figure 8.6 can never be taken to mean that $\overline{AB}$, $\overline{ST}$ are equal because, for example, $A \notin \overline{ST}$. Our third primitive term will provide us with a way of specifying that any loci such as these two line segments have the same size and shape. We shall write $\overline{AB}$ is congruent to $\overline{ST}$ for the situation shown in Figure 8.6.

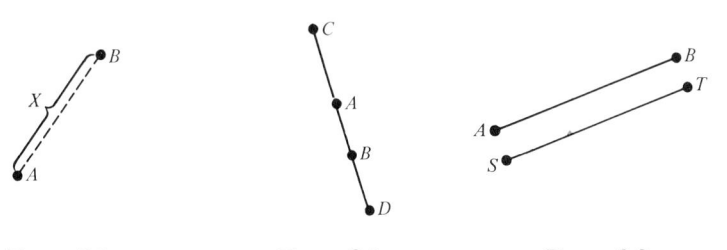

Figure 8.4 **Figure 8.5** **Figure 8.6**

Now returning to Figure 8.4, it ought to seem reasonable that if $\overline{AB}$ is an appropriate name for the line segment, then $\overline{BA}$ would be just as good.

That is, the relative orientation of the two end points is of no importance in naming the locus. This idea is stated in the following proposition.

Theorem (8c-1): For any distinct points A, B, $\overline{AB} = \overline{BA}$.

That is, $\overline{AB}$, $\overline{BA}$ are symbols for the same set of points. A proof of this principle can be made by Axiom G-6.

In discussing sets, we sometimes say that a particular set S is *infinite*. By this we mean that S is nonempty and that for any natural number n, there are at least $n + 1$ elements which belong to S. The following proposition shows that a line segment is such an infinite set.

Theorem (8c-2): Every segment is an infinite set of points.

A proof of this theorem can be constructed using G-1, G-3.

Every set which includes a line segment as a subset is also infinite. In view of Theorem (3e-2), a ray is thus an infinite set according to the following definition.

Definition (8-2): For any two distinct points A, B,

$\overrightarrow{AB} = \overline{AB} \cup \{Y \mid \overline{ABY}\}$ or $\overrightarrow{AB} = \{Y \mid Y \in \overline{AB} \vee \overline{ABY}\}$ is called a **ray**.

The point A of this definition is the *endpoint* of $\overrightarrow{AB}$. Figure 8.7 illustrates the definition. $\overrightarrow{AB}$ has end point A and continues indefinitely in the B direction.

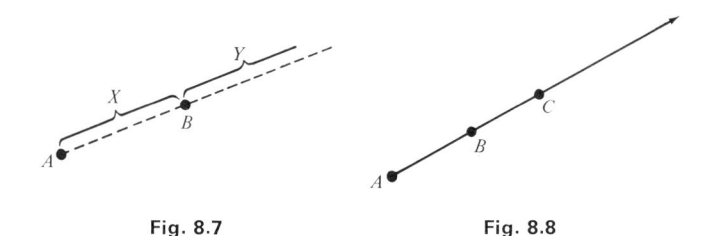

Fig. 8.7   Fig. 8.8

EXAMPLE 1: Write some of the relationships shown in Figure 8.8.
Solution:

$\overline{ABC}$ $\overline{AB} \subset \overrightarrow{AB}$ $\overline{AB} \subset \overline{AC}$ $\{B\} \subset \overline{BC}$ $\overrightarrow{AB} \not\subset \overrightarrow{BC}$

We can now specify the concept of straight line.

Definition (8-3): For any two distinct points A, B, $\overleftrightarrow{AB} = \overrightarrow{AB} \cup \overrightarrow{BA}$ is called a **straight line**.

Another common way of denoting straight lines is by script capitals $\mathscr{L}$, $\mathscr{M}$, etc. Figure 8.9 illustrates this definition where $X \in \overrightarrow{AB}$ and $Y \in \overrightarrow{BA}$.

§8c Straight Lines and Their Subsets

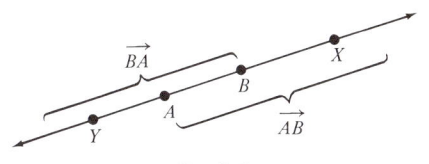

Fig. 8.9

EXAMPLE 2: Write some of the apparent relationships shown in Figure 8.9.

Solution:

$$\overrightarrow{AB} \cap \overrightarrow{BA} = \overline{AB} \qquad \overrightarrow{AB} \cap \overrightarrow{AY} = \{A\} \qquad \overrightarrow{AY} \subset \overrightarrow{XB} \qquad \overrightarrow{AY} \cap \overrightarrow{BX} = \varnothing$$

The next theorem is similar to (8c-1) and its proof should be attempted by everyone in order to gain a better understanding of what Definitions (8-1), (8-2), (8-3) accomplish.

Theorem (8c-3): For any distinct points A, B, $\overleftrightarrow{AB} = \overleftrightarrow{BA}$

A not-so-obvious fact is that if four or more points are known to lie on a line, then the line may be named by any two of them. That is, if A, B, C, D are collinear (belong to the same line) then the line may be named $\overleftrightarrow{AB}$ or $\overleftrightarrow{AC}$ or $\overleftrightarrow{BC}$ or $\overleftrightarrow{BD}$ or $\overleftrightarrow{CD}$. This principle and other properties of straight lines along with properties of the subsets we call line segment and ray will be discussed.

First, however, we must have a mutually agreeable method for representing the primitive and defined terms of the discourse. It is impossible to discuss our geometry adequately on an elementary level without having such a representation.

We have been using a chalkboard or paper to represent our table top. On this surface we have been using a dot to stand for a point. A string of dashes drawn along a ruler has been used to represent $\overline{ABC}$ as in Figure 8.3.

Since a line segment contains *all* of the points between its end points and since it is an infinite set of points by Theorem (8c-2), it is reasonable to represent this locus by an unbroken ruled mark between the endpoints. See Figure 8.10 (i). Starting with a line segment $\overline{AB}$, a ray $\overrightarrow{AB}$ can be represented by continuing the unbroken mark through B in the same direction as illustrated in Figure 8.10 (ii). According to Definition (8-3), we shall draw a straight line $\overleftrightarrow{AB}$ as shown in Part (iii) of the figure. Using these conventions, Figure 8.11 is meant to illustrate $\overline{ABC}$ with the realization that $B \in \overline{AC}, B \in \overrightarrow{AC}, B \in \overleftrightarrow{AC}$ and that the unbroken succession of points extends beyond both A and C from B.

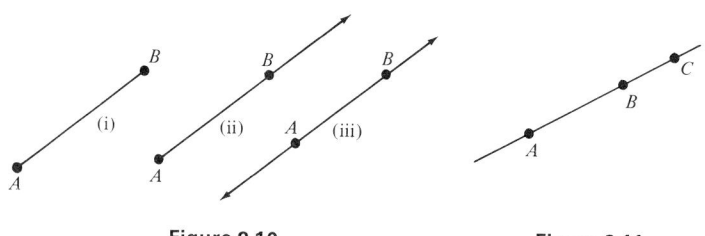

Figure 8.10

Figure 8.11

Returning to consideration of properties of our geometry, recall that we have previously stated that our axioms will allow us to show that at most one of any three points is between the other two. This proposition follows.

Theorem (8c-4): For any three distinct points A, B, C,
$$\overline{ABC} \Rightarrow \sim\overline{ACB} \wedge \sim\overline{BAC}$$

Proof:

1.	$\overline{ACB} \vee \overline{BAC}$	1.	I. P.

(In order to show that this disjunction leads to a contradiction, we show that each component leads separately to a contradiction.)

2a.	$\overline{ACB}$	2a.	I. P. (i)
2b.	$\overline{BCA}$	2b.	G-5
2c.	$\overline{ABC}$	2c.	Hypothesis
2d.	$\overline{ACA}$	2d.	G-6
2e.	Contradiction	2e.	G-2
2f.	$\overline{BAC}$	2f.	I. P. (ii)
2g.	$\overline{CAB}$	2g.	G-5
2h.	$\overline{ABC}$	2h.	Hypothesis
2i.	$\overline{CAC}$	2i.	G-6
2j.	Contradiction	2j.	G-2
3.	$\sim(\overline{ACB} \vee \overline{BAC})$	3.	I. I.
4.	$\sim\overline{ACB} \wedge \sim\overline{BAC}$	4.	DeM.

Q. E. D.

See Figure 8.11.

Proofs of the next few principles tend to be long and involved, and will, therefore, be omitted. Instead, each property will be illustrated in a figure. For example, the following proposition is evident from an examination of Figure 8.12 where the point Y is shown in either of two positions.

Theorem (8c-5): For any points A, X, Y, B,
$$\overline{AXB} \wedge (\overline{AYX} \vee \overline{XYB}) \Rightarrow \overline{AYB}.$$

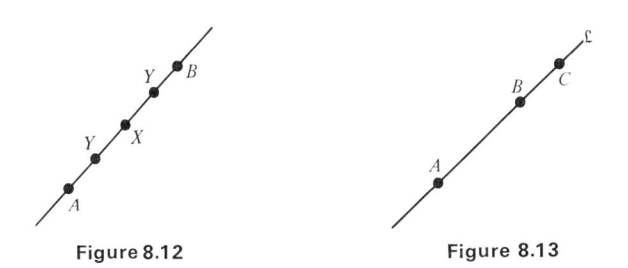

§8c Straight Lines and Their Subsets

Figure 8.12

Figure 8.13

Using this principle we can establish another one which provides us with a convenient alternative way of naming a line segment. Consider a line $\mathscr{L}$ which contains points A, B, C. (See Figure 8.13.) It appears that $\overline{AB} \cup \overline{BC} = \overline{AC}$ provided $B \in \overline{AC}$. This idea can be restated as follows.

Theorem (8c-6): For any points $A, X, B \cdot \ni \cdot \overline{AXB}$, then $\overline{AB} = \overline{AX} \cup \overline{XB}$.

EXAMPLE 3: Write $\overline{AB}$ of Figure 8.14 in three ways as a union of line segments.

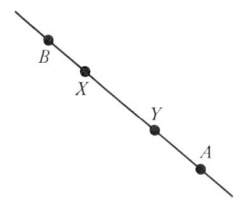

Figure 8.14

Solution:

$$\overline{AB} = \overline{AX} \cup \overline{XB} \qquad \overline{AB} = \overline{AY} \cup \overline{YB} \qquad \overline{AB} = \overline{AY} \cup \overline{YX} \cup \overline{XB}$$

Another principle is illustrated in Figure 8.14. Since we intend that $\overline{AXB}$ and $\overline{AYB}$, then either $\overline{AXY}$ or $\overline{AYX}$ but not both.

Theorem (8c-7): For distinct points A, B, X, Y such that $\overline{AXB} \wedge \overline{AYB}$, exactly one of $\overline{AXY}, \overline{AYX}$ holds.

Theorem (8c-6) allows us to name a line segment in various ways. It is also convenient to have alternative names for rays and straight lines. Do $\overrightarrow{AB}$ and $\overrightarrow{AP}$ represent the same set of points in Figure 8.15, for example? Another way of asking this question is to ask whether there is any point of $\overrightarrow{AP}$ which does not belong to $\overrightarrow{AB}$ and vice versa. If not, $\overrightarrow{AB}, \overrightarrow{AP}$ would be two names for the same ray. The following proposition states that this is the case.

348   *Geometry*

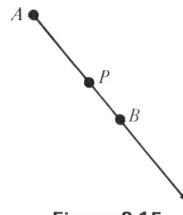

Figure 8.15

Theorem (8c-8): For $\overrightarrow{AB}$ and $P \in \overrightarrow{AB} \cdot \ni \cdot P \neq A$, $\overrightarrow{AB} = \overrightarrow{AP}$.

EXAMPLE 4: Name the ray $\overrightarrow{PS}$ of Figure 8.16 in two other ways.

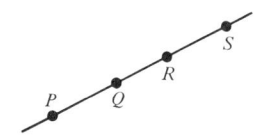

Figure 8.16

Solution: $\qquad \overrightarrow{PS} = \overrightarrow{PQ} \qquad \overrightarrow{PS} = \overrightarrow{PR}$

Just as the preceding theorem provides us with alternative names for rays, the following one gives us a way of renaming straight lines. See Figure 8.17.

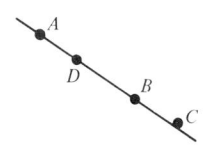

Figure 8.17

Theorem (8c-9): If $C \neq D$ and $C, D \in \overleftrightarrow{AB}$, then $\overleftrightarrow{CD} = \overleftrightarrow{AB}$.

Returning to Euclid's *Postulates* as given in §8a, the first two, taken together, are usually interpreted as specifying that there is exactly one straight line which contains two specified points. We state this as a theorem whose proof is left to the exercises. Note that the proof would be in two parts.

Theorem (8c-10): For any two distinct points, there is one and only one line which contains them.

This preposition states that the situation shown in Figure 8.18 is impossible if $\mathscr{L}, \mathscr{M}$ are both straight lines.

§8c Straight Lines and Their Subsets 349

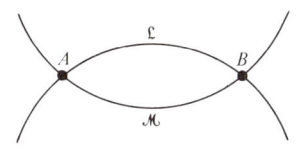

Figure 8.18

A closely related idea is stated in the following theorem. It asserts, for example, that if $\mathscr{L}$, $\mathscr{M}$, A, B are distinct and $\mathscr{L} \cap \mathscr{M} = \{A\}$, then $\mathscr{L} \cap \mathscr{M} = \{B\}$ is impossible.

Theorem (8c-11): Any two distinct lines contain at most one point in common.

EXERCISE SET

1. Using the representation of a line in Figure 8.19, name the line in all possible ways.

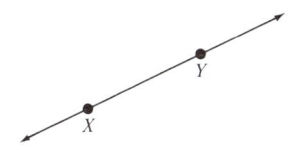

Figure 8.19

2. Given a line $\mathscr{L}$ with points A, B, X identified, list some of the possible names for $\mathscr{L}$.

3. Locate three distinct points X, Y, Z on your paper so that they are elements of the same line $\mathscr{K}$. What is $\overline{XY} \cap \overline{YZ}$? $\overline{XY} \cap \overline{XZ}$? $\overline{XZ} \cap \overline{YZ}$?

4. Given line $\mathscr{M}$ (Figure 8.20) with points A, B, C, D identified, simplify each set if possible.

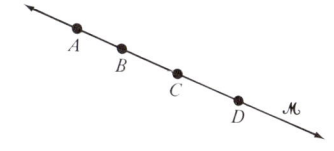

Figure 8.20

a) $\overline{AB} \cup \overline{BD}$ b) $\overrightarrow{AC} \cap \overline{BD}$ c) $\overline{AD} \cap \overline{BC}$
d) $\overleftrightarrow{AB} \cap \overline{BC}$ e) $\overrightarrow{BC} \cup \overrightarrow{CD}$ f) $\overrightarrow{BC} \cap \overrightarrow{BA}$
g) $\overrightarrow{AB} \cap \overline{CD}$ h) $\overrightarrow{AB} \cap \overline{BC}$ i) $\overrightarrow{AD} \cup \overline{BC}$
j) $\overleftrightarrow{AB} \cap \overrightarrow{BC}$ k) $\overrightarrow{BC} \cap \overrightarrow{CD}$ l) $\overrightarrow{DC} \cup \overrightarrow{CA}$

5. Indicate whether the following statements are true or false; if a statement is false, indicate under what circumstances, if any, it might be true.

a) $\overline{EF} = \overline{FE}$
b) $\overline{QR} \cup \overline{RB} = \overline{QR}$
c) $\overrightarrow{AR} = \overrightarrow{RA}$
d) $\overline{EF} \cap \overline{GH} = \overline{FH}$
e) $\overrightarrow{QR} \subset \overrightarrow{RP}$
f) $\overline{GR} \subset \overrightarrow{RP}$
g) $\overrightarrow{RG} \cap \overrightarrow{GH} = \overline{GR}$
h) $\overrightarrow{EF} = \overrightarrow{EG} = \overrightarrow{EH} = \overrightarrow{EI}$
i) $\overrightarrow{RP} \cup \overrightarrow{GH} = \overline{GR}$
j) $\overrightarrow{EF} \cup \overrightarrow{GF} = \overrightarrow{GE}$

6. Using Figure 8.21, complete each sentence.

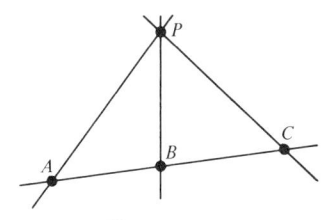

Figure 8.21

a) $(\overline{AB} \cup \overleftrightarrow{BC}) \cap \overrightarrow{CP} = $ _____
b) $\overrightarrow{AB} \cap \_\_ = \overline{AC}$
c) $(\overrightarrow{PA} \cup \overrightarrow{PC}) \cap \overleftrightarrow{AC} = $ _____
d) $(\overline{AB} \cap \overline{BC}) \cap \overleftrightarrow{AP} = $ _____
e) $\{P\} \subset \overrightarrow{AP} \cap \underline{}B$
f) $\overline{AB} = \overline{B}$
g) $(\overrightarrow{AP} \cup \overrightarrow{AC}) \cap (\overrightarrow{BC} \cup \overrightarrow{BP}) = $ _____

7. Prove:
   a) $\overline{AB} \subset \overrightarrow{AB}$
   b) $\overrightarrow{AB} \subset \overleftrightarrow{AB}$
   c) $\overline{AB} \subset \overleftrightarrow{AB}$

8. Prove:
   a) Theorem (8c-1)
   b) Theorem (8c-2)
   c) Theorem (8c-3)
   d) Theorem (8c-10)
   e) Theorem (8c-11)

§8d PLANES AND SPACE

Up to now we have limited ourselves to considerations of Axioms G-1 through G-8. As far as these eight axioms are concerned, the entire set of points under consideration might belong to a single line. Referring to Figure 8.22, for example, we could let A, B be the two points required by G-1. By Theorem (8c-2) we know that there are points such as X between

§8d   **Planes and Space**   351

A, B. Axiom G-4 assures us that there are points such as *Y, Z* beyond *A, B*. According to Definition (8-3), all of those points belong to a single straight line, that is, they are collinear.

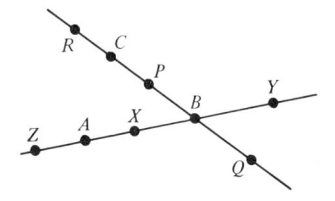

Figure 8.22

This straight line is not enough to represent a table top. Referring to Figure 8.22 again, we note that G-9 does provide us with a point *C* which does not belong to the line $\overleftrightarrow{AB}$. Then, as illustrated, there must also be points *P, Q, R* which belong to $\overleftrightarrow{CB}$ such that $\overline{CBQ}$, $\overline{CPB}$, $\overline{RCB}$. Now by Theorem (8c-10) any two distinct points of $\overleftrightarrow{AB}$, $\overleftrightarrow{BC}$, respectively, except *B* determine straight lines different from $\overleftrightarrow{AB}$ and $\overleftrightarrow{BC}$. Thus G-9 and (8c-10) together guarantee that there are infinitely many distinct straight lines in our geometry. This fact leads to the following definition.

Definition (8-4): For any three distinct non-collinear points *A, B, C*,

$$\{P \mid P \in \overleftrightarrow{AX} \wedge X \in \overline{BC}\} \cup \{P \mid P \in \overleftrightarrow{BY} \wedge Y \in \overline{AC}\} \cup$$
$$\{P \mid P \in \overleftrightarrow{CZ} \wedge Z \in \overline{AB}\}$$

is called a **plane**.

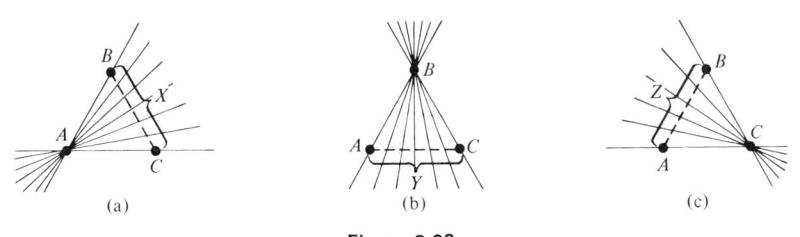

Figure 8.23

According to this definition one might describe a plane as an infinite table top. To see why this is so, we could superimpose the three parts of Figure 8.23.

In Figure 8.24 we illustrate some drawing conventions used to represent a plane or planes. Obviously one cannot actually draw a flat surface of unlimited extent. Instead, we draw a figure to look like a flat piece of cardboard viewed somewhat from the side as in Parts (i) and (ii). In (i) we mean that P belongs to the plane Π whereas P does not belong to plane Π in (ii). In Part (iii) intersecting planes Π, Φ are drawn with those lines dashed which could not be seen if the planes were opaque. Part (iv) is intended to illustrate the case for which the straight line $\mathscr{L}$ is a subset of Π. Since Π can not be drawn as unlimited in extent, $\mathscr{L}$ can not be drawn as endless either. However, a long-standing convention states that if a straight line is not drawn as ending at particular points, it continues endlessly beyond any points which may be shown. Currently, lines are often drawn with arrowheads to indicate indefinite extent.

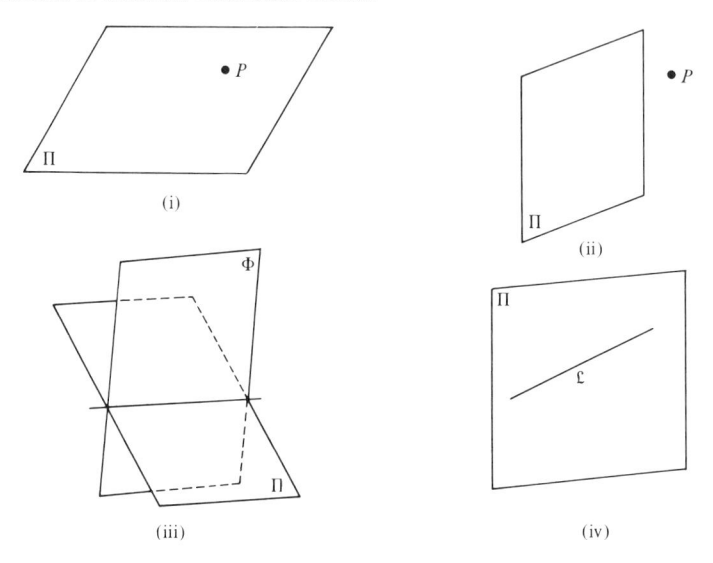

Figure 8.24

We have asked whether more than a single line exists in our geometry and found that at least one plane exists. In trying to make sure that this geometry has other familiar Euclidean properties, we need to ask some further questions. For example, we can ask whether there really is a point P in our geometry which does not belong to a specified plane Π as shown in Figure 8.24 (ii). We cannot guarantee such a point on the basis of axioms G-1 through G-9 and a new axiom is needed for this purpose.

As we have seen, the preceding axioms and theorems do assure us that there will be at least one plane on any three noncollinear points. Can we be certain, however, that three points determine only one plane? Still another axiom is needed to answer this question. Also, in Part (iii) of Figure 8.24, intersecting planes are depicted. Does more than one plane really exist? Is this intersection actually a straight line as drawn? If, then, some points of a straight line belong to a plane, is the entire line a subset of the plane? Still another axiom is needed to answer these three questions.

Further, if we consider *two* distinct lines $\mathscr{L}$, $\mathscr{M}$ in a plane Π as shown in Figure 8.25, what are the possible kinds of intersections? In Figure 8.22

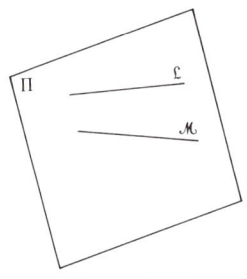

Figure 8.25

we examined a case in which the intersection of two lines contains a single point. We also know by Theorem (8c-11) that no distinct straight lines can have more than one point in common. The question remains: can coplanar lines $\mathscr{L}, \mathscr{M}$ have an empty intersection? It probably seems so intuitively, but under what conditions? These questions too can be answered only by stating an additional axiom.

G-10: For any three distinct non-collinear points, there is no more than one plane that contains them.

G-11: For any plane Π there is at least one point $P \cdot \ni \cdot P \notin \Pi$.

G-12: If the intersection of two planes is nonempty, their intersection contains at least two points.

G-13: For any straight line $\mathscr{L}$ in a plane Π and for any point $P \in \Pi$ $\cdot \ni \cdot P \notin \mathscr{L}$, there is exactly one straight line $\mathscr{M} \subset \Pi \cdot \ni \cdot$ $P \in \mathscr{M} \wedge \mathscr{L} \cap \mathscr{M} = \varnothing$.

These axioms answer some of the questions previously posed. The following three propositions will answer the remaining ones.

Theorem (8d-1): For any two points P, Q belonging to plane Π, $\overleftrightarrow{PQ} \subset \Pi$.

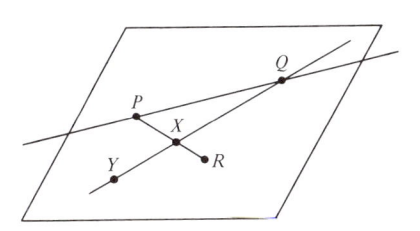

Proof (outline): Let $R \in \Pi$ such that $R \notin \overleftrightarrow{PQ}$. Consider plane PQR. By G-10 this must be the same plane as Π. Now if $X \in \overline{PR} \wedge Y \in \overleftrightarrow{QX}$, we have $Y \in \Pi$ by our definition of plane. Each point of $\overleftrightarrow{PQ}$ is such a point Y as we can see by letting X be P.

The proofs of the following two propositions are left to the exercises.

Theorem (8d-2): For any line and any point not belonging to that line, there is one and only one plane that contains them.

Theorem (8d-3): For any two distinct lines $\mathscr{L}, \mathscr{M}$, if $\mathscr{L} \cap \mathscr{M} \neq \varnothing$, then $\mathscr{L}$ and $\mathscr{M}$ are contained in one and only one plane.

Note that by this theorem two straight lines are guaranteed to determine a plane only if their intersection is nonempty. By Axiom G-13, however, there is a case in which lines having an empty intersection do determine a plane. We call such lines parallel.

Definition (8-5): Two coplanar straight lines are **parallel** if and only if their intersection is empty.

By G-13, then, parallel lines exist.

We now have four ways to specify a plane: by three noncollinear points; by a line and a point which does not belong to the line; by two intersecting lines; by two parallel lines. That is, each of these configurations determines just a single plane. By Axiom G-11 as illustrated by Figure 8.24 (ii), not all points in this geometry belong to the same plane.

Definition (8-6): The set of all possible points is called **space**.

Referring to Figure 8.26, the plane Π and the point P both belong to our set called space. If A, B, Q are noncollinear points of Π, then $\overleftrightarrow{AB} \cap \overleftrightarrow{QP} = \varnothing$. Thus $\overleftrightarrow{AB}, \overleftrightarrow{QP}$ are noncoplanar.

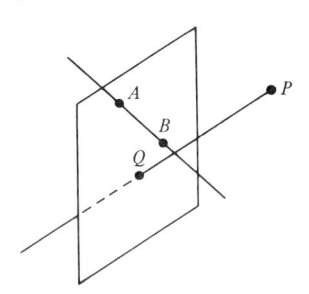

Figure 8.26

Definition (8-7): Two lines are called **skew** if and only if they are nonparallel and noncoplanar.

Proofs of the following two propositions are left to the exercises.

Theorem (8d-4): For any line $\mathscr{L}$ and any plane Π, if $\mathscr{L}$ is not on Π then $\mathscr{L} \cap \Pi$ contains at most one point.

§8d   *Planes and Space*   355

Theorem (8d-5): For any two distinct planes whose intersection is nonempty, their intersection is a line.

EXERCISE SET

1. Answer the following questions based on Figure 8.27.
   a) Is point A an element of Π? an element of $\overleftrightarrow{XY}$?
   b) Is point P an element of Π? of $\overleftrightarrow{XY}$?

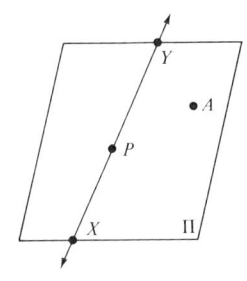

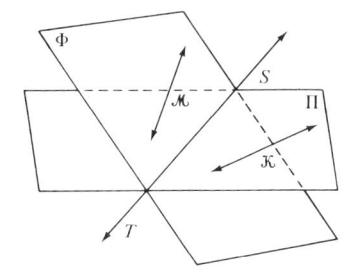

   Figure 8.27   Figure 8.28

2. Consider planes Π, Φ as represented in Figure 8.28.
   a) Is line $\mathcal{M}$ a subset of Π?
   b) Is line $\mathcal{K}$ a subset of Π?
   c) What line is a subset of both planes?
   d) Are there more lines in both planes than are indicated? Explain.

3. Write each intersection for a point P, a line $\mathcal{M}$, a plane Π, and space S as a single set.
   a) $\{P\} \cap \mathcal{M}$ b) $\{P\} \cap \Pi$
   c) $\{P\} \cap S$ d) $\mathcal{M} \cap \Pi$
   e) $\mathcal{M} \cap S$ f) $\Pi \cap S$

4. Describe the possible intersection of planes Π_1, Π_2, Π_3.

5. Explain exactly how Axiom G-9 guarantees the existence of a point C in Figure 8.22 such that $C \notin \overleftrightarrow{AB}$.

6. Is it possible to have skew lines in a plane? Explain.

7. We sometimes say that (a) two line segments or (b) a ray and a line segment are parallel. Definition (8-5) does not, however, state anything about either line segments or rays. Write a definition to cover each of these cases.

8. Two planes are either parallel or their intersection is nonempty exclusively.

a) Is this also true of straight lines in a plane? straight lines in space? Explain.
b) Is this also true of rays in a plane? rays in space? Explain.
c) Is this also true of a ray and a plane? Explain.

9. Draw an illustration of a plane Π containing a straight line $\mathscr{L}$ and a plane Φ containing a straight line $\mathscr{M}$ for each specified case. The intersection of Π, Φ is nonempty in each case.
   a) $\mathscr{L} \cap \Phi \neq \varnothing;\ \mathscr{M} \cap \mathscr{L} = \varnothing$
   b) $\mathscr{L} \cap \Phi = \varnothing;\ \mathscr{M} \cap \Pi = \varnothing$
   c) $\mathscr{L} \cap \mathscr{M} \neq \varnothing$
   d) $\mathscr{L}, \mathscr{M}$ are skew; $\mathscr{L} \cap \Phi \neq \varnothing;\ \mathscr{M} \cap \Pi \neq \varnothing$

10. Prove
    a) (8d-2)   b) (8d-3)
    c) (8d-4) (*Hint:* use I. I.) *d) (8d-5)

§8e PARTITIONS

Returning to considerations of straight lines, let us examine the situation illustrated in Figure 8.29. What subsets of $\mathscr{L}$ relative to the point P of $\mathscr{L}$ can we identify? Certainly $\{P\}$ is one subset. Another is the set S_1 of all points A such that $\overline{APB}$ and another subset of $\mathscr{L}$ is the set S_2 of all points B such that $\overline{APB}$.

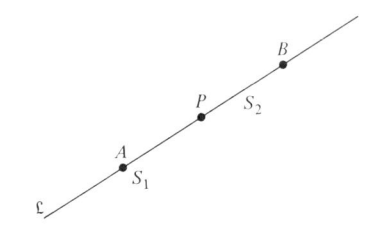

Figure 8.29

Other subsets of $\mathscr{L}$ can be identified, of course, but these three, S_1, S_2, $\{P\}$, are of particular interest. We note for example that these three subsets of $\mathscr{L}$ are disjoint. To see this, recall that Axiom G-3 assures us that $P \notin S_1$ and $P \notin S_2$ because A, P, B are distinct. Likewise S_1, S_2 are disjoint for, if they were not, there would be a point, say C, such that $C \in S_1$ and $C \in S_2$. Then by our definitions of S_1, S_2 we would have $\overline{CPC}$ which is also impossible by G-3. A set of disjoint subsets of this type has a special name.

Definition (8-8): For any nonempty set M and set $\mathscr{P} = \{S_1, S_2, S_3, \ldots, S_n\}$ such that $S_1, \ldots, S_n$ are disjoint nonempty subsets of M and $S_1 \cup S_2 \cup \ldots \cup S_n = M$, the set $\mathscr{P}$ is called a **partition** (or **separation**) of M.

§8e **Partitions**

According to the discussion preceding this definition, then, $\mathscr{P} = \{S_1, \{P\}, S_2\}$ is a partition or separation of $\mathscr{L}$. We sometimes say that $\mathscr{P}$ *separates* $\mathscr{L}$. The subsets S_1, S_2 of $\mathscr{L}$ are sometimes called *half lines;* one must be careful not to confuse this usage of the word *half* with our usual meaning of the term.

In working with geometric loci, it is often important to be sure that all the points of some locus belong to one particuiar set. Knowledge that we have a partition is often helpful in this regard. For example, if $\mathscr{P} = \{S_1, \{P\}, S_2\}$ partitions $\mathscr{L}$ in Figure 8.30 and if $A, B \in S_1$, then we know that *all* points of $\overline{AB}$ belong to S_1. The ability to locate a line segment exactly is so important that we have a special term for a set which contains all the points of a line segment.

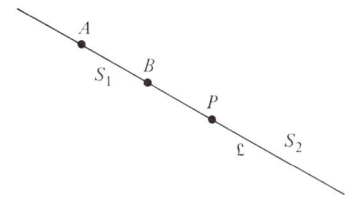

Figure 8.30

Definition (8-9): A set S is a **convex set** iff, for any two points $A, B \in S$, $\overline{AB} \subset S$.

You should note that the sets S_1, S_2, $\{P\}$ of the partition $\mathscr{P}$ of a straight line $\mathscr{L}$ as described previously are each convex sets. In Figure 8.31 the hatched regions of (i) and (iii) are convex sets whereas that of (iv) is not a convex set. The circle of (ii) is not a convex set either since only the points A, B of line segment $\overline{AB}$ belong to it.

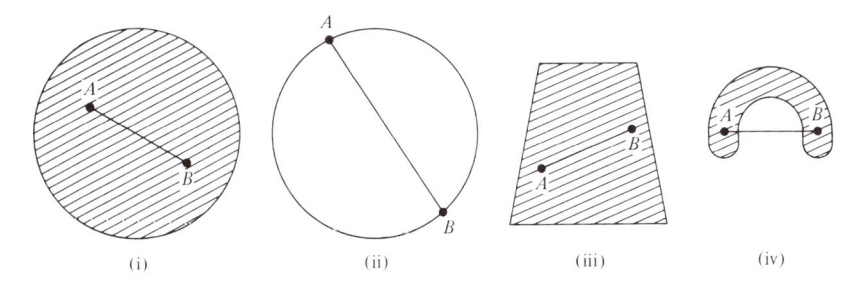

Figure 8.31

Consider the crosshatched region and the line segment $\overline{AB}$ of Figure 8.32 carefully as an illustration of the following proposition about convex sets.

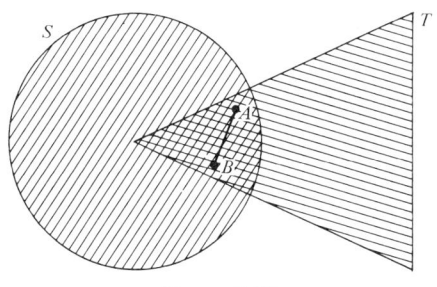

Figure 8.32

Theorem (8e-1): The intersection of any two convex sets is also a convex set.

Proof (outline): By hypothesis, we have convex sets S, T. For any points A, B such that $A, B \in S \cap T$, we have $A \in S \wedge B \in S$ and also $A \in T \wedge B \in T$ by the definition of intersection. Since both S, T are convex, we have $\overline{AB} \subset S \wedge \overline{AB} \subset T$. That is,

$$(X \in \overline{AB} \Rightarrow X \in S) \wedge (X \in \overline{AB} \Rightarrow X \in T)$$

whence

$$X \in \overline{AB} \Rightarrow X \in S \wedge X \in T \quad \text{or} \quad X \in \overline{AB} \Rightarrow X \in (S \cap T)$$

whence

$$\overline{AB} \subset S \cap T$$

Thus $S \cap T$ is convex by the definition. Q. E. D.

Just as a point partitions a line into convex sets so we might expect that a straight line partitions a plane into convex sets. Consider Figure 8.33, for example. If $\mathscr{L} \subset \Pi$, then $\mathscr{L}$, S_1, S_2 will all be disjoint and will contain all the points of Π. In fact, $\{\mathscr{L}, S_1, S_2\}$ is a partition of Π. We shall show that this is indeed the case. The following definition facilitates discussion.

Definition (8-10): For any plane Π and straight line $\mathscr{L} \subset \Pi$ and any point $P \in \Pi \cdot \ni \cdot P \notin \mathscr{L}$, the sets $S_1 = \{P \mid \overline{RQP}\}$ and $S_2 = \{R \mid \overline{PQR}\}$, where Q is a point of $\mathscr{L}$, are called **sides** of $\mathscr{L}$.

Figure 8.34 illustrates these relationships where $\mathscr{L} \subset \Pi$, $Q \in \mathscr{L}$, $\overline{PQR}$, $S_1 = \{P, P', \ldots\}$, $S_2 = \{R, R', \ldots\}$.

The sets S_1, S_2 of this definition are sometimes called *half planes*, and the line $\mathscr{L}$ is then called the *edge* of either half plane.

§8e Partitions

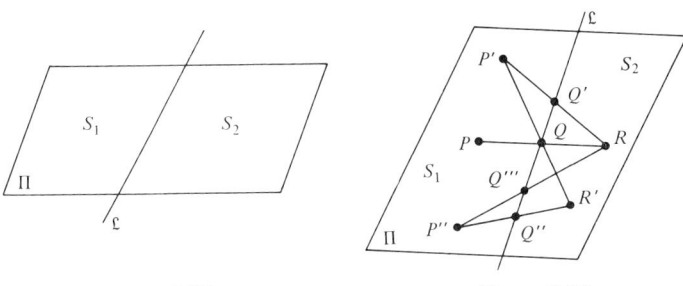

Figure 8.33

Figure 8.34

We shall now outline how to show that $\{S_1, S_2, \mathscr{L}\}$ partitions the plane Π.

Theorem (8e-2): For any plane Π and straight line $\mathscr{L} \subset \Pi$ having sides S_1, S_2, the set $\{S_1, S_2, \mathscr{L}\}$ partitions Π.

Proof (outline): We must show that (i) $S_1, S_2, \mathscr{L}$ are each nonempty; (ii) $S_1 \cup S_2 \cup \mathscr{L} = \Pi$; (iii) $S_1 \cap S_2 = \varnothing$ and $S_1 \cap \mathscr{L} = \varnothing$ and $S_2 \cap \mathscr{L} = \varnothing$.

Part(i): Since a straight line is the union of two rays and since a ray is the union of a line segment and another set of points, we know by (8c-2) that $\mathscr{L}$ is nonempty, say $A \in \mathscr{L}$. By axiom G-9, there is a point P such

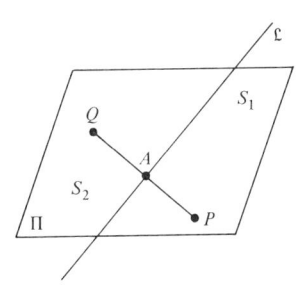

that $P \notin \mathscr{L}$. Then G-4 assures us that there is a point $Q \in \Pi$ such that $\overline{PAQ}$. Now by (8-10) we have $P \in S_1$, and $Q \in S_2$; that is, S_1, S_2 are both nonempty. DONE

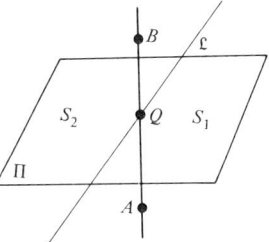

Part (ii): As an indirect premise consider a point $A \in \Pi$ such that $A \notin S_1$ and $A \notin S_2$ and $A \notin \mathscr{L}$. For any point $Q \in \mathscr{L}$, we have $\overleftrightarrow{AQ} \subset \Pi$ by (8d-1). G-4 assures us that a point B exists such that $\overline{AQB}$. Since $B \in \overleftrightarrow{AQ}$, then $B \in \Pi$ also. By (8-10) S_1 contains *all* points P such that $\overline{PQB}$, that is, $A \in S_1 \lor A \in S_2$. This contradiction assures us that there is no point of Π which does not belong to S_1, S_2, or $\mathscr{L}$. That is, $S_1 \cup S_2 \cup \mathscr{L} = \Pi$.

DONE

Part (iii): By the definition, $S_1 = \{P \mid \overline{PQR}\}$ where $Q \in \mathscr{L}$ and $P \notin \mathscr{L}$. Thus $\overleftrightarrow{PR}$ and $\mathscr{L}$ are distinct and Q is their *only* common point by (8c-11). Since P, Q are distinct according to G-3, we have $S_1 \cap \mathscr{L} = \varnothing$. Similarly, we would have $S_2 \cap \mathscr{L} = \varnothing$. (See Figure 8.34.) Now, as an indirect premise, let us consider point $A \cdot \ni \cdot A \in S_1 \land A \in S_2$. Since $A \in S_1$, we

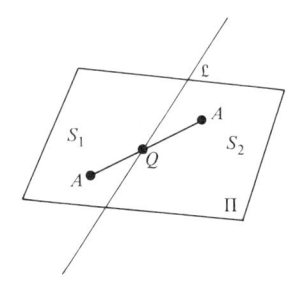

must have $S_2 = \{R \mid \overline{AQR}\}$ for some $Q \in \mathscr{L}$. In particular, we must also have $\overline{AQA}$ because $A \in S_2$. This contradicts G-2. Therefore, no such point A exists. DONE Q. E. D.

Corollary (8e-2a): The sets $\mathscr{L}$, S_1, S_2 consisting of a straight line and each of its sides are each convex sets.

It is also possible to prove that a plane partitions or separates space into three convex sets. These sets are sometimes called *half spaces*, and the plane is called the *face* of each half space.

Returning to the case of a plane, it should be noted that a straight line is not the only locus which partitions the plane. There are, in fact, many different ways to produce a planar partition or separation. Two of the most common loci which have this property are an angle and a triangle.

Definition (8-11): An **angle** is the union of two noncollinear rays having a common endpoint.

The common end point is called the *vertex* of the angle; each ray is called a *side* of the angle. If $\overrightarrow{AB}$, $\overrightarrow{AC}$ are the two sides we denote the angle as

§8e   **Partitions**   361

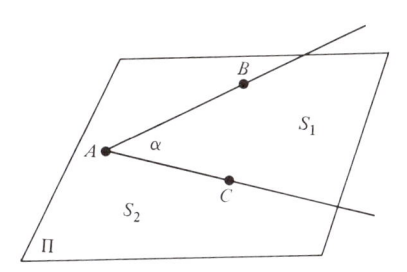

Figure 8.35

∢ *BAC* with the vertex as the middle letter. In cases where no confusion results, an angle may be named by its vertex as ∢ *A*. We sometimes name an angle by a small Greek letter. Thus in Figure 8.35 we have ∢ *BAC* = $\overrightarrow{AB} \cup \overrightarrow{AC}$ = ∢ *A* = α. Figure 8.35 also shows ∢ *BAC* as a subset of a plane Π.

Now it is reasonable to ask whether an angle partitions its plane. The set {∢ *A*, S_1, S_2} shown in Figure 8.35 does indeed partition the plane Π for suitably specified sets S_1, S_2.

Definition (8-12): For any ∢ *ABC* the **interior** of ∢ *ABC* is the set

$$S_1 = \{P \mid \overline{QPR} \wedge Q \in \overrightarrow{BC} \wedge R \in \overrightarrow{BA} \wedge P \neq B\}$$

That is, the interior of an angle is the set of all points between points of its sides. See Part (i) of Figure 8.36 where the interior of ∢ *B* is hatched.

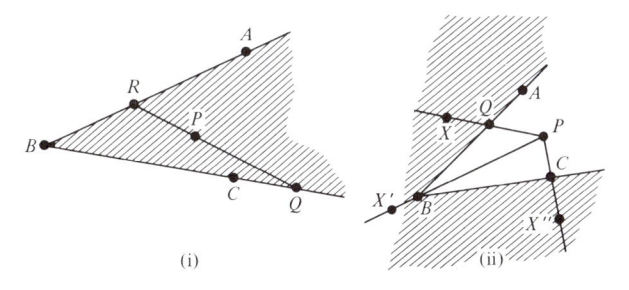

Figure 8.36

Definition (8-13): For any angle *ABC* and any point *P* of its interior, the set $S_2 = \{X \mid \overline{PQX} \wedge Q \in$ ∢ *ABC*} is called the **exterior** of the angle.

The exterior $S_2 = \{X, X', X'', \ldots\}$ is hatched in Part (ii) of Figure 8.36.

The following theorem can be established without too much difficulty.

Theorem (8e-3): For any Π and any angle α ⊂ Π for which S_1 is the interior of α and S_2 its exterior, the set {α, S_1, S_2} is a partition of Π.

362 *Geometry*

Corollary (8e-3a): The interior of an angle is a convex set.

A triangle is still another locus which produces a plane separation, i. e., a partition of the plane.

Definition (8-14): For three noncollinear points A, B, C, the set $\overline{AB} \cup \overline{BC} \cup \overline{CA}$ is called a **triangle**, denoted $\triangle ABC$.

The segments are called the *sides* of the triangle and each point of intersection of any two of the sides is called a *vertex* of the triangle.

We can define the *interior of a triangle* as the intersection of the interiors of its three angles, $\measuredangle ABC$, $\measuredangle CAB$, $\measuredangle BCA$. See Figure 8.37, where the interior of $\measuredangle CAB$ is hatched horizontally, the interior of $\measuredangle CBA$ is hatched vertically, and the interior of $\triangle ABC$ is crosshatched. If W is the interior of $\triangle ABC$ in a plane, the *exterior* can be defined as $(\triangle ABC \cup W)'$. The following proposition then results.

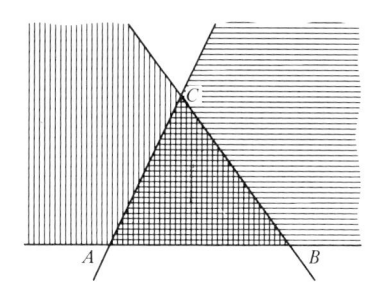

Figure 8.37

Theorem (8e-4): For any plane Π and any $\triangle ABC \subset \Pi$, the set $\{\triangle ABC,$ its interior, its exterior$\}$ is a partition of Π.

EXERCISE SET

1. Which of the hatched regions of Figure 8.38 represent convex sets?

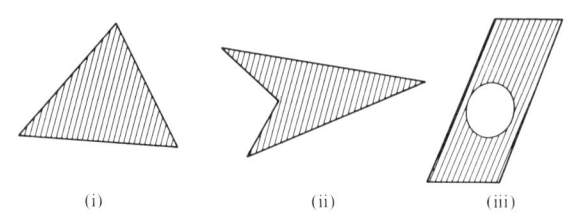

(i) (ii) (iii)

Figure 8.38

2. Explain why a half line is not half of a line in our usual sense of the word *half*.
3. Label each proposition true or false relative to Figure 8.39.

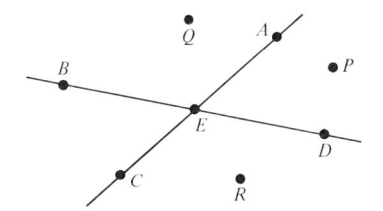

Figure 8.39

   a) *BED* is an angle.
   b) *P* belongs to the interior of $\measuredangle AED$.
   c) *R* belongs to the interior of $\measuredangle BEA$.
   d) *Q, P* are on the same side of $\overleftrightarrow{EC}$.
   e) *A, Q* are on the same side of $\overleftrightarrow{BD}$.
4. Why is the vertex of the angle excluded when defining *interior* in (8-12)?
5. Is each set convex? Explain.
   a) a straight line   b) a triangle
6. Prove or disprove:
   a) The union of two convex sets is convex.
   b) The empty set is convex.
   c) A point set is convex.
   d) A ray partitions a plane.
   e) The interior of a triangle is convex.
   f) An angle is a convex set.
   g) The exterior of a triangle is convex.
   h) The exterior of an angle is convex.
7. Can a line contain the three vertices of a triangle? Explain.
8. Draw three different figures to illustrate that a line may contain exactly one point of an angle.
9. How many triangles are determined by four points in a single plane if no three of them belong to the same line? How many triangles are determined by four points if exactly three of them belong to the same line? Illustrate.
10. Prove:
    a) (8e-2a)
    b) (8e-3a)

364 *Geometry*

11. Prove:
    a) There is one and only one plane which contains an angle.
    b) Either side of a straight line $\mathscr{L}$ contains at least three points.

*12. Prove:
    a) (8e-3)
    b) (8e-4)

§8f CONGRUENCE AND MEASURE

Up to this point, in describing geometric loci and their properties we have restricted ourselves to concepts which can be called *nonmetric*. That is, nonmetric relationships are those which we can specify solely in terms of such notions as between, union, intersection. However, in any extensive discussion of geometric relationships, we find it imperative to make use of concepts such as *size* and *shape* which can be directly related to measurement. Relationships described in these terms are said to be *metric*.

Probably we all have a reasonably correct intuitive idea of what is meant by saying that a pair of loci have the same size or the same shape. You will agree, for example, that the loci of Parts (i), (ii) of Figure 8.40 seem to have the same size and that the two kites of (iii), (iv) seem to have the same shape but not the same size.

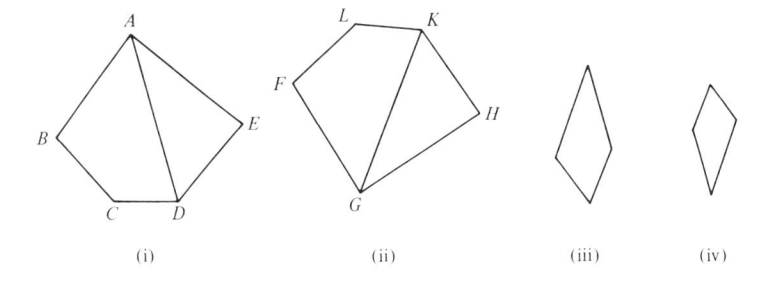

Figure 8.40

As stated previously, we say that loci are *congruent* when they have the same size and shape. However, the terms same size and same shape have not been previously defined. Hence, they can not be used to define congruent. One way to give mathematical meaning to the congruence relation is to recall that we often explain what we mean by "same size" in terms of number. That is, we quote a real number which gives other persons an indication of size relative to some known standard. This technique can be used just as effectively with line segments and angles as with money or herds of sheep.

§8f Congruence and Measure

None of our previous axioms establish any numeric relationship between numbers and loci. The next task in the development of our geometry will be to assure ourselves of such number-locus associations. We must briefly investigate the properties which we shall want to specify in formulating the needed axioms.

The real number we associate with a locus for the purpose of comparing its size with other loci of the same type is called a *measure*. We can use the symbol $m\ (\overline{AB})$ to stand for the real number measure of the line segment $\overline{AB}$. Certainly each possible line segment should have a unique measure if this concept is to be useful to us. Furthermore, if $\overline{ABC}$ as in Figure 8.41, $m(\overline{AB}) + m(\overline{BC})$ ought to be the same as $m(\overline{AC})$. Also it is important to know that whenever the measure of a line segment is known, there are conditions under which other line segments having the same measure can be found as, for example, $m(\overline{A'C'}) = m(\overline{AC})$ when $\overline{A'C'} \subset \overrightarrow{A'P}$. See Figure 8.41. We shall also see that in certain discussions it is very convenient to be sure that *pairs* of parallel lines intersect each other in a particular way. In Figure 8.42, for example, if $\mathscr{L} \parallel \mathscr{L}'$ and if $\mathscr{M} \parallel \mathscr{M}'$, it is important to know that $\overline{AB}$, $\overline{DC}$ are congruent and that $\overline{AD}$, $\overline{BC}$ are congruent.

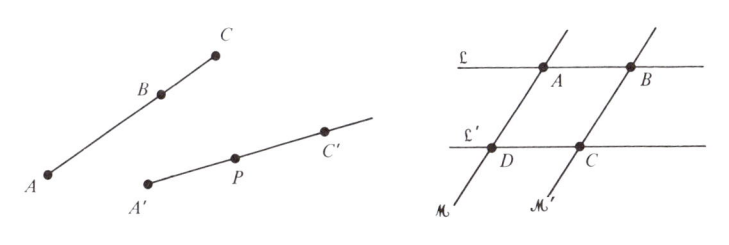

Figure 8.41

Figure 8.42

Measures for angles must also be provided in our axioms. However, it may not be obvious that two angles such as $\measuredangle ABC$ and $\measuredangle A'B'C'$ in Figure 8.43 could be said to have the same size and shape. After all, the sides do extend indefinitely. We must have a way to determine exactly when two angles have the same measure. Further, addition of angle measures must be possible under specified conditions.

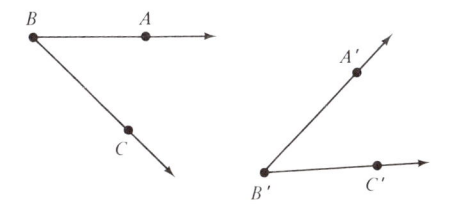

Figure 8.43

One more point about congruence should be noted before we state our axioms. Returning to (i), (ii) of Figure 8.40, we can see that ∡ *ABC* and ∡ *FLK* do not seem to have the same size. Why do we then feel that the two loci *ABCDE*, *FGHKL* are congruent? Obviously, we make this assertion only if we have in mind the point correspondences

$$A \longleftrightarrow G, B \longleftrightarrow F, C \longleftrightarrow L, D \longleftrightarrow K, E \longleftrightarrow H$$

That is, congruence of loci is a particular type of one-to-one correspondence. For our purpose, however, we shall make no note of this principle other than to indicate the requisite point correspondences by using letters in the same alphabetic order for each of two loci. Thus the names △ *ABC*, △ *PQR* for triangles which might be congruent would indicate that the required point correspendence is

$$A \longleftrightarrow P, B \longleftrightarrow Q, C \longleftrightarrow R$$

The following six axioms introduce the measure relationship needed to specify what we mean by congruence of elementary loci. Each measure is a real number, $\mathscr{R}$ is the set of real numbers, and $m(\mathscr{L})$ stands for the measure of locus $\mathscr{L}$.

G-14: For any ray $\overrightarrow{AB}$ and for any point $X \in \overrightarrow{AB}$, there is a one-to-one correspondence between segments $\overline{AX}$ and positive real numbers denoted $m(\overline{AX})$.

G-15: For any points $A, B, C \cdot \ni \cdot \overline{ABC}$, $m(\overline{AB}) + m(\overline{BC}) = m(\overline{AC})$.

G-16: For any pairs of straight lines $\mathscr{L} \parallel \mathscr{L}'$, $\mathscr{M} \parallel \mathscr{M}' \cdot \ni \cdot \mathscr{L} \cap \mathscr{M} = \{A\}$, $\mathscr{L} \cap \mathscr{M}' = \{B\}$, $\mathscr{L}' \cap \mathscr{M}' = \{C\}$, $\mathscr{L}' \cap \mathscr{M} = \{D\}$, $m(\overline{AB}) = m(\overline{CD}) \wedge m(\overline{AD}) = m(\overline{BC})$.

G-17: For any ray $\overrightarrow{AB}$, any plane $\Pi \cdot \ni \cdot \overrightarrow{AB} \subset \Pi$, for points $P \in \Pi$ on one side of $\overleftrightarrow{AB}$, and for any particular positive real number s, there is a one-to-one correspondence between angles *BAP* and positive real numbers $m(\angle BAP) < s$.

G-18: For any ∡ *ABC*, ∡ *DBC* $\cdot \ni \cdot$ *C* is in the interior of ∡ *ABD*, $m(\angle ABC) + m(\angle CBD) = m(\angle ABD)$.

G-19: For any ∡ *ABC*, ∡ *PQR* $\cdot \ni \cdot m(\overline{AB}) = m(\overline{PQ}) \wedge m(\overline{CB}) = m(\overline{RQ})$, $m(\angle ABC) = m(\angle PQR)$ iff $m(\overline{AC}) = m(\overline{PR})$.

Axioms G-14, G-15 together allow us to introduce congruence of line segments.

Definition (8-15): Line segments $\overline{AB}$, $\overline{CD}$ are **congruent** $(\overline{AB} \cong \overline{CD})$ iff $m(\overline{AB}) = m(\overline{CD})$.

Definition (8-16): The measure of a line segment is called its **length**.

§8f   Congruence and Measure

According to these two definitions then, two line segments are congruent iff they have equal lengths. The reader will note that Definition (8-15) specifies only what is meant by congruence of particular loci called line segments. The congruence relation for each other type of locus is stated separately. Based on our axioms we now specify what is meant by angle congruence.

Definition (8-17): Angles ABC, PQR are **congruent** ($\measuredangle ABC \cong \measuredangle PQR$) iff $m(\measuredangle ABC) = m(\measuredangle PQR)$.

Congruence of loci such as circles and triangles can be specified in terms of angle congruence or line segment congruence.

EXAMPLE 1: If the numbers shown in Figure 8.44 are the measures of their associated line segments, what can we conclude about $\overline{AC}$, $\overline{BC}$? About $\measuredangle BAC$ and $\measuredangle ABC$?

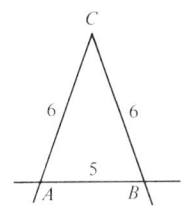

Figure 8.44

Solution: Since $m(\overline{AC}) = 6 = m(\overline{BC})$, we have $\overline{AC} \cong \overline{BC}$ by (8-15).

Since the conditions of G-19 are also fulfilled, we have $m(\measuredangle CBA) = m(\measuredangle CAB)$ whence $\measuredangle CBA \cong \measuredangle CAB$.

We note that since $A \notin \overline{BC} \wedge B \notin \overline{AC}$, $\overline{AC} \neq \overline{BC}$. Compare the definition of equal sets (3-4).

The following theorems about congruence of line segments are not difficult to establish.

Theorem (8f-1): Congruence of line segments is an equivalence relation.

Theorem (8f-2): For any points $A, B, C, A', B', C', \cdot \ni \cdot \overline{ABC} \wedge \overline{A'B'C'}$, $\overline{AB} \cong \overline{A'B'} \wedge \overline{BC} \cong \overline{B'C'} \Rightarrow \overline{AC} \cong \overline{A'C'}$.

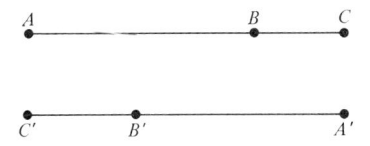

Proof (outline): By G-14 we have real numbers $a = m(\overline{AB})$, $b = m(\overline{AC})$, $c = m(\overline{BC})$, $d = m(\overline{A'B'})$, $e = m(\overline{A'C'})$, $f = m(\overline{B'C'})$ and by G-15 $a + c = b$

368   *Geometry*

and $d + f = e$. Now by (8-15) we have $a = d$ and $c = f$, whence $b = e$ by substitution. This means that $\overline{AC} \cong \overline{A'C'}$.

Q. E. D.

Theorem (8f-3): For any line segment $\overline{AB}$ and ray $\overrightarrow{CD}$, there is a unique point P of $\overrightarrow{CD} \cdot \ni \cdot \overline{AB} \cong \overline{CP}$.

EXERCISE SET I

1. Consider Figure 8.45. What can we conclude about

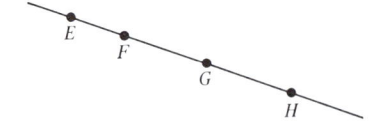

   Figure 8.45

   a) $\overline{FG}, \overline{GH}$ if $m(\overline{FG}) = m(\overline{GH})$?
   b) $m(\overline{EF}) + m(\overline{FG})$?
   c) $m(\overline{EF}), m(\overline{EG})$ if $\overline{EFG}$?
   d) points F, G if $\overline{EG} \cong \overline{GH} \wedge m(\overline{EF}) + m(\overline{GF}) = m(\overline{GH})$?

2. What can we conclude about any distinct points A, B, C of a line if $m(\overline{CA}) + m(\overline{BA}) = m(\overline{BC})$?

3. Consider Figure 8.46 where the numbers shown are the measures of their associated line segments.

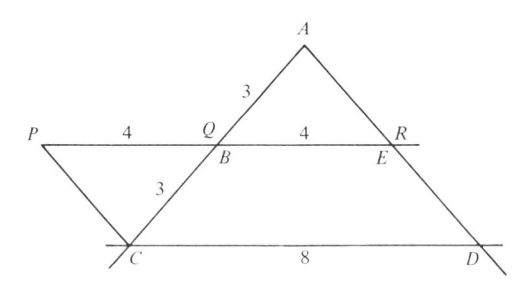

   Figure 8.46

   a) How are $\overline{PB}, \overline{QE}$ related?
   b) How are $\overline{BE}, \overline{QR}$ related?
   c) If $\overline{PC} \cong \overline{AR}$, how are $\measuredangle PQC, \measuredangle RBA$ related?
   d) If $m(\measuredangle PQC) + m(\measuredangle CBE) = 7$, what is the value of s if $s = m(\measuredangle ABE) + m(\measuredangle CBE)$?

e) How do ∡ PQC, ∡ BCD seem to to be related?
4. Discuss the truth value of each assertion.
   a) If two angles are equal, they are congruent.
   b) If two angles are congruent, they are equal.
   c) If two triangles are equal, they are congruent.
   d) If two triangles are congruent, they are equal.
5. Would it be possible to define congruence of rays or straight lines in any manner analogous to that used for line segments? Explain.
6. Prove:
   a) (8f-1)   b) (8f-3)
7. Prove: If $\overline{ABC} \wedge \overline{A'B'C'} \wedge \overline{AB} \cong \overline{A'B'} \wedge \overline{AC} \cong \overline{A'C'}$, then $\overline{BC} \cong \overline{B'C'}$.

As an aid to understanding the three angle measure axioms, we shall make an examination of their provisions in relation to some particular figures. Given a ray $\overrightarrow{AB}$ as in Figure 8.47, for example, G-17 assures us that a real number exists which is the measure of ∡ BAP. Also, for some $s \in \mathscr{R}^+$ and for any $u \in \mathscr{R}$ such that $u < s$, we are assured by the one-to-one correspondence that there is a point Q on one side of $\overleftrightarrow{AB}$ such that $u = m(\angle BAQ)$. Of course, the axiom says nothing about the other side of $\overleftrightarrow{AB}$. There could be a point P' there such that $m(\angle BAP') = m(\angle BAP)$. In fact, if we did not restrict our points to a single plane such as Π, there could be infinitely many angles having one side $\overrightarrow{AB}$ and a particular real number as measure.

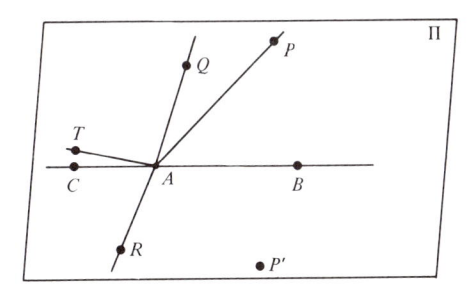

Figure 8.47

Exercise 3c provides us with an illustration of what is intended by G-19. As for G-18, it states particular conditions under which angle measures may be added. However, you must be careful not to read too much into the statement. Returning to Figure 8.47, this axiom says nothing about adding $m(\angle BAQ)$ to $m(\angle QAR)$, for example, because point Q is not interior to ∡ BAR. This is not an important restriction, for in elementary geometry there is little or no reason to want to add measures of such angles.

Another possibility for addition of angle measures is not mentioned in G-18 either. If we understand that *A, B, C* in Figure 8.47 are points of a straight line, what about

(1) $\qquad m(\angle BAP) + m(\angle PAC)?$

As far as the axiom is concerned, we are not told that this particular sum of angle measures is meaningful. Yet angle measures are real numbers according to G-17, and the sum of two real numbers is always another real number. Thus it only remains to try to interpret this real number sum in a useful way.

Now recall that G-17 provides that all angle measures must be numbers less than some particular positive real number *s*. Suppose we let the sum of (1) equal *s*. That is, we let

(2) $\qquad s = m(\angle BAP) + m(\angle PAC)$

where *s* is the positive real number of G-17 and $\overrightarrow{AC} \cup \overrightarrow{AB}$ is a straight line.

How can we pick a specific *s* to use in G-17 and (2)? Associating this number with the measure of an angle called a *right angle* is useful. The following two definitions serve to introduce this concept.

Definition (8-18): A pair of angles are **adjacent** iff they have a common side and a common vertex and the intersection of their interiors is empty.

EXAMPLE 2: Are angles *BAQ, QAT* in Figure 8.47 adjacent? angles *BAT, TAP*?

Solution: Angles *BAQ, QAT* are adjacent according to (8-18). However, angles *BAT, TAP* are not adjacent because the intersection of their interiors is not empty. *Q*, for example, belongs to this intersection.

Now suppose straight lines $\mathscr{L}, \mathscr{M}$ intersect at a point *A* such that the measures of any two adjacent angles are equal. We call such angles right angles.

Definition (8-19): Any angle congruent to either of two congruent adjacent angles whose noncommon sides form a straight line is called a **right angle**.

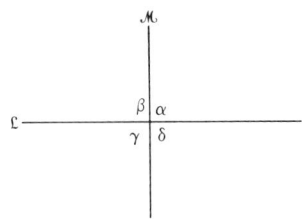

Figure 8.48

§8f   *Congruence and Measure*   371

If the angles α, β of Figure 8.48 are so related that $\alpha \cong \beta$, then according to this definition these angles are right angles. Since the angles are congruent, their measures are equal by (8-17). If we let p be the measure of α, then p is also the measure of β, and we have $m(\alpha) + m(\beta) = 2p$. Then this number $2p$ makes a kind of natural value to use as the s of G-17.

Definition (8-20): For any two angles BAP, $PAC \cdot \ni \cdot \overrightarrow{AB} \cup \overrightarrow{AC}$ is a straight line, $m(\measuredangle BAP) + m(\measuredangle PAC) = 2p$ where p is the measure of a right angle and $2p = s$ for the s of axiom G-17.

In Figure 8.47 we have, for example,

$$m(\measuredangle BAP) + m(\measuredangle PAC) = 2p = m(\measuredangle BAT) + m(\measuredangle TAC)$$

The next three definitions provide us with terms useful in discussing angles and the lines which determine angles.

Definition (8-21): Two lines are said to be **perpendicular** iff they intersect so that the adjacent angles determined are right angles.

If straight lines $\mathscr{L}$, $\mathscr{M}$ are perpendicular as in Figure 8.48 we write $\mathscr{L} \perp \mathscr{M}$. Note that α, β, γ, δ in this figure are all right angles.

Definition (8-22): An angle is said to be **acute** (**obtuse**) according to whether its measure is less than (greater than) the measure p of a right angle.

Definition (8-23): Two angles are said to be **supplementary** iff the sum of their measures is $2p$ where p is the measure of a right angle, and are said to be **complementary** iff the sum of their measures is p.

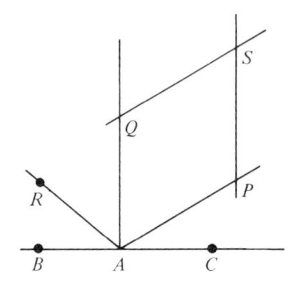

Figure 8.49

In Figure 8.49, $\overleftrightarrow{BC}$ is a straight line, Q is a point of the interior of $\measuredangle CAR$, P is a point of the interior of $\measuredangle CAQ$, $\overleftrightarrow{AQ} \perp \overleftrightarrow{CB}$ at A, and $\measuredangle PAQ \cong \measuredangle PSQ \cong \measuredangle RAB$. Thus $m(\measuredangle CAQ) = p$, $\measuredangle CAQ \cong \measuredangle QAB$, $m(\measuredangle CAP) + m(\measuredangle PAQ) = p$, $m(\measuredangle PAQ) = m(\measuredangle PSQ) = m(\measuredangle RAB)$,

$m(\angle CAR) + m(\angle RAB) = 2p$. As a result, angles PAQ, CAP are complementary; angles PAQ, PSQ are complementary; angles CAQ, QAB are supplementary; angles CAR, RAB are supplementary; angles PSQ, CAR are supplementary.

The following proposition is plausible.

Theorem (8f-4): Two angles are congruent iff their supplements are congruent.

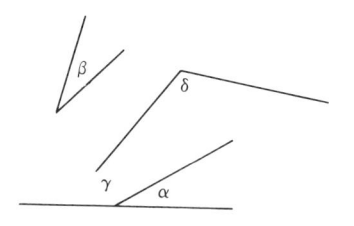

Proof (Outline): (*Only if* part) By hypothesis α, β, γ, δ are angles such that $\alpha \cong \beta$, γ is supplementary to α, δ is supplementary to β. By (8-23) we have

$$m(\alpha) + m(\gamma) = 2p$$
$$m(\beta) + m(\delta) = 2p$$

where p is the measure of a right angle. Now by (8-17) we also have $m(\alpha) = m(\beta)$. Subtracting, we obtain $m(\gamma) = m(\delta)$ whence $\gamma \cong \delta$.   DONE

Proof of the *if* part is left to the exercises.

Returning to consideration of our axioms, it should be reasonable to expect that an angle having a given measure can be shown to exist at a particular point in a plane. Using G-17, the following proposition can easily be established.

Theorem (8f-5): For any $\angle PQR$ and for any ray $\overrightarrow{AB}$ in a plane Π, there is a point $C \in \Pi$ on one side of $\overleftrightarrow{AB} \cdot \ni \cdot \angle BAC \cong \angle PQR$.

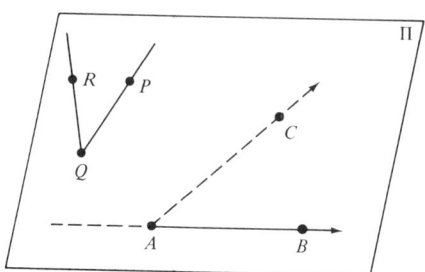

§8f　Congruence and Measure　　373

Proof (Outline): By hypothesis we have $\measuredangle PQR$ and ray $\overrightarrow{AB}$ in plane Π. By G-17 $\measuredangle PQR$ has a measure $m(\measuredangle PQR)$. This axiom also requires that an angle BAC exist having $m(\measuredangle BAC) = m(\measuredangle PQR)$ for some point $C \in \Pi$. By (8-17) we then have $\measuredangle BAC \cong \measuredangle PQR$. 　　Q.E.D.

A few additional theorems about angle congruence as related to straight lines are important. The following one concerns vertical angles which we first define.

Definition (8-24): A pair of angles ABC, DBE are **vertical** iff $\overrightarrow{BA} \cup \overrightarrow{BD}$ is a straight line and $\overrightarrow{BC} \cup \overrightarrow{BE}$ is a straight line or $\overrightarrow{BA} \cup \overrightarrow{BE}$ is a straight line and $\overrightarrow{BC} \cup \overrightarrow{BD}$ is a straight line.

In Figure 8.50 angles ABC, DBE are vertical as are angles ABE, CBD. Angles ABC, ABE are not vertical, they are adjacent.

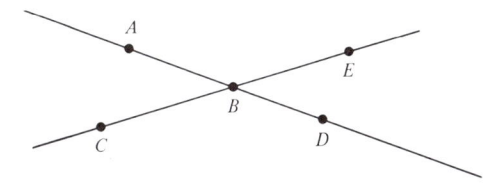

Figure 8.50

Theorem (8f-6): Vertical angles are congruent.

The proof of this proposition is easy using (8f-4). It is left to the exercises.

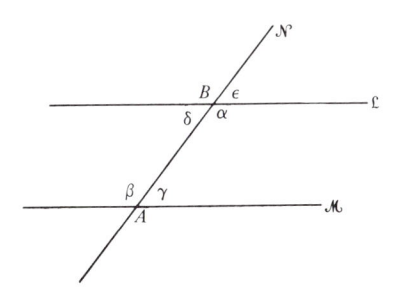

Figure 8.51

The next propositions to be considered make use of the concept of a pair of lines and a transversal. In Figure 8.51 $\mathscr{L}$, $\mathscr{M}$ are a pair of coplanar straight lines. The straight line $\mathscr{N}$ which meets each of them at distinct points such as A, B is called a *transversal* of $\mathscr{L}$, $\mathscr{M}$. A pair of angles such as α, β are said to be *alternate interior angles*. γ, δ are also a pair of alternate interior angles. Angles such as angle ϵ are called *exterior angles*. Note that $\mathscr{L}, \mathscr{M}$ are drawn nearly parallel and that these pairs of alternate interior angles seem congruent.

Theorem (8f-7): If a pair of parallel straight lines $\mathscr{L}, \mathscr{M}$ are met by a transversal $\mathscr{N}$, then the alternate interior angles formed are congruent.

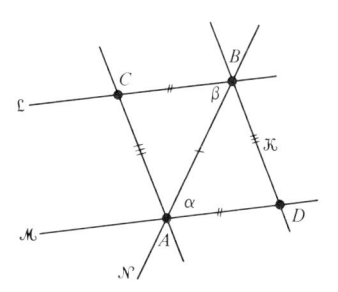

Proof (Outline): By hypothesis, straight lines $\mathscr{L} || \mathscr{M}$ are met by a transversal $\mathscr{N}$ on points A, B in such a way that alternate interior angles α, β are formed. Let C belong to $\mathscr{L}$ distinct from B. By G-13 there is a straight line $\mathscr{K}$ on B such that $\mathscr{K} || \overleftrightarrow{AC}$. Let $\mathscr{K} \cap \mathscr{M} = \{D\}$. By G-16 we now have $\overline{BC} \cong \overline{AD}$ and $\overline{AC} \cong \overline{BD}$. Also $\overline{AB} \cong \overline{AB}$. Thus $m(\alpha) = m(\beta)$ by G-19 whence $\alpha \cong \beta$ by (8-17). Q. E. D.

The converse of this theorem also holds true.

Theorem (8f-8): If straight lines $\mathscr{L}, \mathscr{M}$ are met by a transversal $\mathscr{N}$ in such a way that a pair of alternate interior angles are congruent, then $\mathscr{L} || \mathscr{M}$.

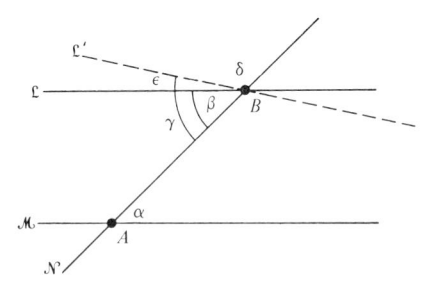

Proof (Outline): By hypothesis, transversal $\mathscr{N}$ meets $\mathscr{L}, \mathscr{M}$ at A, B, respectively, in such a way that $\alpha \cong \beta$. Assume $\mathscr{L} \not{||} \mathscr{M}$. There is a straight line L' on B such that $\mathscr{L}' || \mathscr{M}$ by G-13. This axiom also assures us that $\mathscr{L}'$ is the only such line. That is, $\mathscr{L} \neq \mathscr{L}'$. Thus, in considering angles about B on one side of $\mathscr{N}$, we must have γ, δ, ϵ as shown. By (8f-7) we know

§8f   *Congruence and Measure*   375

that $\alpha \cong \gamma$. Using (8-20) we may write $m(\beta) + m(\epsilon) + m(\delta) = 2\rho$ where ρ is the measure of a right angle. A contradiction of G-17 is now easy to reach. The steps are left to the exercises.

EXERCISE SET II

8. In Figure 8.52, what can we say about the length of $\overline{CD}$ if
   a) $\overleftrightarrow{CD} \not\parallel \overleftrightarrow{AB}$?   b) $\overleftrightarrow{CD} \parallel \overleftrightarrow{AB}$?

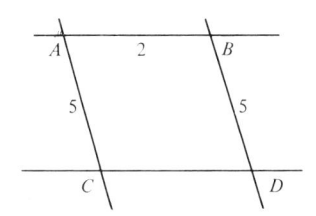

Figure 8.52

   Explain your answers.

9. Consider Figure 8.53.

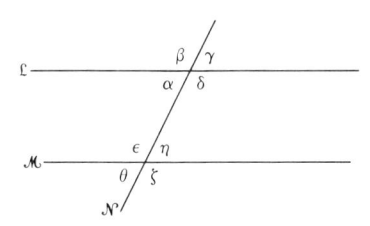

Figure 8.53

   a) List all pairs of alternate interior angles.
   b) List all pairs of alternate exterior angles.
   c) If $\mathscr{L} \parallel \mathscr{M}$ and $m(\eta) = 30°$, what can we conclude about $m(\alpha)$? about $m(\gamma)$?
   d) If $m(\delta) = 1200$, and $\rho = 1000$ is the measure of a right angle, what is $m(\alpha)$? Explain.
   e) If $\mathscr{L} \parallel \mathscr{M}$ and $m(\epsilon) = 110°$ and the measure of a right angle is 90°, what is $m(\alpha)$? Explain.

10. Write out the remaining steps in the proof of (8f-8).

11. Prove:
    a) the *if* part of (8f-4)   b) (8f-6)

12. In general, we say that two loci are *similar* (have the same shape) iff each part of one locus has a measure which is k times the measure of the corresponding part of the second locus. Discuss the proposition "all line segments are similar."

§8g TRIANGLES

Some plane loci are simply sets of line segments. Many of these loci have properties which are interesting and useful enough to be studied informally even in early grades. The most important of these are called polygons.

Definition (8-25): For any set of coplanar points $P_1, P_2, \ldots, P_n \cdot \ni \cdot$ $n \geq 3$ and such that no three consecutive points are collinear,

$$\overline{P_1 P_2} \cup \overline{P_2 P_3} \cup \ldots \cup \overline{P_{n-1} P_n} \cup \overline{P_n P_1}$$

is called a **polygon**.

Each of the points $P_1, P_2, \ldots, P_n$ is called a *vertex* of the polygon and each line segment $\overline{P_1 P_2}, \ldots, \overline{P_n P_1}$ is called a *side* of the polygon. According to this definition, the first three sets of line segments in Figure 8.54 are polygons but the fourth set is not.

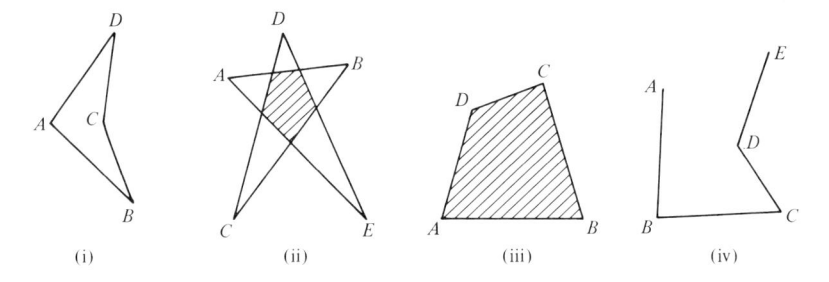

Figure 8.54

Even though the loci of Parts (i), (ii) of this figure are polygons, their properties are difficult to discuss. The following definition provides us with a means of restricting our attention to the type shown in Part (iii).

Definition (8-26): A **polygon** is said to be **convex** iff the interior of each angle of the polygon contains all the points of the polygon not contained in the angle.

By an "angle of a polygon" we mean, unless otherwise specified, an angle having a vertex of the polygon as its vertex and having the sides which intersect at that vertex as subsets. The *interior* of a convex polygon is the intersection of the interiors of its angles. The *exterior* of the polygon is then the set of all points of the plane of the polygon which do not belong to the polygon or to its interior. In view of Theorems (8e-1) and (8e-3a) the interior of a convex polygon is a convex set.

Returning to Part (i) of Figure 8.54, we see that

$$\text{Polygon } ABCD = \overline{AB} \cup \overline{BC} \cup \overline{CD} \cup \overline{DA}$$

is not convex. For one thing, point A is not contained in the interior of $\sphericalangle BCD$ even though this angle is an angle of the polygon. Also note that the intersection of the interiors of its angles is empty. As for

$$\text{Polygon } ABCDE = \overline{AB} \cup \overline{BC} \cup \overline{CD} \cup \overline{DE} \cup \overline{EA}$$

of Part (ii), we may note, for example, that point B does not belong to the interior of $\sphericalangle CDE$ as it must for $ABCDE$ to be convex. Also the hatched intersection of the interiors of the angles of this polygon does not include all of the region which we intuitively expect to be included as part of the interior. Only in the convex case of (iii) does the intersection of the interiors of the angles include all of the region which one would intuitively think of as interior to the polygon. Unless otherwise stated, we normally confine ourselves to study of convex polygons.

Various principles are used to classify polygons for study. Certainly the principle of congruence is useful and important.

Definition (8-27): A polygon whose angles are all congruent and whose sides are all congruent is said to be a **regular polygon**.

Polygons are also usefully classified according to the number of sides. Examination of Definition (8-14) allows us to conclude that a triangle is a polygon having number of sides $n = 3$.

Definition (8-28): According to the number of sides n, polygons are named as follows:

- (i) 3 sides—triangle
- (ii) 4 sides—**quadrilateral**
- (iii) 5 sides—**pentagon**
- (iv) 6 sides—**hexagon**
- (vi) 10 sides—**decagon**
- (vii) 12 sides—**dodecagon**
....
- (viii) n sides—***n*-gon**.

In turn, triangles and quadrilaterals are classified in terms of properties of their angles and sides.

Definition (8-29): A triangle is called

(i) **obtuse** iff one of its angles is an obtuse angle;
(ii) a **right triangle** iff one of its angles is a right angle;
(iii) **acute** iff *all* of its angles are acute angles.

Definition (8-30): A triangle is called

(i) **scalene** iff it has no sides congruent;
(ii) **isosceles** iff it has a pair of sides congruent;
(iii) **equilateral** iff it has all three sides congruent.

Since right triangles are of particular importance, the sides have been given special names. The concept of opposite side is used in defining these terms. A side is said to be *opposite* an angle of a triangle iff it contains points of the interior of the angle.

Definition (8-31): The sides of a right triangle which are subsets of the right angle are called **legs**; the side opposite the right angle is called the **hypotenuse**.

As examples, two easy theorems about triangles can be established.

Theorem (8g-1): If a triangle is isosceles, the angles opposite the congruent sides are also congruent.

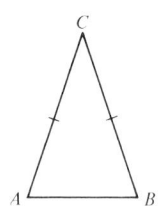

Proof (outline): By hypothesis triangle *ABC* is isosceles; let us say that $\overline{AC} \cong \overline{BC}$. Since $\overline{AB} \cong \overline{AB}$, we have $m(\overline{AC}) = m(\overline{BC})$, $m(\overline{AB}) = m(\overline{AB})$ by (8-15). Now by axiom G-19 we have $m(\measuredangle CAB) = m(\measuredangle CBA)$ whence $\measuredangle CAB \cong \measuredangle CBA$. Q. E. D.

Theorem (8g-2): The sum of the measures of the angles of a triangle equals 2ρ where ρ is the measure of a right angle.

§8g   Triangles

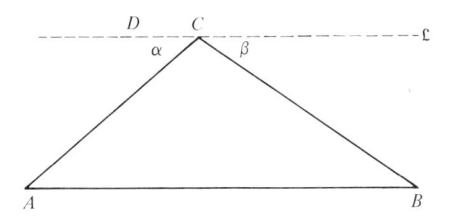

Proof (outline): By hypotheses ABC is a triangle. By axiom G-13 there is a straight line $\mathscr{L}$ containing $C \cdot \ni \cdot \mathscr{L} \parallel \overleftrightarrow{AB}$. $\mathscr{L}$ determines angles α, β with $\overrightarrow{CA}$, $\overrightarrow{CB}$ respectively. Now by (8-20) we have $m(\alpha) + m(\sphericalangle ACB) + m(\beta) = 2p$ where p is the measure of a right angle. We also have $\alpha \cong \sphericalangle CAB$, $\beta \cong \sphericalangle ABC$ according to (8f-8) whence $m(\alpha) = m(\sphericalangle CAB)$, $m(\beta) = m(\sphericalangle ABC)$. Therefore, $m(\sphericalangle CAB) + m(\sphericalangle ACB) + m(\sphericalangle ABC) = 2p$. Q. E. D.

Corollary (8g-2a): If two angles of one triangle are correspondingly congruent to two angles of a second triangle, the third pair of angles is also congruent.

In almost all treatments of geometry, many of the properties of polygons are made to depend on the concept of congruence of triangles.

Definition (8-32): Two **triangles** are **congruent** iff their corresponding angles and sides are congruent.

Recall that the corresponding angles and sides are determined by the *order* in which the vertices are written in the original statement. Thus when $\triangle ABC \cong \triangle DEF$ then A corresponds to D, B to E, and C to F.

EXAMPLE 1: Consider triangles ABC, DEF of Figure 8.55 having side measures as indicated. Is it true that $\triangle ABC \cong \triangle DEF$?

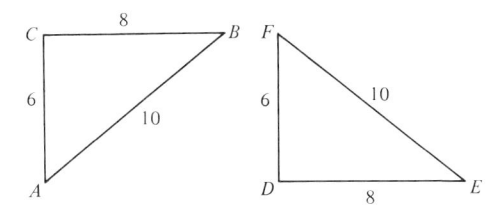

Figure 8.55

Solution: This question involves two considerations; first whether the correct correspondence is specified and then whether the triangles are congruent.

We can see that

(1) $$\triangle ABC \cong \triangle DEF$$

is false because $m(AB) \neq m(DE) \Rightarrow \overline{AB} \not\cong \overline{DE}$. Using a different point correspondence, it might seem that

(2) $$\triangle ABC \cong \triangle FED$$

should hold. However, we cannot logically justify this conclusion at this point, for although (8-15) allows us to conclude that

(3) $\quad\quad\overline{AB} \cong \overline{FE}\quad$ and $\quad\overline{BC} \cong \overline{ED}\quad$ and $\quad\overline{CA} \cong \overline{DF}$

we have no principle which allows us to determine directly anything about the corresponding angles.

Theorem (8g-3): If the corresponding sides of two triangles are congruent, then the triangles are congruent. (SSS)

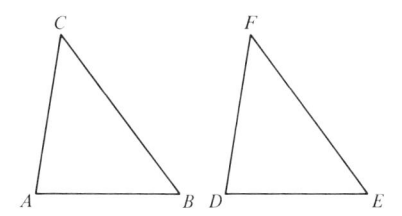

Proof (outline): By hypothesis we have $\overline{AB} \cong \overline{DE}, \overline{BC} \cong \overline{EF}, \overline{AC} \cong \overline{DF}$ for triangles ABC, DEF. Hence $m(\overline{AB}) = m(\overline{DE}), m(\overline{BC}) = m(\overline{EF}), m(\overline{AC}) = m(\overline{DF})$, whence $\angle A \cong \angle D, \angle B \cong \angle E, \angle C \cong \angle F$ by axiom G-19. Now we have $\triangle ABC \cong \triangle DEF$ by (8-32). Q. E. D.

EXAMPLE 2: What can we conclude about the truth value of (2) in Example 1 in the light of this theorem?

Solution: Because we have the congruences of (3), we can immediately conclude from the theorem that the congruence of (2) also holds.

EXAMPLE 3: Can we conclude that $\triangle ADC \cong \triangle EBC$ in Figure 8.56 for the indicated segment measures?

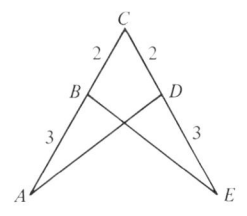

Figure 8.56

Solution: By (8-15) we have $\overline{BC} \cong \overline{DC}$. Also, since $m(\overline{AC}) = 5 = m(\overline{EC})$, we have $\overline{AC} \cong \overline{EC}$. Since this is all that we can conclude about the figure by using principles established to this point, we find it impossible to conclude that the specified triangles are congruent.

Proof of the following congruence theorem is left to the exercises.

Theorem (8g-4): If two pairs of corresponding sides of two triangles are congruent and the angles determined by these sides are congruent, then the triangles are congruent. (SAS)

Corollary (8g-4a): If the hypotenuse and a leg of one right triangle are congruent, respectively, to the hypotenuse and corresponding leg of a second right triangle, the triangles are congruent. (HL)

EXAMPLE 4: By (8g-4) we can now conclude that $\triangle ADC \cong \triangle EBC$ in Example 3 because $\angle C \cong \angle C$.

Theorem (8g-5): If two triangles have two pairs of corresponding angles and the sides common to these angles congruent, then the triangles are congruent. (ASA)

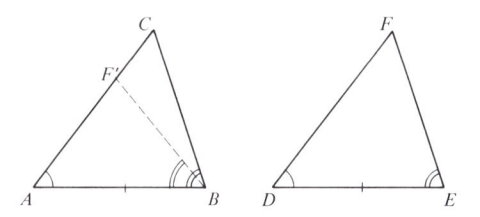

Proof (outline): By hypothesis we have $\triangle ABC$ and $\triangle DEF$ with $\overline{AB} \cong \overline{DE}$, $\angle A \cong \angle D$, $\angle B \cong \angle E$. By (8f-3) there is a segment $\overline{AF'} \subset \overrightarrow{AC}$ such that $\overline{AF'} \cong \overline{DF}$. Thus, according to (8g-4), we have $\triangle ABF' \cong \triangle DEF$, whence $\angle ABF' \cong \angle DEF$ by (8-32). Also $\angle ABF' \cong \angle ABC$ because congruence of angles is an equivalence relation. Now we have $m(\angle ABF') = m(\angle ABC)$, but according to the one-to-one correspondence requirement of axiom G-17, this means that $\overrightarrow{BF'}, \overrightarrow{BC}$ are the same segment. Since $\overline{BF'} \cong \overline{EF}$ by (8-32) it must be true that $\overline{BC} \cong \overline{EF}$. Now (8g-4) requires that $\triangle ABC \cong \triangle DEF$. Q. E. D.

Using the SSS, SAS, and ASA congruence principles, an extensive assortment of propositions about triangles and other polygons can be established. A few examples follow.

Theorem (8g-6): If two angles of a triangle are congruent, then the sides opposite those angles are congruent.

Corollary (8g-6a): $\triangle ABC$ is equilateral iff $\angle ACB \cong \angle ABC \cong \angle BAC$.

Definition (8-33): A **bisector** separates a geometric locus into two congruent loci.

Theorem (8g-7): In isosceles $\triangle ABC$ ($\overline{AB} \cong \overline{AC}$), the bisector of $\angle BAC$ (the vertex angle) bisects $\overline{CB}$ and is perpendicular to it.

Theorem (8g-8): The line that connects the vertex of an isosceles $\triangle$ with the middle point of the opposite side (the base) bisects the vertex angle and is perpendicular to the base.

EXERCISE SET

1. Discuss the proposition "Some polygons have empty interiors."
2. Explain why the locus of Part (iv) of Figure 8.54 is not a polygon according to (8-25).
3. Make a drawing to show that each polygon is not convex.
   a) Part (i), Figure 8.54
   b) Part (ii), Figure 8.54
4. Which sets are convex? Explain.
   a) A polygon
   b) The interior of a polygon
   c) The exterior of a polygon
5. Is it possible to have:
   a) an obtuse triangle which is
      1) scalene
      2) isosceles
      3) equilateral
   b) a right triangle which is
      1) scalene
      2) isosceles
      3) equilateral
   c) an acute triangle which is
      1) scalene
      2) isosceles
      3) equilateral
6. Is it true that $\triangle ABC \cong \triangle PQR$ in Figure 8.57 for the indicated measures if we know that $\angle B \cong \angle P$?

§8g   *Triangles*   383

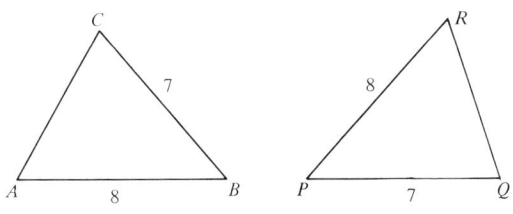

Figure 8.57

7. What conclusions can we reach for the locus of Figure 8.58? Explain each.

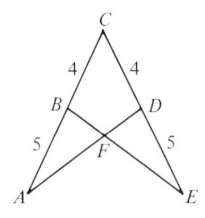

Figure 8.58

8. Which of these congruence conditions are sufficient to assure us that two triangles are congruent? Illustrate each.
   a) SAA   b) SSA
   c) AAA   d) ASA
   e) SSS   f) SAS

9. Write the converse of each theorem from (8g-1) to (8g-6). Does each hold? Explain.

10. If each angle of one triangle is congruent to the corresponding angle of a second triangle, are the triangles congruent? Show that, in fact, the two triangles are only similar.

11. Prove that if $\overleftrightarrow{AC} \cap \overleftrightarrow{DE} = \{B\}$ and $\overline{ABC}$ and $\overline{DBE}$ and $\overline{AB} \cong \overline{BC}$ and $\overline{EB} \cong \overline{BD}$, then $\triangle ABE \cong \triangle CBD$.

12. For the quadrilateral of Figure 8.59, prove that if $\overline{AB} \cong \overline{DC}$ and $\overline{AD} \cong \overline{BC}$, then $\angle A \cong \angle C$ and $\angle B \cong \angle D$.

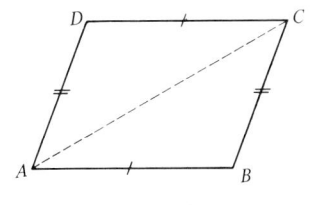

Figure 8.59

13. Prove each proposition:
    a) (8g-2a)   b) (8g-4)
    c) (8g-4a)   d) (8g-6)
    e) (8g-6a)   f) (8g-7)
    g) (8g-8)

§8h QUADRILATERALS

The next several definitions provide a classification scheme for quadrilaterals. You will note that the basic properties of each of the various types are given in terms of parallelism and congruence.

Definition (8-34): A quadrilateral is called a **trapezium** iff it has no pair of sides parallel.

Part (iii) of Figure 8.54 illustrates a trapezium.

Definition (8-35): A quadrilateral is called a **trapezoid** iff at least one pair of its nonconsecutive sides are parallel.

Definition (8-36): A quadrilateral is called a **parallelogram** iff both pairs of its nonconsecutive sides are parallel.

Definition (8-37): A quadrilateral is called a **kite** iff both pairs of distinct consecutive sides are congruent.

The parts of Figure 8.60 illustrate a trapezoid, a parallelogram, and a kite in order.

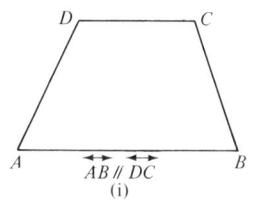

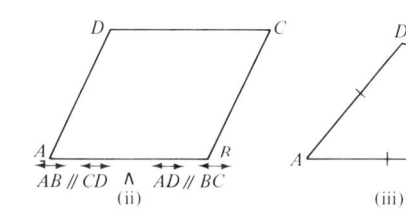

Figure 8.60

Various subsets of the set of parallelograms are traditionally singled out for special study according to the following definitions.

Definition (8-38): A parallelogram is called a **rectangle** iff it has one right angle.

§8h Quadrilaterals

Definition (8-39): A rectangle is called a **square** iff a pair of consecutive sides are congruent.

Definition (8-40): A parallelogram is called a **rhombus** iff it has a pair of consecutive sides congruent.

Additional important properties of these various quadrilaterals are stated in the following theorems. Proofs of all but the first are left to the exercises.

Theorem (8h-1): A quadrilateral is a parallelogram iff opposite pairs of sides are congruent.

Proof (outline): (*Only if* part) By hypothesis, $ABCD$ is a parallelogram, whence $\overleftrightarrow{AB} // \overleftrightarrow{DC}$ and $\overleftrightarrow{AD} // \overleftrightarrow{BC}$ by definition. Then by axiom G-16, we have $m(\overline{AD}) = m(\overline{BC})$ and $m(\overline{AB}) = m(\overline{DC})$ whence $\overline{AD} \cong \overline{BC}$ and $\overline{AB} \cong \overline{DC}$ by (8-15). DONE

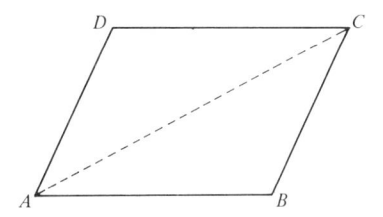

(*If* part) By hypothesis $ABCD$ is a quadrilateral having $\overline{AB} \cong \overline{DC}$ and $\overline{AD} \cong \overline{BC}$. Since $\overline{AC} \cong \overline{AC}$, we have $\triangle ABC \cong \triangle CDA$ by SSS (8g-3), whence $\measuredangle CAB \cong \measuredangle DCA$ and $\measuredangle BCA \cong \measuredangle DAC$ by (8-32). Now by (8f-9) we have $\overleftrightarrow{AB} // \overleftrightarrow{DC}$ and $\overleftrightarrow{AD} // \overleftrightarrow{BC}$. Therefore, $ABCD$ is a parallelogram by definition. DONE Q. E. D.

EXAMPLE 1: Consider the relationships of Figure 8.61. Classify quadrilaterals $ACDE$, $ABCE$, $ACDF$.

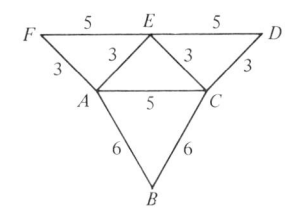

Figure 8.61

Solution: By the theorem just proved, *ACDE* is a parallelogram. By (8-37), *ABCE* is a kite. Since *ACEF* is also a parallelogram, we have $\overleftrightarrow{FD}//\overleftrightarrow{AC}$. Thus *ACDF* is a trapezoid.

Corollary (8h-1a): A quadrilateral is a parallelogram iff nonconsecutive pairs of angles are congruent.

Corollary (8h-1b): Each angle of a rectangle is a right angle.

Corollary (8h-1c): The diagonals of a rectangle are congruent.

Theorem (8h-2): A quadrilateral is a parallelogram iff a pair of non-consecutive sides are parallel and congruent.

Theorem (8h-3): The diagonals of a parallelogram bisect each other.

Theorem (8h-4): The diagonals of a kite are perpendicular to each other.

Corollary (8h-4a): The diagonals of a rhombus are perpendicular to each other.

Corollary (8h-4b): The diagonals of a square are perpendicular to each other.

From consideration of principles stated in the preceding definitions and theorems, it is not difficult to see that the set of squares is a subset of the set of rhombi which in turn is a subset of the set of kites. The set diagram of Figure 8.62 illustrates these and other relationships among quadrilaterals.

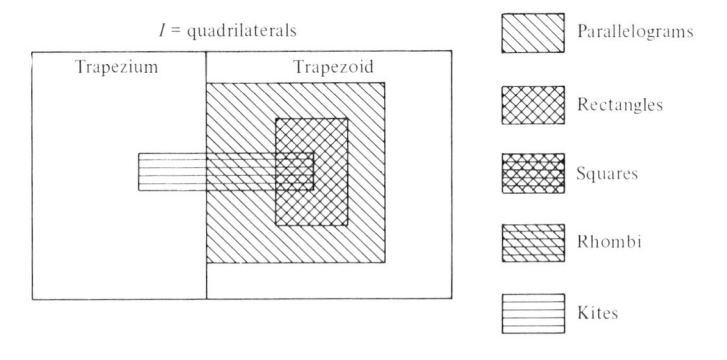

Figure 8.62

EXERCISE SET

1. Consider a quadrilateral *ABCD*.
   a) Can *ABCD* be a trapezium if $\overline{AB} \cong \overline{CD}$? Explain.
   b) Can *ABCD* be a trapezium if $\overline{AB} \cong \overline{CD}$ and $\overline{BC} \cong \overline{AD}$? Explain.
   c) Can *ABCD* be a trapezoid if known to be a rhombus?
   d) Can *ABCD* be a trapezoid if known to be a kite?

2. Consider Figure 8.63 in which $\overline{AD} \cong \overline{BC}$ and $\overleftrightarrow{AD} // \overleftrightarrow{BC}$ in quadrilateral *ABCD*.

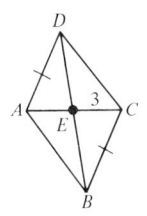

Figure 8.63

   a) How can *ABCD* be classified if we know that $\overline{AB} \not\cong \overline{BC}$?
   b) What is the length of $\overline{AE}$?
   c) How can quadrilateral *ABCD* be classified if $\overline{AB} \cong \overline{BC}$?

3. Prove or disprove:
   a) All trapezoids are parallelograms.
   b) All squares are rhombi.
   c) All rhombi are kites.
   d) All squares are kites.
   e) All rectangles are kites.
   f) All parallelograms are trapezoids.

4. What is the sum of the measures of the angles of a quadrilateral? Prove your result.

5. Prove each proposition.
   a) (8h-1a)   b) (8h-1b)
   c) (8h-1c)   d) (8h-2)
   e) (8h-3)   f) (8h-4)
   g) (8h-4a)   h) (8h-4b)

6. Prove (8h-4a) independently of (8h-4).

§8i PLANE CURVES

In general, it is not easy to define the term curve. Rather, it is easier to define each particular curve as we wish to make use of it. Descriptively, however, we can say that a *curve* is any continuous locus, that is, any locus which can be drawn without lifting the pencil.

The concept of circle is useful in further describing curves.

Definition (8-41): For any point O in plane Π and for any segment $\overline{AB}$, the set of *all* points $X \in \Pi$ such that $\overline{OX} \cong \overline{AB}$ is called a **circle**.

The point O is called the **center** of the circle and any $\overline{OX}$ such that X belongs to the circle is called a **radius** of the circle. We note that the center is not a point of the circle.

Continuing our classification of curves, we can say that a **simple curve** is one which has the property that for any point A of the curve and for a circle having center A and *small enough* radius, each circle $\mathcal{K}$ having center A and *smaller* radius has at most two points of intersection with the curve. A curve is a **closed curve** iff each such circle $\mathcal{K}$ has at least two points of intersection with the curve. If each such $\mathcal{K}$ intersects the curve exactly twice, we have a *simple closed curve*.

Figure 8.64

In Figure 8.64 (i) the point A and most of the points of the curve have the property that for a small enough circle with center A each such circle $\mathcal{K}$ with smaller radius would have two points of intersection with the curve. However, each circle $\mathcal{K}$ at B would have at most one point of intersection with the curve. Thus this curve is simple but not closed. For 8.64(ii), each requisite circle $\mathcal{K}$ with center not at A would have two points of intersection with the curve, but circle $\mathcal{K}$ with center at A would have four points of intersection. Hence, this curve is closed but not simple. Figure 8.64 (iii) illustrates a simple closed curve while 8.64 (iv) illustrates a curve that is neither simple nor closed.

Line segments, straight lines, and angles associated in particular ways with circles have special names.

Definition (8-42): A segment both of whose endpoints belong to a circle is called a **chord** of the circle. If a chord contains the center it is called a **diameter**.

Definition (8-43): A line in the plane of a circle which contains one and only one point of the circle is called a **tangent** and is said to be tangent to the circle.

Definition (8-44): An angle is said to be a **central angle** of a circle iff its vertex is the center of the circle and each of its sides has a nonempty intersection with the circle.

Just as particular subsets of straight lines are useful, so it is advantageous to define particular subsets of circles.

Definition (8-45): For any circle and any central angle of the circle, an **arc** of the circle is a subset of the circle which contains the two points of intersection of the angle and the circle and either all of the points of the circle interior to the angle or all the points of the circle exterior to the angle.

In the interior case the arc is said to be a *minor arc*; in the exterior case a *major arc*. A **semicircle** is an arc which contains the two endpoints of a diameter of a circle and all of the points of the circle on one side of the diameter. In Figure 8.65 (i), $\overline{BC}$ is a chord, $\overleftrightarrow{AG}$ is a tangent, $\overline{DE}$ is a diameter, and $\angle FOE$ is a central angle. The subset of the circle using endpoints B and C is arc BC (denoted $\widehat{BC}$). Either arc $\widehat{DE}$ is a semicircle. $\widehat{FAD}$ is a minor arc subtended by the central angle FOD, and $\widehat{FBD}$ is the related major arc.

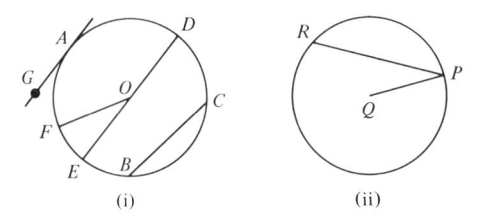

Figure 8.65

As with other loci, we define a congruence relation for circles.

Definition (8-46): Two **circles** are **congruent** iff a radius of one is congruent to a radius of the other.

Thus circle $(Q, \overline{QP})$ of Figure 8.65 (ii) is congruent to circle $(O, \overline{OD})$ of (i) iff $\overline{QP} \cong \overline{OD}$ or $\overline{QP} \cong \overline{OF}$ or $\overline{QP} \cong \overline{OE}$. Congruence of subsets of a circle is also useful.

Definition (8-47): **Minor arcs** of the same or congruent circles are **congruent** iff they are subtended by congruent chords.

390   Geometry

The following propositions can now be proved.

Theorem (8i-1): In the same or congruent circles, central angles are congruent iff they intercept congruent minor arcs.

Theorem (8i-2): If a line which contains the center of a circle is perpendicular to a chord, it bisects the chord and its arc.

Corollary (8i-2a): A line which contains the center of a circle and which bisects a chord is perpendicular to it.

EXERCISE SET

1. Which geometric figures defined thus far are curves and which are not?
2. Classify the curves in Figure 8.66.

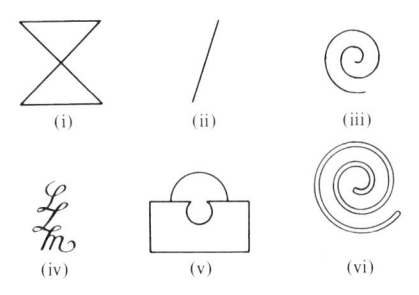

Figure 8.66

3. Show that congruence of (a) circles, (b) arcs is an equivalence relation.
4. How could one define congruence of major arcs?
5. Discuss the truth value of each proposition.
    a) A circle is a simple closed curve.
    b) All circles are similar to one another.
6. Prove each proposition
    a) (8i-1)
    b) (8i-2)
    c) (8i-2a)

§8j   SPACE LOCI

The loci we have considered to this point have been plane loci because their points have all been confined to a single plane. Loci whose points are not necessarily confined to a single plane are called space loci.

Familiar plane relationships may not always hold in space. Recall, for example, that in a plane two straight lines must either have a nonempty intersection or be parallel whereas in space there is the third possibility that they may be skew.

It will be sufficient simply to define some of the more common loci, for any attempt to develop the subject extensively enough to arrive at meaningful theorems would produce a treatment much too cumbersome for our purpose here. Any reader who is disappointed by this approach to space geometry can find plentiful material for further study.

A common classification system for space loci is to place them in four subsets: polyhedra, cones, cylinders, and spheres. For this purpose we need to specify what we mean by a *region* of a plane. For any plane locus having an interior, we say that the union of the locus and its interior is a **region**. Thus the union of a convex polygon and its interior would be called a **polygonal region**. If we consider $\mathscr{D}$ as represented by vertical, $\mathscr{F}$ by horizontal, and $\mathscr{E}$ by cross hatching in Figure 8.67, then $\mathscr{K} \cup \mathscr{D} \cup \mathscr{E}$ is the circular region or disc where $\mathscr{K}$ stands for the circle and $\alpha \cup (\mathscr{E} \cup \mathscr{F})$ is the angular region or wedge where α is the angle.

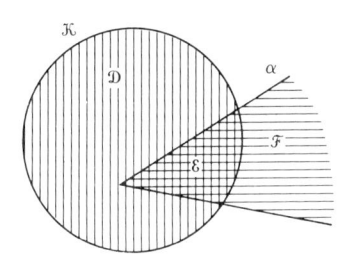

Figure 8.67

The sphere is perhaps the simplest of the four types of space loci mentioned.

Definition (8-48): For any point O and for any segment $\overline{AB}$, the set of all points X such that $\overline{OX} \cong \overline{AB}$ is called a **sphere**.

The fixed point O is called the **center** of the sphere and any $\overline{OX}$ is called a **radius** of the sphere.

Definition (8-49): A **polyhedron** is the union of a finite number of polygonal regions such that:
(i) the interiors of any two of the polygonal regions have an empty intersection:
(ii) Each side of any of the polygonal regions is also a side of exactly one of the other regions.

Definition (8-50): In a polyhedron each polygonal region is called a **face**. The intersection of any two faces is called an **edge**. The intersection of any two edges is called a **vertex**.

EXAMPLE 1: The polyhedron $\mathscr{P}$ of Figure 8.68 can be expressed as the union of the four triangular regions ABC, BDC, DAC, ABD.

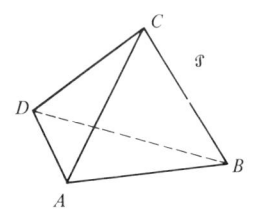

Figure 8.68

Each of these regions is a face; $\overline{AB}$, $\overline{AC}$, $\overline{AD}$, $\overline{BD}$, $\overline{BC}$, $\overline{DC}$ are each edges; C, D, A, B are each vertices of the polyhedron.

We now define three of the most familiar polyhedra.

Definition (8-51): A **prism** is a polyhedron having two parallel congruent faces called bases with the remaining faces being parallelograms.

Definition (8-52): A **parallelepiped** is a prism whose bases are parallelograms.

Definition (8-53): A **pyramid** is a polyhedron such that one face called the base is a polygon of any number of sides and the other faces are triangles whose nonbasal sides meet in a common point called the vertex.

The locus illustrated in Fig. 8.68 is a triangular pyramid.

EXAMPLE 2: In Figure 8.69 parallel planes Π and Φ provide the re-

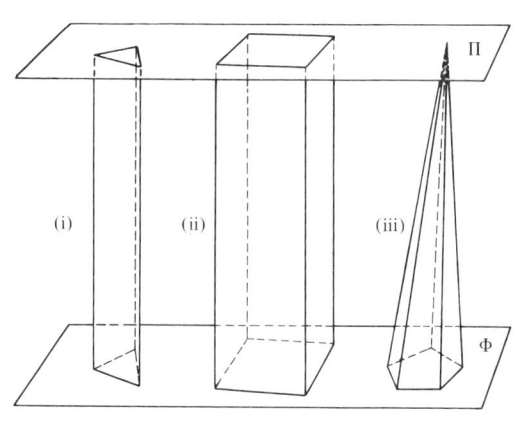

Figure 8.69

§8j Space Loci 393

quired parallel bases for the two prisms (i), (ii). The first is a triangular prism, the second a parallelepiped. A **cube** (not shown) is a parallelepiped having all edges congruent. The polyhedron of (iii) is a pentagonal pyramid.

Definition (8-54): For any simple closed curve $\mathscr{C}$ in a plane Π and for any point $P \notin \Pi$, the set of all straight lines $\overleftrightarrow{XP}$, where $X \in \mathscr{C}$, is called a **conical surface**.

Definition (8-55): For any simple closed curve $\mathscr{C}$ in a plane Π, and for any straight line $\mathscr{L}$ having a single point of intersection with Π, the set of all straight lines $\mathscr{M} \cdot \ni \cdot \mathscr{M} \parallel \mathscr{L}$ and $\mathscr{M}$ has a point in common with $\mathscr{C}$ is called a **cylindrical surface**.

EXAMPLE 3: In Figure 8.70 a plane Π, a point $P \notin \Pi$, and a straight line $\mathscr{L}$ are illustrated. Part (i) illustrates a few of the elements $\overleftrightarrow{XP}$ of a conical surface determined by P and a simple closed curve $\mathscr{C} \subset \Pi$. Part (ii) illustrates a few of the elements $\mathscr{M}$ determined by $\mathscr{L}$ and a simple closed curve $\mathscr{F} \subset \Pi$. The point P is called the **vertex** or **apex** of the conical surface. The straight line $\mathscr{L}$ is called a **generator** of the cylindrical surface.

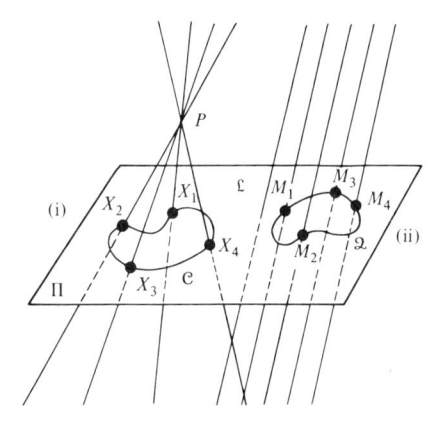

Figure 8.70

Definition (8-56): A **cone** is that subset of a conical surface which is bounded by the vertex and a plane which does not contain the vertex but does have a point in common with each element of the surface.

Definition (8-57): A **cylinder** is that subset of a cylindrical surface which is bounded between two parallel planes which are not parallel to a generator of the surface.

EXAMPLE 4: In Figure 8.71 parallel planes Π, Φ are illustrated. In Part (i) Π contains the vertex of a conical surface so that with Φ a cone is determined. In Part (ii) the planes and a cylindrical surface intersect so as to determine a cylinder.

394   *Geometry*

Figure 8.71

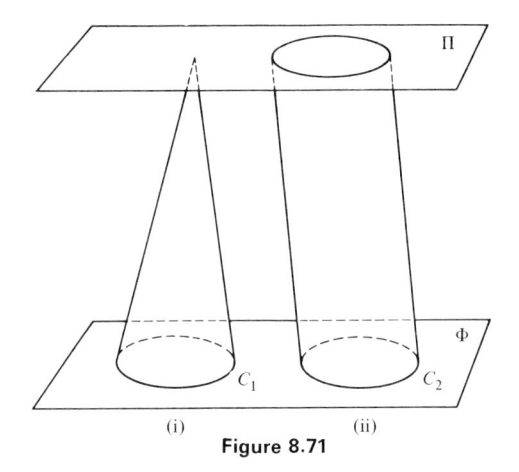

Since the elements of the cone and cylinder in this figure were determined by a simple closed curve, each of the intersections $\mathscr{C}_1$, $\mathscr{C}_2$ with Φ must be a simple closed curve in turn. If $\mathscr{C}_1$ is a circle, the cone is called a **circular cone**. If the straight line determined by the center of the circle and the vertex of the cone is perpendicular to Π, the cone is said to be a **right circular cone**. It is this locus which is very often called a cone for short. If $\mathscr{C}_2$ is a circle, the cylinder is said to be a **circular cylinder**. If, then, a generator $\mathscr{L}$ is perpendicular to Π, the cylinder is called a **right circular cylinder**. This term is often shortened to cylinder.

EXERCISE SET

1. What three kinds of intersections can a sphere and a plane have? Explain.
2. Compare the definitions of a sphere (8-48) and circle (8-41). How do they differ?
3. Is the actual shape of a stick of butter a cube? Explain.
4. What kind of a pyramid is the standard Egyptian pyramid?
5. Geometrically, how would we describe the shape of a cone for ice cream?
6. Disregarding the ends, what is the precise geometric name for the shape of the usual tin can?

§8k   CONSTRUCTIONS

In the later grades, certain elementary constructions are attempted. The only tools used are compass, unmarked ruler, and a sharp number three or number four pencil. The first of these is used to construct circles or arcs; the unmarked ruler, to construct line segments. The better the instruments used, the better will be the work.

The constructions here all depend on having previously mastered the art of constructing straight line segments and circles. One value in learning such constructions is, of course, to use them in applying geometry to various other studies. More important, perhaps, is the fact that making the effort to argue why the construction does, indeed, give the desired result can provide an excellent review of fundamental principles and thereby increase one's understanding. For this purpose, you should make certain that you can justify each step logically.

Construction I: At a point on a line construct an angle congruent to a given angle.

Given: $\measuredangle ABC$, $\overleftrightarrow{PQ}$ as in Fig. 8.72.

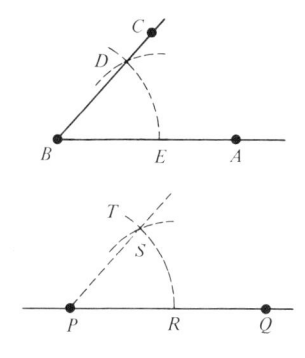

Figure 8.72

Construction:

1. With B as center and any fixed segment as radius on the compass, draw an arc intersecting $\overrightarrow{BA}$ at E and $\overrightarrow{BC}$ at D.
2. With P as center and the same fixed radius, draw arc RT intersecting $\overleftrightarrow{PQ}$ at R.
3. With E as center, fix radius $\overline{ED}$ on compass.
4. With R as center and the same radius as in step 3, draw an arc intersecting arc RT at S.
5. Draw $\overrightarrow{PS}$.
6. $\measuredangle SPR \cong \measuredangle ABC$.

Construction II: Construct the bisector of an angle.
Given: $\measuredangle ABC$ as in Fig. 8.73.

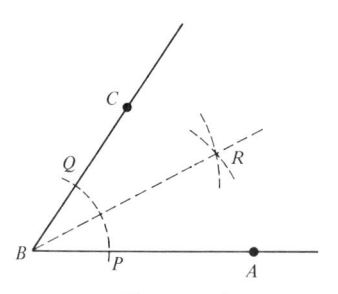

Figure 8.73

Construction:

1. With B as center and any fixed radius, draw an arc intersecting $\overrightarrow{BA}$ at P, $\overrightarrow{BC}$ at Q.
2. With P and Q as centers, fix a radius more than half of the segment $\overline{PQ}$, draw arcs interior to the angle intersecting at R.
3. Draw $\overrightarrow{BR}$.
4. $\overrightarrow{BR}$ is the angle bisector.

Construction III: Construct a line perpendicular to a line such that the perpendicular line contains an arbitrary point P.

(a) Given: Line $\mathscr{L}$ with P on the line as in Fig. 8.74.

Construction:

1. With P as center, take any radius and intersect $\mathscr{L}$ at A and B.
2. Using A and B as centers, fix a radius more than half $\overline{AB}$.
3. Draw arcs intersecting at Q.
4. Draw $\overleftrightarrow{PQ}$.
5. $\overleftrightarrow{PQ}$ is perpendicular to $\mathscr{L}$ at P.

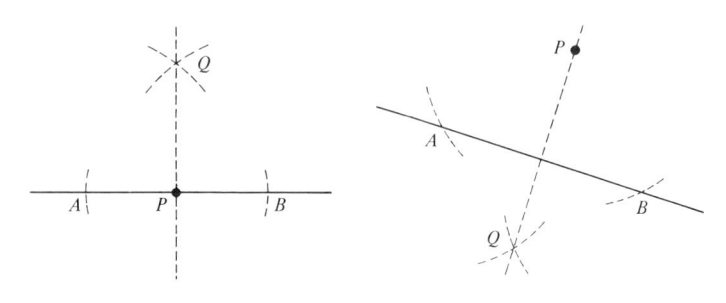

Figure 8.74

Figure 8.75

(b) Given: Line $\mathscr{L}$ with P not on the line as in Fig. 8.75.

Construction:

1. With P as center, fix a radius that will make an arc that intersects $\mathscr{L}$ in two points A, B.
2. Using A, B as centers, fix a radius more than half $\overline{AB}$.
3. Draw arcs that intersect at Q.
4. Draw $\overleftrightarrow{PQ}$.
5. $\overleftrightarrow{PQ}$ is perpendicular to $\mathscr{L}$.

Construction IV: Construct the perpendicular bisector of a given line segment.

Given: Line segment $\overline{AB}$ as in Fig. 8.76.

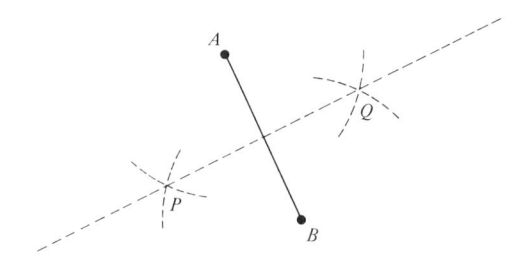

Figure 8.76

Construction:

1. Using A, B as centers, fix a radius greater than half $\overline{AB}$ and draw arcs intersecting each other at P and Q.
2. Draw $\overleftrightarrow{PQ}$.
3. $\overleftrightarrow{PQ}$ is the perpendicular bisector of $\overline{AB}$.

Construction V: Construct through a given point a line parallel to a given line.

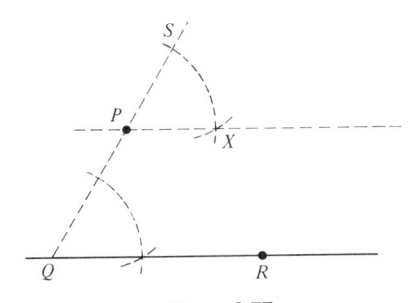

Figure 8.77

Given: Line $\overleftrightarrow{QR}$ and point P not on the line (Fig. 8.77).

Construction:

1. Through P, draw any line $\overleftrightarrow{PQ}$.
2. With P as vertex and $\overrightarrow{PS}$ where S is not on $\overline{PQ}$, as side, construct $\measuredangle SPX \cong \measuredangle PQR$ such that X and R lie on the same side of $\overleftrightarrow{PQ}$.
3. Draw $\overleftrightarrow{PX}$.
4. $\overleftrightarrow{PX}$ is parallel to $\overleftrightarrow{QR}$ at P.

Construction VI: Divide any given line segment into n congruent segments, where n is an arbitrary natural number.

Given: Line segment $\overline{PQ}$ as in Figure 8.78.

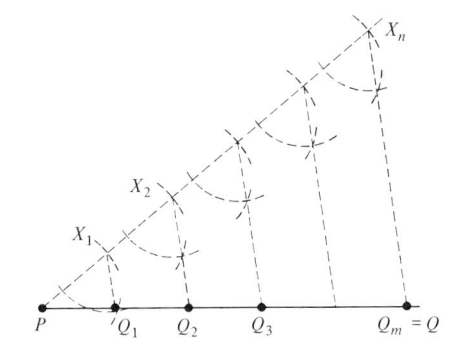

Figure 8.78

Construction:

1. Draw ray $\overrightarrow{PX}$ through P making $\measuredangle XPQ$ acute.
2. On $\overrightarrow{PX}$ mark off n successive congruent segments with compass of fixed radius using P as the first center.
3. Join the end point X_n that has all the other end points between it and P to Q making $\measuredangle PX_nQ$.
4. At each other end point X_i construct angles PX_iQ_i congruent to $\measuredangle PX_nQ$ so that the Q_i will be on $\overline{PQ}$.
5. The desired number of congruent segments are the $\overline{PQ_1}, \overline{Q_1Q_2}, \ldots, \overline{Q_{n-1}Q}$, where $Q_n = Q$.

EXERCISE SET

1. Why is the "more than half" requirement made in Step 2 of Construction II and elsewhere?
2. Why is an acute angle called for in Step 1 of Construction VI?

§8i MISCELLANEOUS EXERCISES

1. According to the standard pattern for naming polygons given in §8h, what should a triangle and a quadrilateral be called?
2. Discuss the truth values of the following propositions:
    a) Any line segment is congruent to any other line segment.
    b) Any line segment is similar to any other line segment.
    c) The set of rectangles is a subset of the set of kites.
    d) The set of squares is a subset of the set of kites.
    e) The set of squares is a subset of the set of rhombi.
3. In Figure 8.79 how many rays have end point A' and determine with $A'P$ an angle congruent to $\measuredangle A$?

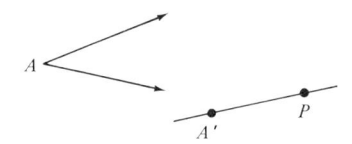

Figure 8.79

For each of the loci described in Exercises 4 through 21, try to decide whether the set described is a plane figure, that is, whether some plane contains all of its points. If it is impossible to decide exactly, indicate with a figure showing a counterpossibility.

4. A set containing exactly three distinct points
5. A line
6. A set of exactly two points
7. A set consisting of five distinct points
8. Three lines through a single point such that every angle determined is a right angle
9. Two lines which are not parallel and have no common points
10. A set of four points
11. Two intersecting lines
12. A triangle and a line which contains two distinct points of the triangle
13. A triangle and a point not belonging to the triangle
14. A line and a point not belonging to the line
15. Two triangles having a common angle
16. Two triangles having a common side
17. Two triangles with a common side and no other points in common
18. Four points A, B, C, D such that B is between A and C

19. Three rays from a single point which determine three congruent acute angles

20. Three rays from a single point which determine a pair of congruent angles

21. A point A and all points B such that all the segments $\overline{AB}$ are congruent to each other

22. Is the intersection of two planes always nonempty? Explain. Give a geometric description of the nonempty intersection of two planes.

23. What do these mean?

    a) "A point P is *on* a line."
    b) "A line L is *on* a plane."

24. How many line segments are determined by four distinct points A, B, C, D in a line?

25. Suppose a line coplanar with a triangle contains one vertex of the triangle. Is it possible that the line contains another point of the triangle? Is it possible that it doesn't? Illustrate.

26. Answer the two questions of exercise 25 if the line contains a point of the triangle which is not a vertex. Illustrate.

27. What conclusion can you draw if you know that a line contains three points of a triangle?

28. Is it true that lines which are not in the same plane are not parallel?

29. How would you define parallel line segments? Would you want to call $\overline{AB}$ and $\overline{CD}$ of Figure 8.80 parallel?

Figure 8.80

Figure 8.81

30. How many angles are determined by two intersecting lines? List all the angles determined by the lines $\mathscr{L}$ and $\mathscr{M}$ of Figure 8.81.

31. Name all the adjacent pairs of angles determined by $\mathscr{L}$, $\mathscr{M}$ of Figure 8.81.

32. If we consider that a triangle determines three disjoint sets in the plane and if we know that points P and Q are not points of the triangle but are in different disjoint sets, what can we say about line segment $\overline{PQ}$?

33. Complete the statement. "If $\overline{AC} \cong \overline{PR}$ and $\overline{BC} \cong \overline{QR}$ then _____," where A, B, C are points on one line and P, Q, R are points on another.

34. Illustrate the following:
    a) a line containing exactly one point of a simple closed curve.
    b) a line containing exactly two points of a simple closed curve.
    c) a line containing exactly three points of a simple closed curve.
    d) a line containing exactly four points of a simple closed curve.

35. If the intersection of simple closed curves $\mathcal{K}$ and $\mathcal{M}$ is also a simple closed curve, what can be said about sets $\mathcal{K}$ and $\mathcal{M}$?

36. Draw figures to illustrate the following possibilities for the intersection of a triangle T and a circle C:

    a) the empty set
    b) exactly one point
    c) exactly two points
    d) exactly three points
    e) exactly four points
    f) exactly five points
    g) exactly six points.

37. State why an angle is not a simple closed curve.

38. How is the set of all triangles T related to the set of all polygons P?

39. What is the intersection of the set of all triangles and the set of all quadrilaterals?

40. Referring to Figure 8.82, label the sets of points below as coplanar or non-coplanar.
    a) A, B, C
    b) A, B, E
    c) A, B, C, D
    d) B, C, D, E.

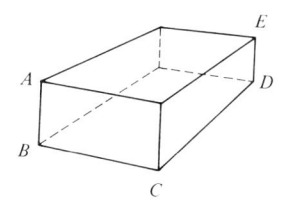

Figure 8.82

SELECTED ANSWERS AND HINTS

§1a

1. The abstract property shared is roundness.
3. Greeks felt that most mathematical principles should be proved carefully. Sumerians and Egyptians were content to rest their conclusions on observation.

§1b

1. On inductive reasoning. The earth seems flat if we look about us from a point on land.
3. A quick conclusion might be "He will reach the cheese in about two minutes on the 1,000,001st try." Is this conclusion affected by the fact that a mouse lives only three or four years in captivity?

§1c

1. (a) Grocery store arithmetic: true; Clock arithmetic: false.
   (c) Base-five numeration: true; Base-ten: false.
3. All x's are z's.
5. Mathematics: The primary statements are not checked against any outside reference, i. e., they are arbitrary. Inductive reasoning is used only to show plausible conjectures. Natural Science: Inductive reasoning is used to arrive at primary statements which may be changed by further observation.

§1d

1. (a) no variable   (c) x, y are variables.

2. (a) A *trapezoid* is a polygon of four line segments two of which are everywhere equidistant.
   (c) A *parallelogram* is a polygon of four line segments, nonconsecutive pairs of which are everywhere equidistant.
3. A *term* is the name for one particular idea and may consist of one word or a phrase. A *statement* is a complete sentence.
4. (a) As it is given: true; reversed: also true. Thus the statement might be a definition.
   (c) Given form: true; reversed: false. Thus, the statement is not reversible and cannot, therefore, be a definition.
5. (a) integer, factor, 2

§1e

1.

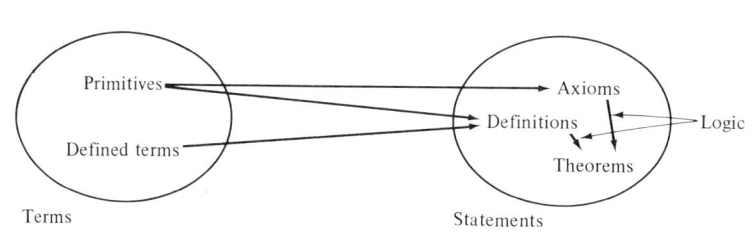

§1f

1. $=$, real numbers, $+$, $\times$, 0, 1, $<$.

3. The relationships in the axioms can be illustrated as follows:

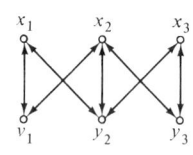

A-1 is satisfied certainly. There is exactly one y associated with each pair of x's as required by A-2. There is one x not associated with any particular y as required by A-3. The illustration shows that there are three x's as required by Theorem 1. Theorems 2, 3 are also satisfied. None of the axioms or theorems is contradicted by this illustration.

5. This illustration shows three points as endpoints and three line segments. Theorem 1 tells us that there are at least three points. Theorem 2 states that any two segments have at most one point in common. Theorem 3 states that there are at least three line segments in the system.

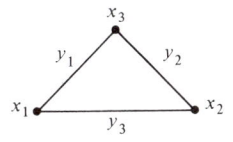

x_1 — y_1 — x_3 — y_2 — x_2, y_3

7. The theorems tell us that there are at least three tigers and three tanks and that any two distinct tanks have one tiger in them. This is hard to visualize for our everyday meaning of these terms; so to be reasonable the application must be to some very special kinds of tanks and ticklish tigers.

§2a

1. No. *Truth* refers to labels for propositions relative to a given truth standard. *Validity* refers to the way in which an argument is constructed, that is, to the reasoning process.

3. P(i) false; P(ii) true; conclusion true; argument valid.

5. P(i) false; P(ii) false; conclusion false; argument valid.

7. P(i) false; P(ii) false; conclusion false; argument invalid.

9. The conclusion "This pink panther passes a prelim" will form a valid argument.

§2b

1.

p	$\sim p$	$\sim(\sim p)$
T	F	T
F	T	F

Therefore, $\sim(\sim p) \equiv p$.

3. No. Valid and invalid are labels for the structure of *arguments*. *Propositions* are labeled true or false.

Selected Answers and Hints

5. a) a proposition provided Bertha is one particular creature or thing
   c) a proposition
   e) not a proposition, not a declarative sentence
   g) not a proposition, not a sentence
   i) a proposition

7. a) Bertha is not a bird.
   c) This lion is not lazy.
   e) not possible; logical negation applies only to propositions
   g) not possible; logical negation applies only to propositions
   i) $3 \not< 8$

§2c

1. Tree diagram

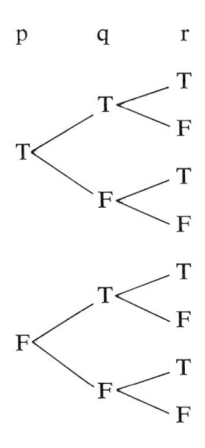

   Truth table

p	q	r
T	T	T
T	T	F
T	F	T
T	F	F
F	T	T
F	T	F
F	F	T
F	F	F

2. a) H: The earth is a planet.
      C: The moon is a satellite.
   c) H: Manfred mangles mastodons.
      C: Bertha is a bird.
   e) Not possible because the sentence is not an implication.

3. a) $p \wedge (\sim q \vee p)$ c) $p \Longleftrightarrow q$
4. a) $p \Rightarrow q \vee r$ c) $q \Rightarrow [q \wedge (q \Rightarrow p)]$
5. a) $(p \Rightarrow q) \vee r$ c) $\sim s \vee (x \Rightarrow \sim y)$
6. a) True, $f \Rightarrow t \underline{e} t$
   c) False, if Mary *is* merry
   e) True

Selected Answers and Hints

7. a) If Roy is not a boy, then Pearl is a girl.
   c) Roy is a boy iff Pearl is not a girl.
   e) It is not the case both that Roy is a boy only if Pearl is a girl and that Roy is a boy.

8. a) T c) F e) F

9. a) T-F; $p \vee p \equiv p$ c) T-F-T-T; $\sim p \vee q \equiv p \Rightarrow q$
   e) F-F; $p \wedge \sim p \equiv f$ g) T-T; $p \Rightarrow p \equiv t$
   i) T-T; $p \Rightarrow t \equiv t$ k) T-F; $\sim p \Rightarrow p \equiv p$
   m) T-T-T-T; $p \vee (p \Rightarrow q) \equiv t$
   o) T-F-T-T; $[p \vee (\sim p \wedge q) \Rightarrow (q \vee \sim p)] \equiv p \Rightarrow q$
   q) T-T-F-T; $p \vee q \Rightarrow p \equiv q \Rightarrow p$
   s) T-T-T-T; $p \Rightarrow p \vee q \equiv t$
   u) F-T-T-F; $\sim(p \Longleftrightarrow q) \equiv p \vee q$
   w) T-F-F-T; $\sim(p \vee q) \equiv p \Longleftrightarrow q$
   y) F-T-T-T; $(q \wedge \sim r) \vee \sim q \equiv \sim(q \wedge r)$

10. a) T-F-F-F-T-F-T-T   c) F-T-T-T-T-T-F-T

11. a) T-T-F-T   c) T-F-T-T

12. a) $p \Leftarrow q$ c) $p \vee q$
    e) $\sim(p \wedge q)$

§2d

2. a) argument; premise is p
   c) not an argument; a conjunction

3. a) Let the assignment of symbols be: p: a zosterops is a bird and q: this animal is two legged. In symbolic form the argument becomes: $(p \Rightarrow q) \wedge \sim q \Rightarrow \sim p$.

p	q	(i) $p \Rightarrow q$	(ii) $\sim q$	∴ $\sim p$
~~T~~	~~T~~	~~T~~	~~F~~	~~F~~
~~T~~	~~F~~	~~F~~	~~T~~	~~F~~
~~F~~	~~T~~	~~T~~	~~F~~	~~T~~
F	F	T	T	boxed T

The conclusion is true in the only case that each premise is true. The argument is valid.

Selected Answers and Hints

5.

p	q	r ‖ $p \Leftrightarrow q$ (a)	$\sim r$	$p \vee \sim r$ (b)	$\sim q$ (c)	∴ $\sim p$
~~T~~	~~T~~	~~T~~	~~T~~	~~F~~	~~T~~	~~F~~	~~F~~
~~T~~	~~T~~	~~F~~	~~T~~	~~T~~	~~T~~	~~F~~	~~F~~
~~T~~	~~F~~	~~T~~	~~F~~	~~F~~	~~T~~	~~T~~	~~F~~
~~T~~	~~F~~	~~F~~	~~F~~	~~T~~	~~T~~	~~T~~	~~F~~
~~F~~	~~T~~	~~T~~	~~F~~	~~F~~	~~F~~	~~F~~	~~T~~
~~F~~	~~T~~	~~F~~	~~F~~	~~T~~	~~T~~	~~F~~	~~T~~
~~F~~	~~F~~	~~T~~	~~T~~	~~F~~	~~F~~	~~T~~	~~T~~
F	F	F	T	T	T	T	boxed T

The argument is valid, for there is no case that each premise is true and the conclusion false.

7. Let the assignment of symbols be *p:* This is Ace polish and *q:* This is good polish. The symbolic translation is: $[(p \Rightarrow q) \wedge q] \Rightarrow p$. The argument is invalid and consistent.

9. Let *p:* $\mathscr{L}_1 // \mathscr{L}_2$ and *q:* $\mathscr{L}_3 // \mathscr{L}_2$. The symbolic translation is: $[(p \vee q) \wedge \sim q] \Rightarrow p$. The argument is valid and consistent.

11. Let *p:* not eat spinach and *q:* not drink milk. The symbolic translation is: $p \wedge q \Rightarrow \sim(\sim p \vee \sim q)$. The argument is valid and consistent.

13. Let *p:* major is math, *q:* required to take physics, and *r:* required to take logic. The symbolic translation is: $(p \Rightarrow q \wedge r) \wedge \sim r \Rightarrow \sim p$. The argument is valid and consistent.

15. Let *p:* rained yesterday, *q:* rains today, and *r:* will rain tomorrow. The symbolic translation is: $[(p \Rightarrow q) \wedge (q \vee r) \wedge \sim(p \wedge r) \wedge (q \Rightarrow r)] \Rightarrow \sim q$. The argument is invalid and consistent.

17. No. That conclusion forms an invalid argument.

19. a) Cork will float on water. Valid.
    c) A needle is less dense than water. Invalid.

21. a) Rows 3 and 4 can be crossed off. Nothing can be concluded about the truth value of *q*.

§2e

1. Construct a truth table for $(p \Rightarrow q) \wedge \sim q \wedge p$.
3. Show that $(p \vee \sim q) \wedge \sim p \Rightarrow \sim q$ is a tautology.

			(a)		(b)	
p	q	$\sim q$	$p \vee \sim q$	$\sim p$	$a \wedge \sim p$	$b \Rightarrow \sim q$
T	T	F	T	F	F	T
T	F	T	T	F	F	T
F	T	F	F	T	F	T
F	F	T	T	T	T	T

The compound is a tautology.

5. a) $\mathscr{L}_1$ is not $\perp \mathscr{L}_3 \Rightarrow \mathscr{L}_1 \perp \mathscr{L}_2$.
   b) $\mathscr{L}_1$ is not $\perp \mathscr{L}_2 \Rightarrow \mathscr{L}_1 \perp \mathscr{L}_3$.
   c) $\mathscr{L}_1 \perp \mathscr{L}_3 \Rightarrow \mathscr{L}_1$ is not $\perp \mathscr{L}_2$.
7. a) Mickey is sticky if Portia is portly.
   b) Portia is not portly if Mickey is not sticky.
   c) Mickey is not sticky if Portia is not portly.
9. If q, then p.
11. Construct their truth tables.
12. a) F-T-T-T: $\sim(p \wedge q) \underset{e}{=} \sim p \vee \sim q$.
    c) F-T-F-F: ?
    e) T-T-T-T: logically equivalent to any true proposition.
    g) T-T-T-F: $(\sim p \Rightarrow q) \vee q \underset{e}{=} p \vee q$.
    i) F-T-F-F: $p \wedge \sim q \underset{e}{=} \sim(p \Rightarrow q)$. See Part (c).
13. a) True
    c) False
    e) False
    g) True
    i) False
15. The propositions are logically equivalent.
17. The argument is valid.
19. Since the top argument is valid and is logically equivalent to the bottom by the result of Exercise 22, the latter must also be valid.
21. If $f'(c) \neq 0$ and $f'(c)$ exists, then $f(c)$ is not an extremum of a function f in some neighborhood of c.

22. a) $\sim p \vee q$ c) $q \wedge s$
23. a) $\sim q \Rightarrow \sim s$ c) $\sim p \Rightarrow q$
24. a) $\sim p \vee \sim q$
    c) $\sim(p \vee q) \vee (p \wedge q) \stackrel{e}{=} (\sim p \wedge \sim q) \vee (p \wedge q)$
25. a) Spiros does not eat spinach or Millie does not drink milk.
    c) $\mathscr{L}_1$ is not parallel to $\mathscr{L}_2$ and $\mathscr{L}_3$ is not parallel to $\mathscr{L}_2$.
    e) $x = y$ and $2x \neq 2y$.
    g) A natural number is not even and it is not odd.
    i) $3 \not< 8 \wedge 3 \neq 8$, i.e., $3 > 8$
26. a) In (2e-11) we are to prove that

$$\sim(p \vee q) \stackrel{e}{=} \sim p \wedge \sim q$$

p	q	$p \vee q$	$\sim(p \vee q)$	$\sim p$	$\sim q$	$\sim p \wedge \sim q$
T	T	T	F	F	F	F
T	F	T	F	F	T	F
F	T	T	F	T	F	F
F	F	F	T	T	T	T

The possible truth values of the two compounds are the same; that is, they are logically equivalent. Q. E. D.

27. a) True, that is, can be proved.
    c) Can be proved.
    e) We are to prove or disprove this conjecture:

$$p \vee (q \wedge r) \stackrel{e}{=} (p \vee q) \wedge r$$

p	q	r	$q \wedge r$	$p \vee (q \wedge r)$	$p \vee q$	$(p \vee q) \wedge r$
T	T	T	T	T	T	T
T	T	F	F	T	T	F
T	F	T	F	T	T	T
T	F	F	F	T	T	F
F	T	T	T	T	T	T
F	T	F	F	F	T	F
F	F	T	F	F	F	F
F	F	F	F	F	F	F

The possible truth values of the two compounds are not the same; that is, they are not logically equivalent. The conjecture is disproved.

28. 2^n

§2f

1. valid form

3. (a)

p	q	$p \wedge q$
T	T	T
~~T~~	~~F~~	~~F~~
~~F~~	~~T~~	~~F~~
~~F~~	~~F~~	~~F~~

The conclusion is true in the one case that the premises are simultaneously true. The argument is valid.

5.

p	q	$p \vee q$	$\sim q$	$(p \vee q) \wedge \sim q$	$[(p \vee q) \wedge \sim q] \Rightarrow p$
T	T	T	F	F	T
T	F	T	T	T	T
F	T	T	F	F	T
F	F	F	T	F	T

The implication is a tautology; hence, the argument is valid.

7. Construct the truth table.

9. Construct the truth table.

11. Valid, D. S.

13. Valid, C. S.

15. Valid, R. C.

17. Invalid, Inverse Argument.

19. Valid, R. D.

21. Valid, R. C.

23. Invalid, Converse Argument

25. Valid, D. A.

27. Valid, R. D.

29. Invalid, Converse Argument

31. Valid, C. R.

33. Invalid. Construct truth table.

35. Valid, R. C.

37. Valid, R. C.

39. There is no joy in Mudville, by R. D.

41. No conclusion
43. Bertha is a bird, by D. S.
45. Harry is hairy and Mary is merry by C. A.
47. Let p: A shfnexi is a glacho; q: An ootster is a glacho; and r: A shfnexi exists. The argument in symbolic form is

$$[(p \Rightarrow q) \land (r \lor p) \land \sim r] \Rightarrow ?$$

Our conclusions are: p: "A shfnexi is a glacho" from the second and third premises using D. S.; and q: "An ootster is a glacho" from the first conclusion and the first premise using R. D.

48. a) $p \subseteq f, q \subseteq f$.
    c) not possible
    e) $p \subseteq t, r \subseteq t$.
49. a) $r \subseteq f, q \subseteq f, p \subseteq t$.
    c) $q \subseteq t, p \subseteq f$.

§2g

1. a) premises inconsistent
   c) premises consistent: $p \subseteq t, q \subseteq t$.
3. Step 3 R. C.
   Step 5 D. S.
   Step 6 D. A.
5. Step 2 NIMP.
   Step 3 L-4
   Step 4 C. S.
   Step 5 C. S. (3)
   Step 7 R. D.
   Step 9 D. S.
   Step 10 C. A. (4,9)
7. Step 2 Defininition of $\Longleftrightarrow$.
   Step 3 L-4
   Step 4 C. S.
   Step 6 R. C.
9. 1) $\sim(p \lor r)$ P(a)
   2) $\sim(p \lor r) \subseteq \sim p \land \sim r$ De M. ($\lor$)
   3) $\sim p \land \sim r$ L-4
   4) $\sim r$ C. S.
   5) $q \Rightarrow r$ P(b)

		6)	$\sim q$	R. C.
		7)	$q \vee \sim s$	P(c)
		8)	$\sim s$	D. S.

Argument Valid Q. E. D.

11.	1)	$\sim r$	P(b)
	2)	$p \wedge s \Longleftrightarrow r$	P(a)
	3)	$p \wedge s \Longleftrightarrow r \underline{e}$	
		$(p \wedge s \Rightarrow r) \wedge (r \Rightarrow p \wedge s)$	Definition of $\Longleftrightarrow$.
	4)	$(p \wedge s \Rightarrow r) \wedge (r \Rightarrow p \wedge s)$	L-4
	5)	$p \wedge s \Rightarrow r$	C. S.
	6)	$\sim (p \wedge s)$	R. C. (1,4)
	7)	$\sim (p \wedge s) \underline{e} \sim p \vee \sim s$	DeM.($\wedge$)
	8)	$\sim p \vee \sim s$	L-4
	9)	p	P(c)
	10)	$\sim s$	D. S.

Argument Valid Q. E. D.

13.	1)	$\sim p$	P(d)
	2)	$\sim s \Rightarrow p$	P(c)
	3)	$\sim \sim s$	R. C.
	4)	$\sim s \vee \sim r$	P(a)
	5)	$\sim r$	D. S.
	6)	$\sim r \Rightarrow \sim q$	P(b)
	7)	$\sim q$	R. D.
	8)	$\sim q \wedge \sim p$	C. A. (7,1)

Argument Valid Q. E. D.

15.	1)	$\sim (p \vee \sim r)$	P(a)
	2)	$\sim (p \vee \sim r) \underline{e}$	
		$\sim p \wedge \sim \sim r$	DeM ($\vee$)
	3)	$\sim p \wedge \sim \sim r$	L-4
	4)	$\sim \sim r \underline{e} r$	D. N.
	5)	$\sim p \wedge r$	L-4
	6)	$\sim p$	C. S.
	7)	r	C. S. (5)
	8)	$q \vee p$	P(b)
	9)	q	D. S. (6, 8)
	10)	$r \Rightarrow s$	P(c)
	11)	s	R. D. (7, 10)
	12)	$q \wedge s$	C. A. (9, 11)
	13)	$q \wedge s \Rightarrow u \wedge s$	P(d)
	14)	$u \wedge s$	R. D.

	15)	$u \wedge s \underline{e} s \wedge u$	Commutative Property
	16)	$s \wedge u$	L-4

Argument Valid Q. E. D.

17.	1)	$\sim r$	P(e)
	2)	$q \Rightarrow r$	P(a)
	3)	$\sim q$	R. C.
	4)	$p \Rightarrow q$	P(b)
	5)	$\sim p$	R. C.
	6)	$p \vee u$	P(c)
	7)	u	D. S.
	8)	$u \Rightarrow s$	P(d)
	9)	s	R. D.

Argument Valid Q. E. D.

19)	1)	$\sim(p \wedge q)$	P(a)
	2)	$\sim(p \wedge q) \underline{e} \sim p \vee \sim q$	DeM ($\wedge$)
	3)	$\sim p \vee \sim q$	L-4
	4)	$\sim p \vee \sim q \underline{e} p \Rightarrow \sim q$	LEDI
	5)	$p \Rightarrow \sim q$	L-4
	6)	$\sim q \Rightarrow u$	P(b)
	7)	$p \Rightarrow u$	C. R.
	8)	$(p \Rightarrow u) \underline{e} (\sim u \Rightarrow \sim p)$	CLEP
	9)	$\sim u \Rightarrow \sim p$	L-4
	10)	$\sim p \Rightarrow u$	P(c)
	11)	$\sim u \Rightarrow u$	C. R.
	12)	$(\sim u \Rightarrow u) \underline{e} (u \vee u)$	LEDI
	13)	$u \vee u$	L-4
	14)	$(u \vee u) \underline{e} u$	Idempotent Property
	15)	u	L-4
	16)	$s \Rightarrow \sim u$	P(d)
	17)	$\sim s$	R. C.

Argument Valid Q. E. D.

21.	1)	$p \vee \sim r$	P(c)
	2)	$p \vee \sim r \underline{e} \sim p \Rightarrow \sim r$	LEDI
	3)	$\sim p \Rightarrow \sim r$	L-4
	4)	$\sim p \Rightarrow \sim r \underline{e} r \Rightarrow p$	CLEP
	5)	$r \Rightarrow p$	L-4
	6)	$p \Rightarrow q$	P(a)

Selected Answers and Hints 415

7) $r \Rightarrow q$ C. R.
8) $q \lor r$ P(b)
9) $q \lor r \equiv \sim q \Rightarrow r$ LEDI
10) $\sim q \Rightarrow r$ L-4
11) $\sim q \Rightarrow q$ C. R. (10, 7)
12) $\sim q \Rightarrow q \equiv q \lor q$ LEDI
13) $q \lor q$ L-4
14) $q \lor q \equiv q$ Idempotent Property
15) q L-4
16) $q \lor \sim r$ D. A.

           Argument Valid Q. E. D.

23. p: Flipper is a mammal   1) $p \Rightarrow q$ P(a)
    q: gets oxygen from the air   2) $q \Rightarrow r$ P(b)
    r: has no need of gills   3) $p \Rightarrow r$ C. R.
    s: habitat is the ocean   4) $p \land s$ P(c)
     (a) $p \Rightarrow q$ 5) p C. S.
     (b) $q \Rightarrow r$ 6) r R. D. (3,5)
     (c) $p \land s$ Valid
        r

25. p: Tom is seven
    q: Tom is same age as June $p \Rightarrow q$
    r: John not as old as Tom $r \Rightarrow s$
    s: John not as old as June $p \land \sim s$
             $\sim r \land q$ Valid

27. p: John is shorter than Bob $p \Rightarrow q$
    q: Mary is taller than Jean $\sim q$
    r: John and Bill same height $r \Rightarrow p$
             $\sim r$ Valid

29. p: clock is slow $p \Rightarrow q \land r$
    q: arrived before 12:00 $s \Rightarrow \sim r$
    r: saw horse leave $s \lor u$
    s: is telling truth p
    u: was in building u Valid

31. p: contract not approved $p \Rightarrow q$ let $p \equiv t$
    q: total remains as it is $q \Rightarrow r$ $q \equiv t$
    r: not add new staff $r \lor s$ $r \equiv t$
    s: production will be delayed $\sim s$ $s \equiv f$
             $\sim p$ Invalid

Selected Answers and Hints

33. $p:$ a watched pot never boils $\quad \sim(p \Rightarrow q) \Rightarrow r$
    $q:$ it is not worth a bird bush $\quad \sim r$
    $r:$ can tickle tigers $\quad\quad\quad\quad p \Rightarrow q$
    $\quad\quad\quad\quad\quad\quad\quad\quad\quad\quad\quad\quad \sim q \Rightarrow p$
    $\quad\quad\quad\quad\quad\quad\quad\quad\quad\quad\quad\quad\overline{p \Longleftrightarrow q}$ $\quad\quad$ Invalid

35. $p:$ rule approved $\quad\quad\quad\quad p \Longleftrightarrow q$
    $q:$ supported by boss $\quad\quad q \lor r$
    $r:$ wife opposed $\quad\quad\quad\quad r \Rightarrow s$
    $s:$ family deliberations $\quad\overline{p \lor s}$ $\quad\quad$ Valid

37. Step 3 R. D.
    Step 5 D. S.
    Step 6 D. A.
    Step 7 C. I.

39. 1) $\sim s$ $\quad\quad\quad\quad\quad$ I. P.
    2) $\underline{p \lor s}$ $\quad\quad\quad\quad$ P(b)
    3) p $\quad\quad\quad\quad\quad\quad$ D. S.
    4) $\underline{p \Rightarrow u \land s}$ $\quad\quad$ P(a)
    5) $\underline{u \land s}$ $\quad\quad\quad\quad$ R. D.
    6) s $\quad\quad\quad\quad\quad\quad$ C. S.
    7) $\underline{s \land \sim s}$ $\quad\quad\quad$ C. A. (1,6)
    8) s $\quad\quad\quad\quad\quad\quad$ I. I.
    $\quad\quad\quad\quad\quad\quad\quad\quad\quad\quad$ Argument Valid. Q. E. D.

41. 1) $\sim p$ $\quad\quad\quad\quad\quad$ C. P.
    2) $\underline{\sim p \Rightarrow q \land r}$ $\quad$ P(a)
    3) $q \land r$ $\quad\quad\quad\quad$ R. D.
    4) q $\quad\quad\quad\quad\quad\quad$ C. S.
    5) $\underline{q \Rightarrow s}$ $\quad\quad\quad$ P(b)
    6) s $\quad\quad\quad\quad\quad\quad$ R. D.
    7) $\sim p \Rightarrow s$ $\quad\quad\quad$ C. I.
    $\quad\quad\quad\quad\quad\quad\quad\quad\quad\quad$ Argument Valid. Q. E. D.

43. 1) $\sim w$ $\quad\quad\quad\quad\quad$ I. P.
    2) $\sim w \lor s$ $\quad\quad\quad$ D. A.
    3) $\underline{\sim w \lor s \Rightarrow w}$ $\quad$ P
    4) w $\quad\quad\quad\quad\quad\quad$ R. D.
    5) $\underline{w \land \sim w}$ $\quad\quad$ C. A. (1,4)
    6) w $\quad\quad\quad\quad\quad\quad$ I. I.
    $\quad\quad\quad\quad\quad\quad\quad\quad\quad\quad$ Argument Valid. Q. E. D.

Selected Answers and Hints 417

45. Let $u \doteq t$
    $s \doteq f$
    $q \doteq t$ Argument Invalid

47. Let $u \doteq f$
    $s \doteq f$
    $w \doteq t$ or $w \doteq f$ Argument Invalid

49.
1)	s	C. P.
2)	$\sim p \Rightarrow \sim s$	P(a)
3)	$\sim(\sim p)$	R. C.
4)	$\sim p \vee r$	P(b)
5)	r	D. S.
6)	$r \Rightarrow \sim u$	P(c)
7)	$\sim u$	R. D.
8)	$s \Rightarrow \sim u$	C. I.

Argument Valid. Q. E. D.

51.
1)	$\sim \sim u$	I. P.
2)	$\sim \sim u \doteq u$	D. N.
3)	u	L-4
4)	$u \Rightarrow \sim p$	P(b)
5)	$\sim p$	R. D.
6)	$p \vee q$	P(a)
7)	q	D. S.
8)	$\sim(q \vee r)$	P(c)
9)	$\sim(q \vee r) \doteq \sim q \wedge \sim r$	DeM($\vee$)
10)	$\sim q \wedge \sim r$	L-4
11)	$\sim q$	C. S.
12)	$q \wedge \sim q$	C. A. (7,11)
13)	$\sim u$	I. I.

Argument Valid. Q. E. D.

53.
1)	$\sim \sim (p \wedge s)$	I. P.
2)	$\sim \sim (p \wedge s) \doteq p \wedge s$	D. N.
3)	$p \wedge s$	L-4
4)	p, s	C. S.
5)	$p \Rightarrow q \vee r$	P(a)
6)	$q \vee r$	R. D.
7)	$s \Rightarrow \sim r$	P(c)
8)	$\sim r$	R. D. (4, 7)

Selected Answers and Hints

	9)	q	D. S. (6, 8)
	10)	$q \Rightarrow \sim p$	P(b)
	11)	$\sim p$	R. D.
	12)	$p \wedge \sim p$	C. A. (4, 11)
	13)	$\sim (p \wedge s)$	I. I.

Argument Valid. Q. E. D.

55.	1)	p	C. P.
	2)	$s \Rightarrow \sim p$	P(b)
	3)	$\sim s$	R. C.
	4)	$s \vee r$	P(a)
	5)	r	D. S.
	6)	$r \Rightarrow q$	P(c)
	7)	q	R. D.
	8)	$p \Rightarrow q$	C. I.
	9)	$p \Rightarrow q \equiv \sim p \vee q$	LEDI
	10)	$\sim p \vee q$	L-4

Argument Valid. Q. E. D.

57. Invalid; e.g., let $s \equiv t$, $p \equiv t$, $q \equiv f$, $r \equiv t$.

59.	1)	$\sim s$	C. P.
	2)	$p \Rightarrow q$	P(a)
	3)	$\sim q$	P(d)
	4)	$\sim p$	R. C.
	5)	$\sim s \wedge \sim p$	C. A. (1, 4)
	6)	$\sim s \wedge \sim p \equiv \sim (s \vee p)$	DeM ($\vee$)
	7)	$\sim (s \vee p)$	L-4
	8)	$r \Rightarrow s \vee p$	P(b)
	9)	$\sim r$	R. C.
	10)	$u \vee v \Rightarrow r$	P(c)
	11)	$\sim (u \vee v)$	R. C.
	12)	$\sim (u \vee v) \equiv \sim u \wedge \sim v$	DeM ($\vee$)
	13)	$\sim u \wedge \sim v$	L-4
	14)	$\sim u$	C. S.
	15)	$\sim s \Rightarrow \sim u$	C. I.

61. $p:$ eaten 13 bananas (a) $(p \wedge q) \vee r$
 $q:$ eaten a cheeseburger (b) $q \vee s \Rightarrow \sim r$
 $r:$ is hungry $p \Rightarrow \sim s$
 $s:$ in kitchen Argument Invalid
 For example, let $p \equiv t$, $s \equiv t$, $q \equiv t$, $r \equiv f$

§2h

1. a) Some integers are not even. This is an integer and it is not even.
   c) No book is on the shelf. If this is a book, it is not on the shelf.
   e) Some integers are neither even nor odd.
   g) $\exists x, x^2 + 1 \neq 0$.
   i) $\exists x, x \geq 0$.
   k) $\forall p, p$ is not false.

3. The collection may not be a sentence; if a sentence, it may not have required grammatical form; a suitable truth standard may not be specified; the collection may be an open sentence.

6. (a) False (c) False (e) not a proposition because y is not quantified (g) True (i) False

7. (a) propositional form (c) propositional form (e) proposition (g) proposition (i) proposition

9. Let $p:$ this is a parallelogram
       $q:$ this is a quadrilateral
       $r:$ this is a rectangle
       $s:$ this is a square

   The symbolic translation is   a) $r \Rightarrow p$
                                  b) $s \Rightarrow r$
                                  c) $p \Rightarrow q$
                                     $q \Rightarrow s$ Invalid

11. Invalid
13. Valid
15. Valid

§2i

2. a) Column 5
3. a) $p \Rightarrow (q \Rightarrow r)$
5. Valid
7. True. If two triangles have congruent bases and equal areas, then they have congruent altitudes. If two triangles have equal areas and congruent altitudes, then they have congruent bases.

§2j

7. A-1, A-2, A-3, A-4, A-7, A-8, A-9, A-10, A-11, A-12, A-13, A-14, A-19 are properties which correspond to properties of the algebra of numbers.

§2k

3. Inductive
7. a) $\sim(p \land q)$ c) $\sim(p \Rightarrow q)$ e) $\sim(p \lor q)$ g) $p \lor q$
   i) $q \lor f$
8. a) converse: If it is cold, then it snows.
   inverse: If it does not snow, then it is not cold.
   contrapositive: If it is not cold, then it does not snow.
9. No. The positive is true but the converse is false.
10. a) T-F-T-F-T-F-F-T   c) F-F-F-F
11. a) False   c) False   e) False   g) True   i) True
13. a) True   c) True   e) False   g) not possible, open sentence.
14. a) No dogs are lazy. If it is a dog, it is not lazy.
    c) No Berthas are birds.
    e) Some burbot fish don't equal 100,000 or no astonished men equal 1,000,000.
    g) $8 > 2$
    i) $\forall\, x,\ x \leq 3 \lor x \geq 5$
15. a) proposition if Cindy and Sid are known.
    c) proposition
    e) not a declarative sentence
    g) proposition
17. Symbolic form $\dfrac{\begin{array}{l} p \Rightarrow q \\ p \Longleftrightarrow \sim q \end{array}}{\sim p}$ Valid

19. Valid
21. Invalid, Converse Argument
23. Invalid
25. Valid
27. Valid
29. Invalid
31. Invalid

Selected Answers and Hints

33. Valid
35. Valid
37. Invalid
39. Invalid
41. Valid
43. Valid
45. Valid
47. Valid
49. Symbolic Form $p \vee q$
    $\underline{\sim q \vee r}$
    $\sim r \Rightarrow p$ Valid
51. Symbolic Form $p \vee q$
    $r \Rightarrow \sim q$
    $\underline{\sim r \Rightarrow s}$
    $p \Rightarrow s$ Invalid
53. a) Nothing is known.   c) It makes contact with kids.
55. One conclusion is "Babies are despised." Another is "Illogical persons cannot manage a crocodile." Optimal is "Babies cannot manage a crocodile."
57. "No jug in this cupboard will hold water."

§3a

1. troop — scouts, pack — hounds, class — students
2. a) True   c) False
3. a) $\{a, e, i, o, u\}$
   c) {Maine, New Hampshire, Vermont, Massachusetts, Connecticut, Rhode Island}
4. a) the elements of nature according to some Greek philosophers
   c) unit fractions 1/1 through 1/5
   e) the set of writing implements
   g) the set of oil companies
5. The dog likes to play in the sunshine near the maple tree.
6. a) {Washington, Lincoln, Roosevelt, Kennedy $\cdots$}
   c) {Washington, Oregon, California, Alaska, Hawaii}

e) {Stars and Stripes, Tricolor, Union Jack, ···}
7. a) nonempty   c) empty   e) empty
8. a) {2}   c) {1, 2, 3, 4, 5, 6, 7}   e) { }   g) $\left\{\dfrac{\pi - 28}{3}\right\}$
   i) {3}
9. a) ∅   c) nonempty; ∅ is an element.

§3b

1. a) Two sets A, B are disjoint iff $x \in A \Rightarrow x \notin B$.
2. a) equal
   c) equal if we accept "family" to mean a married couple and their children
3. a) {1, 9, 16, 17}   c) {1, 9, 17}   e) ∅
4. a) {Left, Center, Right}   c) {First, Second, Third}
5. a) True   c) True   e) False   g) False   i) False
   k) True   m) True
6. a) $\Rightarrow$ c) $\not\Rightarrow$
7. a) No   b) Yes   c) Yes   d) No   e) No   f) No
9. a) Yes   b) No   c) Yes   d) No   e) Yes   f) Yes

§3c

1. a)

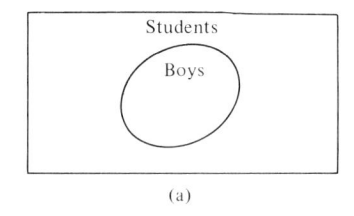

(a)

c)

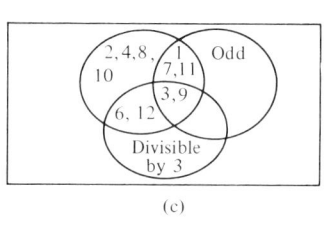

(c)

2. a) empty   c) empty
3. a) $\{x\}, \{y\}, \{x, y\}, \varnothing$ c) ∅
4. a) Yes   c) No   e) No
5. a) Yes   c) Yes   e) Yes

Selected Answers and Hints

6. (1) C. P.   (5) R. D.   (6) C. I.
7. (2) S-3   (5) (3-3)   (6) R. D.   (7) C. I.
9. True
11. False
19. a) $x = 0$ c) nothing
21. Yes; sometimes.
23. Yes; sometimes; yes; sometimes; yes; sometimes.
25.
   1. $a = b$ 1. C. P.
   2. $a = b \Rightarrow \{a\} = \{b\}$ 2. (3-7)
   3. $\{a\} = \{b\}$ 3. R. D.
   4. $\{a\} = \{b\} \Rightarrow \{b\} = \{a\}$ 4. (3c-4)
   5. $\{b\} = \{a\}$ 5. R. D.
   6. $\{b\} = \{a\} \Rightarrow b = a$ 6. (3-7)
   7. $b = a$ 7. R. D.
   8. $a = b \Rightarrow b = a$ 8. C. I.

Q. E. D.

§3d

1. a) $A' = \{1, 10, 15, 28, 36\}$ c) $A' = \{1, 3, 6, 10, 15, 21, 28, 36, 45\}$
2. a) $A' = \{3, 9, 27\}$ c) $C' = \{8, 14, 27\}$
3. a) $S' = \{x \mid x > 0 \land x \in \mathscr{Z}\}$ b) $S' = \mathscr{N}$
4. a) $S' = \varnothing$
5. If I were empty, we would have $x \notin \varnothing \Longleftrightarrow x \in \varnothing$, which would contradict the definition.

§3e

1. a) negation, absolute value
2. a) addition   c) division   e) raising to the power
    g) $a \circ b$ could be (a times b) plus b.
3. Two sets are overlapping iff $A \cap B \neq \varnothing \land A \not\subset B \land B \not\subset A$.
5. a) Real numbers
7. a) $\{2,5\}$ c) $\{7\}$ e) $\{2,4,5\}$

8. a) S c) $S \cup T$ e) $S \cap T'$ g) $(S \cap T') \cup (S' \cap T)$
   i) $(S \cap T)'$

9. a)   c)   e)   g)

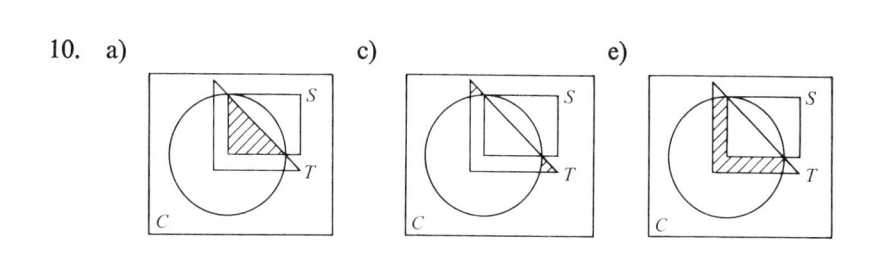

10. a)   c)   e)

11. a) $\{-, \dot{c}, \Delta\}$ c) Yes

12. a) $\{a, b, c, d, e, f, i, o, u\}$ c) $\{a, b, c, d, e, f, i, m, n, o, p, q, r, s, t, u\}$
    e) $\emptyset$ g) $\{o, u\}$

13. $\{a, b, c, d, \triangle, \dot{c}, \theta, \#, \$\}$

14. a) $\{3, 6\}$ c) $\{2, 3, 4, 6, 8, 10\}$ e) $\{6, 8, 10\}$

15) a) $\{c, d, f, i, k, m\}$ c) $\{c, d, f\}$
    e) Not defined in this case. Definition 3-7 provides only for a complement relative to a *nonempty I*.

17. a) 16 b) 10

19. a) 29 c) 2 e) 3 g) 7

20. a) False c) False

21. a) red, blue, green c) It results in the original set.

22. a) $\{e, d\}$
    c)

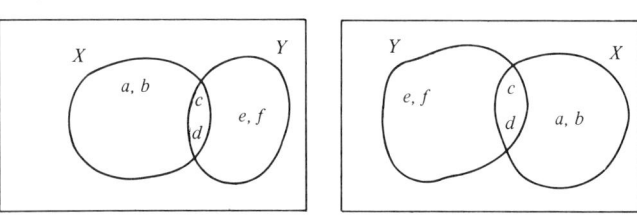

23. a) E is a subset of S.
24. a) $\{e\}$ c) They are equal.
25. a) $\{g\}$ c) $\{g\}$
26. a) Closure under union   c) Idempotent law for union
    e) Inverse laws   g) Associativity   i) DeMorgan's Laws
27. a) $(\varnothing \cap M) \cup S = \varnothing \cup S = S$
    c) $M \cap (M \cup S) = (M \cup \varnothing) \cap (M \cup S) = M \cup (\varnothing \cap S) = M \cup \varnothing = M$
28. a) $\{1, 2, -2, -3\}$ c) $\varnothing$ e) $\mathscr{R}$ g) $\mathscr{W}$
29. a) Commutative property   c) Distributivity of $\cap$ over $\cup$
    e) Identity for $\cup$ g) Identity for $\cap$
30. a) $M' \cup S$
31. a) $A \cap (B \cap B')$
32. a) $(R \cap S) \cup (R \cap M)$ c) $(X \cap Z) \cup (X \cap Y)$
    e) $M \cup (R \cap S)$
33. a) $(C \cup B) \cup C' = C' \cup (C \cup B)$ (3e-5)
    $= (C' \cup C) \cup B$ (3e-6)
    $= (C \cup C') \cup B$ (3e-5)
    $= I \cup B$ (3e-4)
    $= B \cup I$ (3e-5)
    $= I$ (3e-3c)
    c) $(C \cup B') \cap (C \cup B') = C \cup B'$ (3e-3a)
    e) $[(C \cup \varnothing)' \cup C]' = (C' \cup C)'$ (3e-3b)
    $= (C \cup C')'$ (3e-5)
    $= I'$ (3e-4)
    $= \varnothing$ (3d-1b)
    g) $S \cap (S \cap B) = (S \cap S) \cap B$ (3e-6)
    $= S \cap B$ (3e-3a)
    i) $(S' \cap M')' = S'' \cup M''$ (3e-8)
    $= S \cup M$ (3d-2)
34. a) False   c) True   e) False   g) False

§3f

1. a) $\sim(f \wedge \sim p) \subseteq \sim f \vee p$ b) $\sim p \wedge (t \vee \sim p) \subseteq \sim p$
2. a) I c) $C \cup B'$ e) $\varnothing$ g) $S \cap B$ i) $S \cup M$
3. a) $I \cap A' = A'$ c) $\varnothing \cap A' = \varnothing$ e) $A \cap A' = \varnothing$
    g) $A \cap (B \cap C)' = A \cap (B' \cup C')$

4. a) $(S \cup \emptyset) \cap (S \cap \emptyset)' = S \cap I = S$ c) $(S \cup I) \cap (S \cap I)' = I \cap S' = S'$

5. a) True   c) True   e) True   g) False   i) False

§3g

1. a) $Z = \{x \mid x \text{ is a zoy}\}$ P(a): $Z \subset G$
$G = \{x \mid x \text{ is a glacho}\}$ P(b): $G \subset O$
$O = \{x \mid x \text{ is an ootster}\}$ P(c): $S \cap O = \emptyset$
$S = \{x \mid x \text{ is a shfnexi}\}$ C : $S \cap Z = \emptyset$

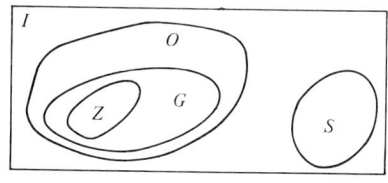

Valid

c) $T = \{x \mid x \text{ is a tired tiger}\}$ P(a): $T \subset L$
$L = \{x \mid x \text{ is ticklish}\}$ P(b): $M \subset L$
$M = \{x \mid x \text{ is a man}\}$ C : $M \subset T$

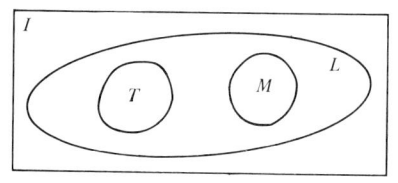

Invalid

3. $H = \{x \mid x \text{ has no arms}\}$ P(a): $B \subset H$
$B = \{x \mid x \text{ is a normal human being}\}$ P(b): $S \subset B$
$S = \{x \mid x \text{ is a snail}\}$ C : $S \subset H$

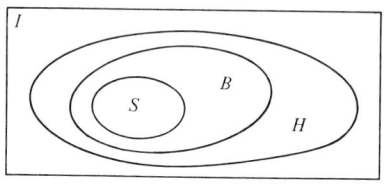

Valid

5. $R = \{x \mid x \text{ is a rose}\}$ P(a): $R \subset E \wedge V \subset B$
$V = \{x \mid x \text{ is a violet}\}$ P(b): $B \subset E$
$B = \{x \mid x \text{ is blue}\}$ P(c): $\exists x \in E \wedge x \in B$
$E = \{x \mid x \text{ is red}\}$ C : $\exists x \in V \wedge x \in R$

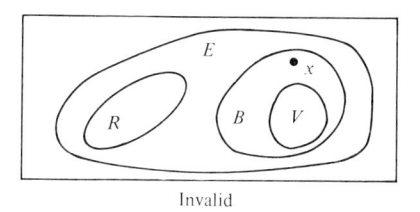

Selected Answers and Hints 427

[Diagram: Box labeled I containing oval E, with R inside E, and B inside E containing V, with point x inside B but outside V.] Invalid

7.

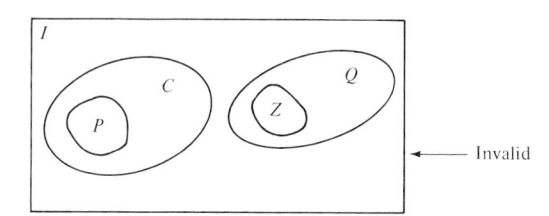

[Diagram: Box labeled I containing oval C with P inside, and oval Q with Z inside.] ← Invalid

9.

[Diagram: Box labeled I with overlapping ovals B and P, and oval R overlapping P, with point x in the intersection of P and R.] Valid →

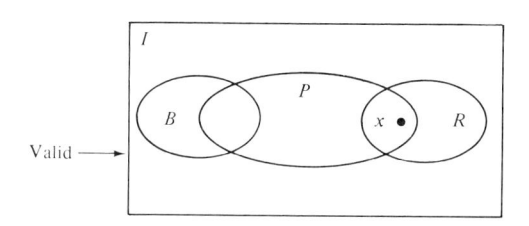

11.

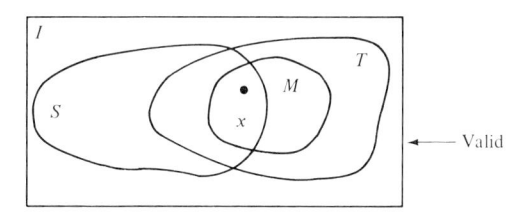

[Diagram: Box labeled I containing oval S overlapping oval T, with M inside the overlap, and point x inside M.] ← Valid

13.

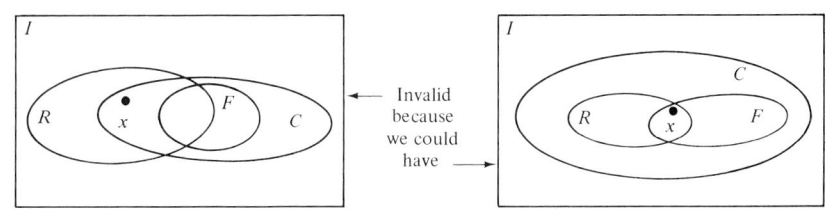

[Two diagrams: Left — Box labeled I with overlapping ovals R, F, C, point x in R near F. Right — Box labeled I with C containing R and F overlapping, point x in the intersection.] ← Invalid because we could have →

428 **Selected Answers and Hints**

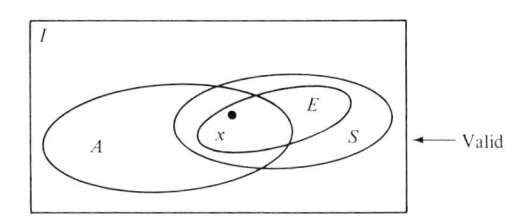

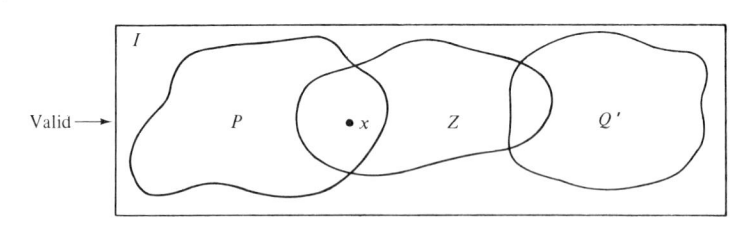

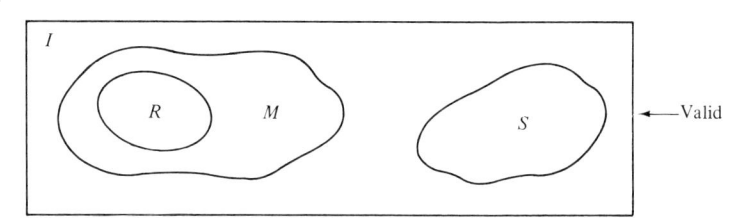

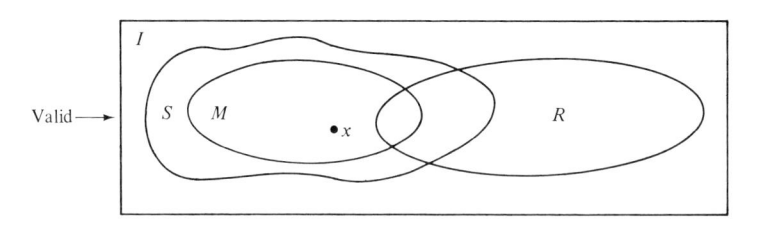

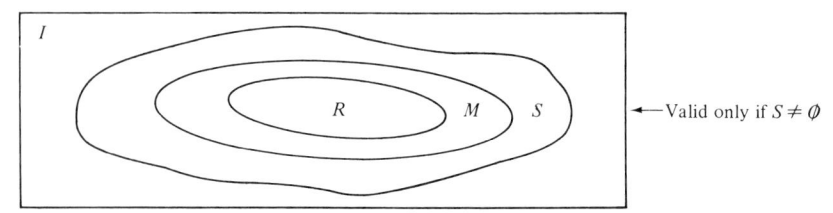

15.

Valid

17.

Valid

19.

Valid

21.

Valid

23.

Valid only if $S \neq \emptyset$

25.

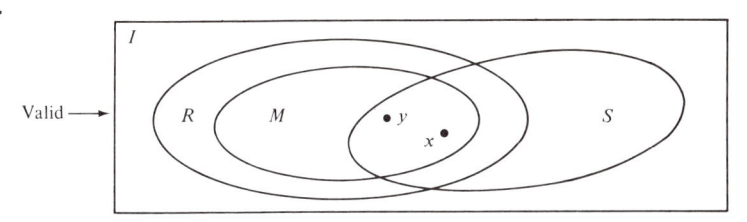

27. Valid
29. Valid
31. Invalid
33. Valid
35. Invalid

§3h

1. c) Let I be the set of teachers you have during the day, $\{x \mid x \in I \wedge x \text{ is a male}\}$.

2) a) {Nixon, Johnson, Kennedy, Eisenhower, Truman}
   c) {Kennedy}
   e) $\{x \mid x \text{ is a president} \wedge x \text{ is not L. Johnson, Kennedy, Truman, F. Roosevelt, Wilson}\}$

3. a) $\{x \mid x \in I \wedge x \text{ is not eight feet tall}\}$, I = teachers in school
   c) $\{x \mid x \in I \wedge x \text{ is not a male teacher}\}$, I = teacher during day

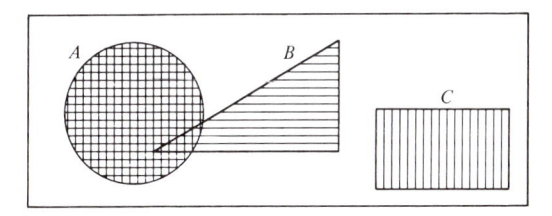

4. a) See figure   c) the crosshatched region of B e) C
5. a) $\{a, b, c, d\}$ c) ∅   e) ∅   g) disjoint sets
6. a) {1}   c) {1}
7. Yes, same elements
8. a) $A = B$ c) $A = \emptyset$ e) $A = B = \emptyset$
9. a) (v)   c) (ii)   e) (i), (ii)   g) (iv)   j) (i), (ii), (iii), (v)
   l) (i)

Selected Answers and Hints

10. a) $=, =$ c) A e) A g) overlapping
11. a) Yes, a set with one element
    c) Yes, $\emptyset \subset A$ for all sets A.
13. a) Yes c) Yes
14. a) False c) True e) False
15. a) $\{x \mid e \leq x < 9 \wedge x \text{ is even}\}$
    c) $\{x \mid x = y^2 \wedge 1 \leq y \leq 4\}$
    e) $\{x \mid -1 < x < 2\}$
16. a) $\{1, 2, 3\}$ c) $\{0\}$
17. a) Four c) Two
19. a) Closure c) Identity e) Commutative g) Uniqueness
21.

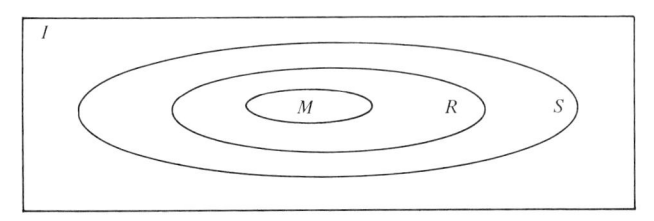

Valid Provided $M \neq \emptyset$

23. Valid

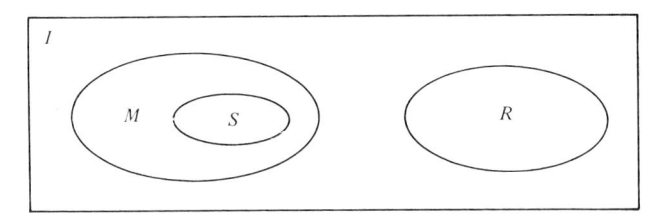

25. Invalid

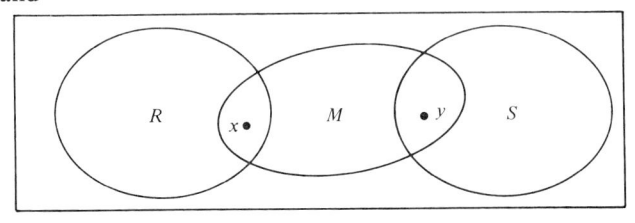

§4a

1. a) True c) True
2. a) $\{(1, 0), (1, 3), (2, 0), (2, 3)\}$
3. $a = b = c = d$
4. a) $x = 0 \wedge y = 0$ c) $y = 0 \wedge y = \frac{3}{4}$ e) $x = 5 \wedge y = 2$
5. a) $\{(0, 0), (1, 2), (2, 4)\}$
6. a) $\mathscr{D}_R = \{3, 2, 1\}$
7. Yes

§4b

1. a) $\{(\text{cat}, \wedge), (\text{cat}, \vee), (\text{cat}, \Rightarrow), (\text{rat}, \wedge), (\text{rat}, \vee), (\text{rat}, \Rightarrow),$
      $(\text{bat}, \wedge), (\text{bat}, \vee), (\text{bat}, \Rightarrow)\}$
   c) $\{(\text{big}, \wedge), (\text{big}, \vee), (\text{big}, \Rightarrow), (\text{bird}, \wedge), (\text{bird}, \vee), (\text{bird}, \Rightarrow)\}$
2. a) $\{(1, 1), (1, 5), (2, 1), (2, 5)\}$ c) $\{(2, 3), (2, 4), (2, 5)\}$
3. a) 30   c) 18   e) 90   g) 8   i) 14
4. a) 12   c) 6   e) 24   g) 4   i) 6
5. 84
6. a) $A = \{1, 2, 3\}$ c) $\mathscr{D}_R \subset A, \mathscr{E}_R \subset B$
7. a) $\mathscr{D}_R = \{3, 2, 1\}$
   c) No, $\{3, 2, 1\} \times \{2, 1, 3\}$ would have 9 elements and R has only 6.
9. a) $\{\ \}$ c) $\{(\varnothing, \varnothing)\}$
11. $A = \varnothing$, or $A = \{1\}$, etc.

§4c

1. a) $R = \{(1, 1), (1, 2), (1, 3), (1, 4), (1, 5), (2, 2), (2, 4), (3, 3), (4, 4),$
      $(5, 5)\}$
   c) $R = \{(1, 2), (1, 3), (1, 4), (1, 5), (2, 3), (2, 4), (2, 5), (3, 4), (3, 5),$
      $(4, 5)\}$
   e) $R = \{\ \}$
3. a) Not reflexive: A yard is not longer than a yard. Not symmetric: A yard is longer than a foot but a foot is not longer than a yard.

c) Not reflexive: 3 is not different from itself by 5. Not transitive, $(1, 6) \in R$ and $(6, 11) \in R \Rightarrow (1, 11) \in R$ is false.
e) Not reflexive, not symmetric: $(2, \sqrt{4}) \in R \Rightarrow (\sqrt{4}, 2) \in R$ is false, not transitive
g) Not reflexive, not symmetric, not transitive
i) Not transitive
k) Not reflexive, not transitive

4. a) Less than on $\mathscr{R}$ c) perpendicular on coplanar lines.
   e) Impossible: It can be shown that any relation which is symmetric and transitive is also reflexive.
5. a) {(cat, cat), (cat, mouse)}   c) {(mouse, cat)}, { }
6. a) Neither: $\perp$ is neither reflexive nor transitive; // is not reflexive.
7. a) (cat, cat), (dog, cat), (dog, dog)
   c) (2, 2), (3, 3), (8, 2), (8, 8), (5, 3), (5, 5)

§4d

1. a) Not a function   c) function   e) function   g) not a function
3. a) $f + g = \{(5, 10,), (7, 2), (2, 2), (0, 6)\}$
   c) $f/g = \{(5, -6), (7, 0), (0, 1)\}$
   e) $f \circ g = \{(7, 2), (2, 3), (4, 3), (-6, 2)\}$
   g) $f \circ f = \{(7, 3), (2, 2)\}$
   i) $g - f = \{(5, -14), (7, 2), (2, 2), (0, 0)\}$
   k) $g/f = \{(5, -1/6), (2, 0), (0, 1)\}$
   m) $h/f = \{(0, 0)\}$
   o) $h/g = \{(0, 0)\}$
   q) $h \circ g = \{(2, 0), (4, 0)\}$
   s) $h \circ h = \{(0, 0), (-5, -5), (4, 4)\}$
4. a) -2 c) $\sqrt{3} - 2$ e) $(2x + 2) - 2$ or $2x$ g) 10
   i) $3x$ k) -2 m) $2x$
5. a) $x^2 + x + 1$ c) 6   e) 17   g) $x^2 - 2x + 3$
7. a) $\{(a, b), (b, c)\}$ c) $\{(a, b), (b, a)\}$ e) $\{(a, a)\}$

§4e

1. a) $\{(1, 1), (2, 2), (0, 3)\}$ c) $\{(x, y) \mid y$ weighs less than $x\}$
   e) $\{(1, 1), (2, 1), (2, 2), (1, 2)\}$

Selected Answers and Hints   433

g) $\{(x, y) \mid x = 2y + 1\} = \{(y, x) \mid y = 2x + 1\}$
i) $\{(x, y) \mid x = y^2 + 5\} = \{(y, x) \mid y = x^2 + 5\}$
k) $\{(x, y) \mid x = 0 \lor y = 0\} = \{(y, x) \mid y = 0 \lor x = 0\}$

3. a) function, reverse function   b) function, not a reverse function
   d) not a function, reverse function   e) not a function, not a reverse function

5. $\bar{F} = \{(3, 2), (7, 4), (8, 9), (\sin 12, \pi)\}$
   $F \circ \bar{F} = \{(3, 3), (7, 7), (8, 8), (\sin 12, \sin 12)\}$
   $\bar{F} \circ F = \{(2, 2), (4, 4), (9, 9), (\pi, \pi)\}$

§4f

1. a)

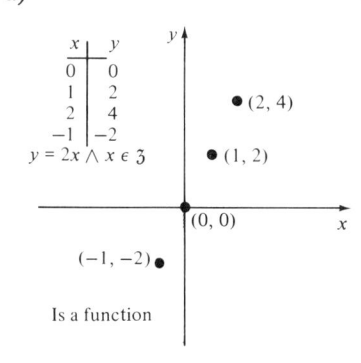

x	y
0	0
1	2
2	4
−1	−2

$y = 2x \land x \in \mathbb{Z}$

points: $(2, 4)$, $(1, 2)$, $(0, 0)$, $(-1, -2)$

Is a function

c) $y > 2x$
Not a function
$y = 2x$

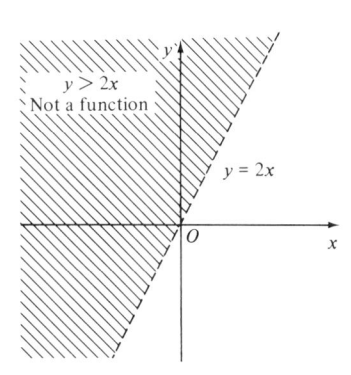

e)

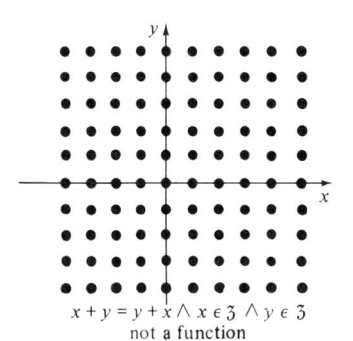

$x + y = y + x \land x \in \mathbb{Z} \land y \in \mathbb{Z}$
not a function

g) $y = x^3$

x	y
0	0
1	1
2	8
−1	−1
−2	8

Is a function

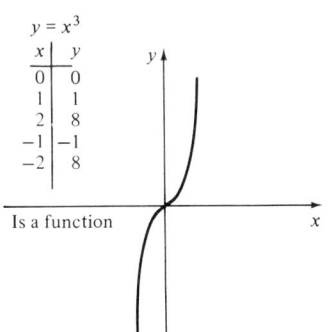

Selected Answers and Hints

2. a)

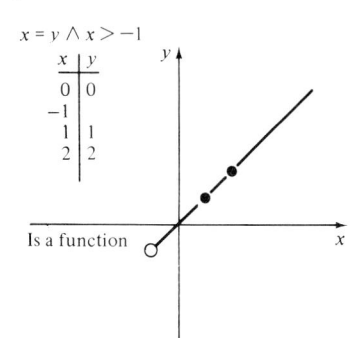

$x = y \wedge x > -1$

x	y
0	0
-1	
1	1
2	2

Is a function

c)

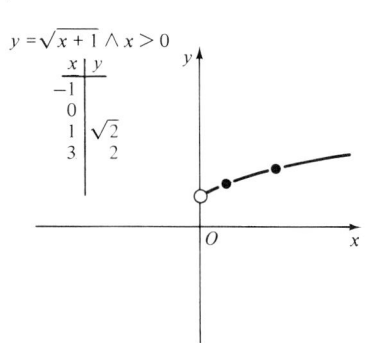

$y = \sqrt{x+1} \wedge x > 0$

x	y
-1	
0	
1	$\sqrt{2}$
3	2

e)

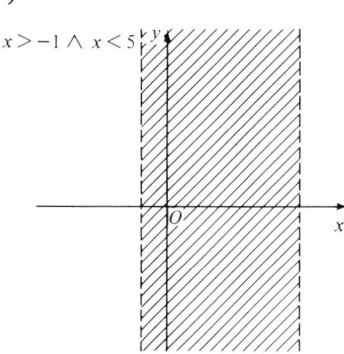

$x > -1 \wedge x < 5$

g)

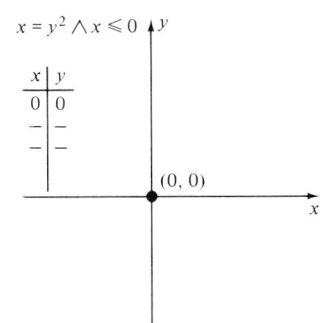

$x = y^2 \wedge x \leq 0$

x	y
0	0
—	—
—	—

(0, 0)

3. a)

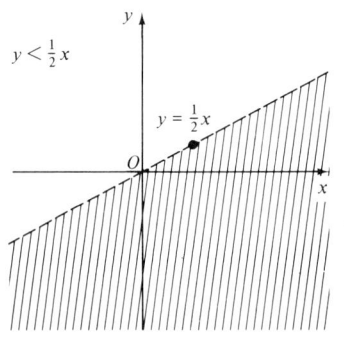

$y < \frac{1}{2}x$

$y = \frac{1}{2}x$

c)

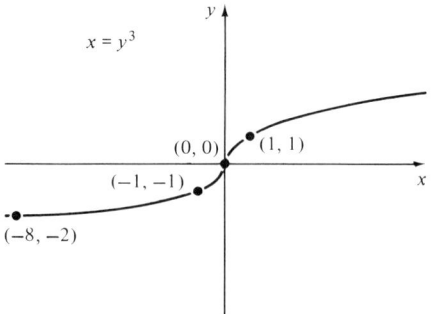

$x = y^3$

(0, 0)
(1, 1)
(-1, -1)
(-8, -2)

Selected Answers and Hints

4. a)

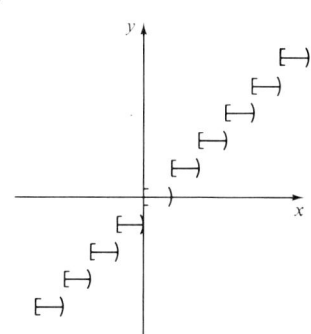

   c)

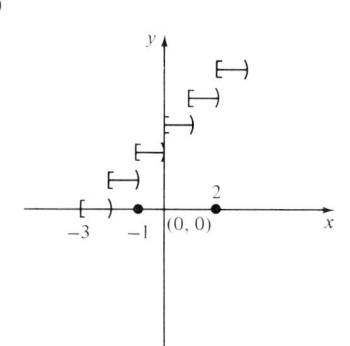

5. a)

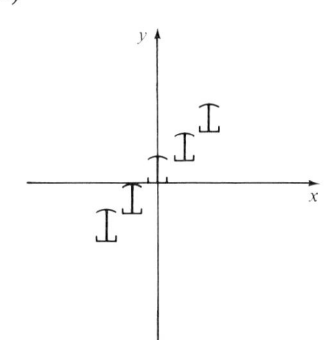

   c)

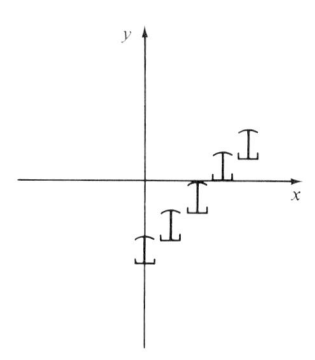

6. a)

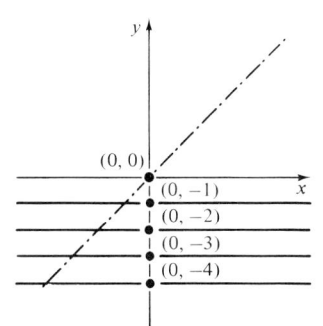

   b)

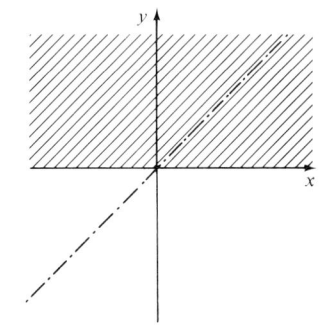

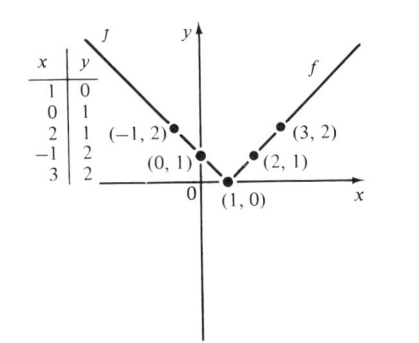

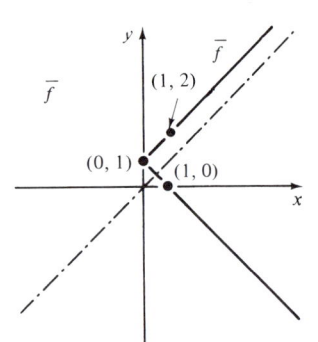

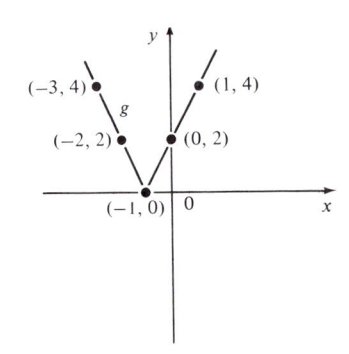

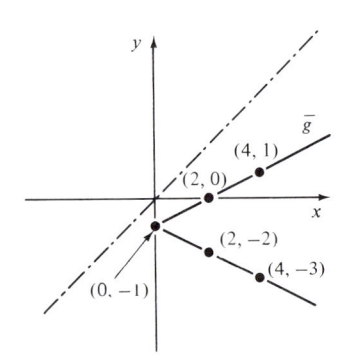

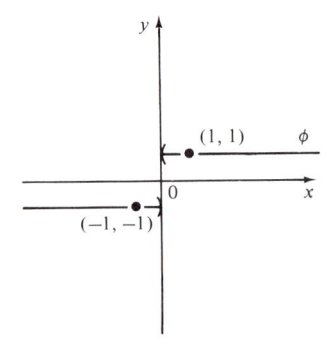

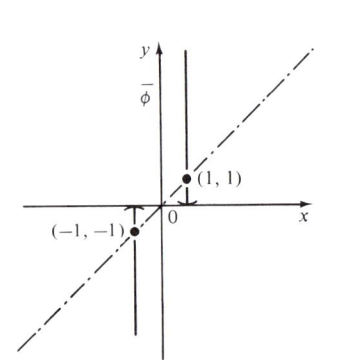

Selected Answers and Hints

7. a)

c)

e)

Selected Answers and Hints

e)

$y = x^3$

x	y
0	0
1	1
−1	−1

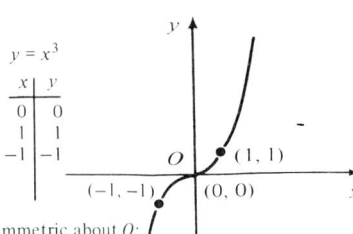

Symmetric about O;
$a = 0$, $b = 0$;
no exclusions

g)

$x^2 + y^2 = 4$

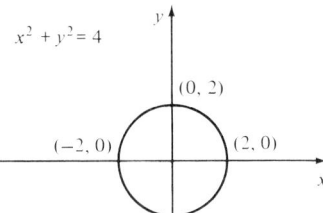

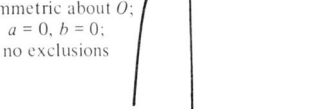

Symmetric about both axes and O; $a = 2$, $a = -2$, $b = 2$, $b = -2$; exclude $x > 2$, $x < -2$, $y > 2$, $y < -2$.

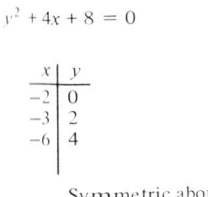

i)

$y^2 + 4x + 8 = 0$

x	y
−2	0
−3	2
−6	4

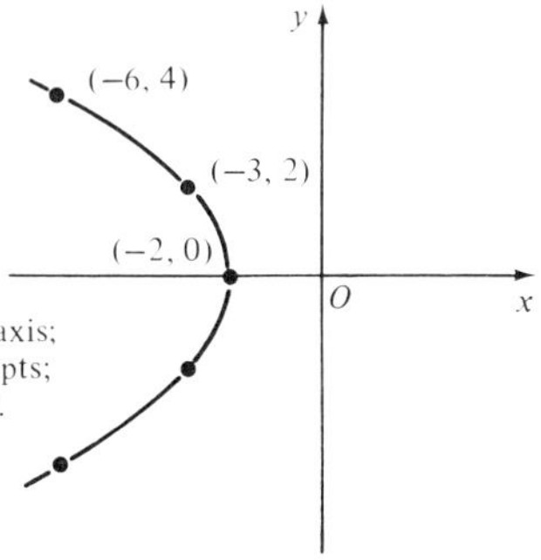

Symmetric about x-axis; $a = -2$, no y-intercepts; exclude $x > -2$.

k)

$y = -\sqrt{x + 4}$

x	y
0	−2
−4	0
−3	−1

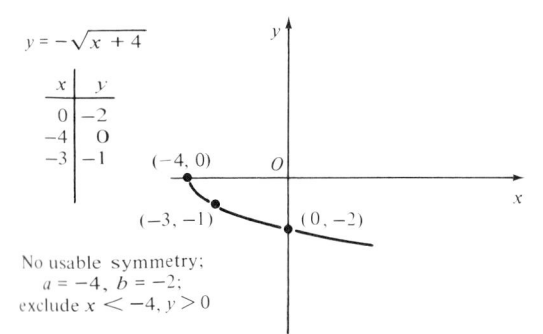

No usable symmetry; $a = -4$, $b = -2$; exclude $x < -4$, $y > 0$

Selected Answers and Hints

g)

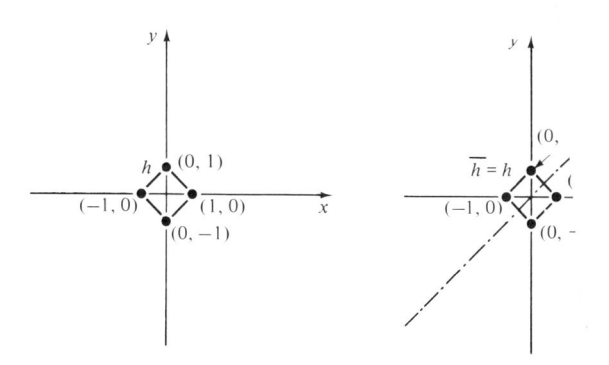

§4g

1. a) no usable symmetry; $a = -2$; no y-intercept
   c) symmetric about the origin; no intercepts
   e) no usable symmetry; $a = 2, a = 0$; $b = 0, b = -1$
   g) symmetric about y-axis; no x-intercepts; $b = -7$

2. a) Exclude $x > 2, x < -2, y > 2, y < -2$; $a = 2, a = b = -2$
   c) no exclusions; $a = 0$; $b = 0$
   e) exclude $-2 < x < 2$; $a = 2, a = -2$; no y-intercept
   g) exclude $y < 0$; $a = 2$; $b = 2$
   i) exclude $x < -4, y > 0$; $a = -4, b = -2$

3. a)

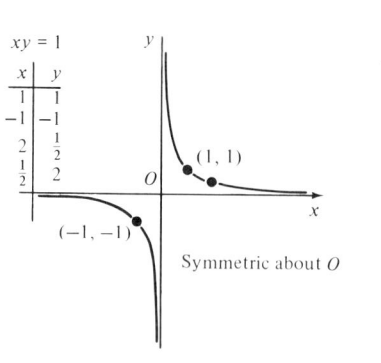

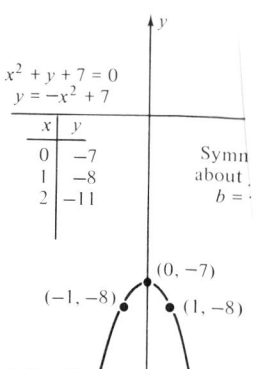

   $xy = 1$

x	y
1	1
-1	-1
2	$\frac{1}{2}$
$\frac{1}{2}$	2

   (1, 1)
   (-1, -1)
   Symmetric about O

   c)

   $x^2 + y + 7 = 0$
   $y = -x^2 + 7$

x	y
0	-7
1	-8
2	-11

   Symm about $b =$

   (0, -7)
   (-1, -8) (1, -8)
   (-2, -11) (2, -11)

§5b

1. $\{(H, 1), (H, 2), (H, 3), (H, 4), (H, 5), (H, 6), (T, 1), (T, 2), (T, 3), (T, 4), (T, 5), (T, 6)\}$

3. a) $\{R_1, R_2, R_3, B_1, B_2\}$
   b) $\{(R_1, R_2), (R_1, R_3), (R_1, B_1), (R_1, B_2), (R_2, R_3), (R_2, B_1), (R_2, B_2),$
      $(B_3, B_1), (R_3, B_2), (B_1, B_2), (R_2, R_1), (R_3, R_1), (B_1, R_1), (B_2, R_1),$
      $(R_3, R_2), (B_1, R_2), (B_2, R_2), (B_1, R_3), (B_2, R_3), (B_2, B_1)\}$

5. a) $\{\ \}$
   c) $\{(1, 1), (1, 2), (1, 3), (1, 4), (1, 5), (2, 1), (2, 2), (2, 3), (2, 4), (3, 1),$
      $(3, 2), (3, 3), (4, 1), (4, 2), (5, 1)\}$

6. a) $\{x \mid x$ is a card and x is a heart or a diamond$\}$
   c) $\{x \mid x$ is a card and x is a king or queen or jack$\}$
   e) {Ace of spades, Ace of clubs}

7. a) The order of apples is not important.
      $\{(v, w, x), (v, w, y), (v, w, z), (v, x, y), (v, x, z), (v, y, z)\}$
   b) $\{v, w, x), (v, w, y), (v, w, z)\}$
   c) $\{(x, y, z)\}'$
   d) $\{v, x, y), (v, y, z), (x, y, z)\}$
   e) $\{(x, y, z)\}$

8. a) $\{(H, H, 3), (H, T, 3), (T, H, 3)\}$
   b) $\{(H, H), (H, T), (T, H)\} \times \{4, 5, 6\}$
   c) $\{(H, H)\} \times \{1, 2, 3, 4, 5, 6\} \cup \{(H, H), (T, H), (H, T), (T, T)\} \times \{5\}$
   d) S

9. $S = \{(R_1, R_2), (R_1, B), (R_2, B)\}$
   a) $\{(R_1, B), (R_2, B)\}$ b) $\{(R_1, B), (R_2, B)\}$ c) S d) $\emptyset$

§5c

1. a) $26/52 = 1/2$ b) $13/52 = 1/4$ c) $12/52 = 3/13$
   d) $4/52 = 1/13$ e) $2/52 = 1/26$ f) $1/52$

2. $S = \{(H, H, H), (H, H, T), (H, T, H), (H, T, T), (T, H, H), (T, H, T),$
   $(T, T, H), (T, T, T)\}$
   a) $7/8$ b) 1 c) $7/8$

3. $S = \{(E, E), (O, E), (E, O), (O, O)\}$
   a) $1/2$ b) $1/2$ c) $10/36 = 5/18$ d) $5/36$
   e) $21/36 = 7/12$

4. $S = \{(b, b, b), (b, b, g), (b, g, b), (b, g, g), (g, g, g), (g, g, b), (g, b, g),$
   $(g, b, b)\}$
   a) 1 b) $1/2$ c) $3/4$ d) $1/4$

5. S will contain all the combinations of 9 things taken two at a time: $9!/(2! \ 7!) = (9)(8)/2 = 36$. To find the number of those of the same color, add the combinations of three things taken two at a time, four things taken two at a time and two things taken two at a time.
   a) 10/36 b) 26/36

§5d

1. a) 5/14 b) 9/14 c) 12/14 d) 1
2. a) 6/10 b) 4/10 c) 8/10 d) 3/10 e) 7/10
3. a) 1/2 b) 1/2 c) 1/3 d) 2/3
4. $P(A|B) = n(A \cap B)/n(B) = 3/10$
5. a) 6/36 or 1/6 b) 1/6 c) 1/6
6. a) 3/4 b) 1/4 c) 1/8 d) 0
7. a) 20/35 b) 15/21 c) 6/21 d) 1

§5e

1. $(1/2)(1/2) = 1/4$
2. $(4/52)(3/51)$
3. a) $(2/10)(1/9) = 2/90$ b) $(8/10)(7/9) = 56/90$
   c) $P(D \cap G) + P(G \cap D) = P(D) \cdot P(G|D) + P(G) \cdot P(D|G)$
   $= (2/10) \cdot (8/9) + (8/10) \cdot (2/9)$
   $= 32/90$
4. a) 3/5 b) $(2/5)(1/6) = 2/30$ c) $(3/5)(2/4) = 6/20$
   d) $(3/5)(2/4) = 6/20$
5. a) $.2 + .3 - .05 = .45$ b) $(.05)/.30 = 1/6$ c) .95
   d) 1/4 e) .8 f) .7 g) .55/.7 h) .95 i) .55
6. a) $((3/7)(2/6)) \cdot ((2/7)(1/6))$
   b) $P(RR, WW) + P(RW, RW) + P(RW, WR) + P(WR, RW) + P(WR, WR) + P(WW, RR) = ((3/7)(2/6) \cdot (5/7)(4/6) + 4((3/7) \cdot (4/6)) \cdot ((2/7)(5/6)) + ((4/7)(3/6)) \cdot ((2/7)(1/6))$
   c) $((4/7)(3/6)) \cdot ((5/7)(4/6))$

§6a

1. It is circular.

3. {I, II, III, IV, ...} ⇋ {1, 2, 3, 4, ...} ⇋ {one, two, three, four, ...}, etc. Yes, each set is ordered.

7. Use eighteen pebbles or cut off two of the chief's toes.

§6b

1. a) not meaningful   c) not meaningful   e) all right
   For example, $1 + 1 = 2$ by the definition of 2.
4. a) $x = 5$ c) $y = 12$ e) $x = 4$ g) no solution
5. a) $2(x + 2)$ c) $3[(7 + 5) + 4]$ e) $17b(3a + 7)$
   g) $(a + 2)(x - y)$
7. (1) C. P.   (2) N-9   (3) Transitivity of =   (4) N-1
   (6) C. I.
9. (1) C. P.   (2) N-5   (3) Transitivity   (4) $(6-2)$
   (5) $(6b-7)$ (6) Transitivity   (7) C. I.

§6c

1. a) $\{1, 2\}$ c) $\{1, 2, 3, 4, 5\}$ e) $\emptyset$

§6d

1. a) $y^2 + 6y + 9$ c) $(2z)(2z) = (2 \cdot 2)(z \cdot z) = 4z^2$
2. a) $\gcd = 2$
3. a) $\gcd = 42$
5. (1) $(6-1)$ (2) N-4   (3) N-9   (4) N-10   (5) (6-1)
   (6) N-3   (7) N-7   (8) Example 5, Section 6b   (9) N-3
   (10) (6-1)   (11) Transitivity (6, 7, 9, 10)
   (12) Transitivity (2, 3, 4, 11)

§6e

1. $1 \in \mathscr{L}$ Z-12
   $1 + 1 = 2$ (6-1)
   $2 \in \mathscr{L}$ Z-1
   $2 + 1 = 3$ (6-1)
   $3 \in \mathscr{L}$ Z-1

                                                                         Q. E. D.

2. a) $\{4\}$ c) $\{1\}$ e) $\{-5\}$ g) $\{2\}$ i) $\{1\}$ k) $\{3\}$
   m) $\{-1\}$

§6f

1. a) $\{^-3\}$ c) $\{3\}$ e) $\{^-14\}$
2. a) $\{x \mid x \in \mathscr{Z} \wedge x < {}^-1\}$ c) $\{x \mid x \in \mathscr{Z} \wedge x > 1\}$
   e) $\{x \mid x \in \mathscr{Z} \wedge x < 1\}$ g) $\{x \mid x \in \mathscr{Z} \wedge x > {}^-1\}$
3. a) $\{1\}$ c) $\varnothing$ e) $\{3\}$

§6g

1. a) $\{0\}$ c) $\varnothing$ e) $\{^-3, 3\}$
2. a) 7 c) 4 e) $|{}^-\pi - 5|$
3. a)

   c)

§6h

2. a) $x = 4$ c) $x = 16/3$ e) $x = 4$ g) $x = 1/2$
3. a) $\{0, 5\}$ c) $\{2, 3\}$ e) $\{3/2, 2/3\}$ f) $\{-4, 0, 4\}$
   h) $\{5/2, -13/2\}$
4. a) 2 c) 2 e) not possible g) 1
5. a) 1 c) 8 e) $25/1029$ g) $a + 2(ab)^{1/2} + b$
   i) $(b^2 + 2ab + a^2)/a^2 b^2$

§6i

1. a) $17/12$ c) $-1/7$ f) $137/60$ g) $57/64$ i) $8/7$
2. a) $<$ c) $<$ e) $=$ g) $=$ i) $=$
3. a) $-2, 4/(-7), 0, 5/8$
4. a) $2/3$ c) $7/8$ e) $13/17$

§6j

1. a) .14   c) .0001   e) .005
2. a) $.\overline{142857}$ c) $.1\overline{6}$ e) .0625
3. a) 33/100   c) 1/3   e) 73/99
4. a) rational   c) rational   e) irrational

§6k

1. a) $3i\sqrt{2}$ c) $-3\sqrt[3]{2}$ e) $5\sqrt{7}$ g) $-i$
2. a) null set—bounded by all positive numbers and zero and zero is the lub
   c) upper bounds are all numbers greater than or equal to $-1+\sqrt{5}$ and $-1+\sqrt{5}$ is the lub
3. a) $\{(-1+i\sqrt{3})/2, (-1-i\sqrt{3})/2\}$ c) $\{7, -4\}$
   e) $\{(-5+\sqrt{17})/2, (-5-\sqrt{17})/2\}$

§7a

1. a) 1112   c) 31,222
2. a) ∩∩∩∩ ||||| c) 𝒫𝒫𝒫𝒫 |||||||
3. a) 472,323   c) 216,660
4. a) [cuneiform]   c) [cuneiform]
5. 𝒫𝒫𝒫𝒫𝒫𝒫 |
7. a) [Chinese numerals]   c) [Chinese numerals]   e) [Chinese numerals]
8. a) 36   c) 200
9. a) [Mayan numerals]   c) [Mayan numerals]
10. a) 4087   c) 8051
11. a) φξβ   c) ψϛ
12. a) 21   c) 583
14. a) 57   c) 2004

§7b

1. a) $5(10)^3 + 2(10)^2 + 7(10)^1 + 4(10)^0$

2. a) 7532
3. aaa where a is the $(b-1)$st digit
5. a) no

§7c

1. a) 9   c) 15
2. a) 48_{twelve} c) 1001100_{two}
4. a) 1011110_{two}
5. a) 80292_{twelve}
6. a) 4743_{eight}
7. a) $507 + 25/\cap \text{\Pitchfork}_{twelve}$
9. a) 2123_{four} c) 1123_{four}
11. last digit ends in one—odd; zero—even
13. the base b

§7e

1. a) 3494
3. a) $315 \to 0, 2784 \to 3, 365 \to 5, 22 \to 4, 8 \to 8, 3494 \to 2$;
   $0 + 3 + 5 + 4 + 8 \to 2$: it checks
4. a)
$$\begin{array}{r} 4\ 3\ ^{1}2\ ^{1}1 \\ -1\ \overset{3}{\cancel{2}}\ \overset{4}{\cancel{3}}\ 4 \\ \hline 3\ 0\ 8\ 7 \end{array}$$
5. a)
$$\begin{array}{r} 4321 \\ +8766 \\ \hline \cancel{1}3087 \end{array}$$
7. a)

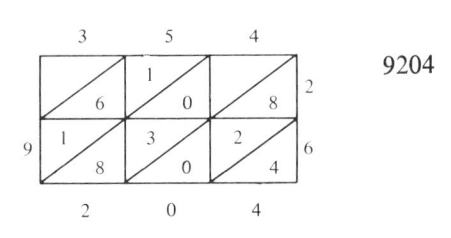

9204

§8b

8. a)   ~~354~~   ~~26~~
        708   13
        ~~1416~~ ~~6~~
        2832   3
        +5664   1
        ─────
        9204

§8b

3. opposite

§8c

1. $\overleftrightarrow{XY}, \overleftrightarrow{YX}, \overrightarrow{XY} \cup \overrightarrow{YX}$, etc.

3. $\{Y\}$ or $\overline{XY}$ or $\overline{YZ}$; $\{X\}$ or $\overline{XY}$ or $\overline{XZ}$; $\{Z\}$ or $\overline{XZ}$ or $\overline{YZ}$

4. a) $\overline{AD}$ c) $\overline{BC}$ e) $\overline{BC}$ g) ∅   i) $\overline{AD}$ k) $\overrightarrow{CD}$

5. a) True   c) False   e) False   g) True provided $\overline{GHR} \lor \overline{GRH}$
   i) False

6. a) $\{C\}$ c) $\{A, C\}$ e) P g) $\overrightarrow{BC} \cup \{P\}$

§8d

1. a) Yes   No

2. a) No   c) $\overleftrightarrow{ST}$

3. a) $\{P\}$ or ∅   c) $\{P\}$ e) $\mathcal{M}$

9. a)   c)

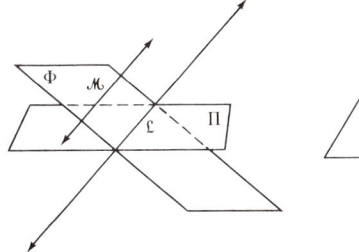

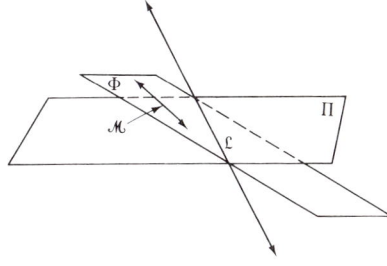

§8e

1. (i) only
3. a) False, $\overrightarrow{EB}, \overrightarrow{ED}$ are collinear rays c) False e) True
5. a) Yes b) No
6. a) False c) True e) True g) False
7. No, the definition of triangle calls for noncollinear vertices.
9. four, three

§8f

1. a) $\overline{FG} \cong \overline{GH}$ c) $m(\overline{EF}) < m(\overline{EG})$
3. a) $\overline{PB} \cong \overline{QE}$ c) $\measuredangle PQC \cong \measuredangle RBA$
   e) They seem to be congruent.
4. a) True c) True
5. No, these loci extend endlessly and our axioms do not provide us with a measure to use in defining congruent.
8. a) nothing
9. a) $\alpha - \eta, \delta - \varepsilon$ c) $m(\alpha) = 30° = m(\gamma)$ e) $m(\alpha) = 70°$

§8g

1. True; see Figure 8.54 (i).
4. a) never c) never
5. a) Yes, yes, no c) Yes, yes, yes.
7. $\triangle ACD \cong \triangle ECB$ (SAS), $\overline{BE} \cong \overline{DA}$, $\measuredangle A \cong \measuredangle E$, $\measuredangle CDA \cong \measuredangle CBE$. It is also possible to show $\triangle ABF \cong \triangle EDF$.
8. a) sufficient c) not sufficient e) sufficient
9. Converse of 8g-1: If two of the angles of a triangle are congruent, then the triangle is isosceles. Holds.

   Converse of 8g-2: If the sum of the measures of the angles of a convex polygon is 2 p, then the polygon is a triangle. Holds.

   Converse of 8g-3: If two triangles are congruent then the corresponding sides are congruent. Holds.

Converse of 8g-5: If two triangles are congruent, then two pairs of corresponding angles and the side common to these angles are congruent. Holds.

13. e) *Hint*: Consider $\triangle ABC$ and $\triangle ACB$ produced by correspondences $A \leftrightarrow A, B \leftrightarrow C$.

§8h

1. a) Yes   c) Yes
2. a) parallelogram   c) rhombus
3. a) False   c) True   e) False

§8i

1. single point, line segment, ray, line, polygons, circle
2. (i) closed   (iii) simple   v) closed
5. a) True

§8j

1. empty, single point, circle
3. No
5. right circular cone

§8k

1. The arcs won't intersect if we don't.

§8l

1. trigon, tetragon
2. a) False   c) False   e) True
3. infinite—two in each plane that contains $\overrightarrow{A'P}$
5. plane figure
7. not necessarily a plane figure

9. not a plane figure
11. plane figure
13. not necessarily a plane figure
15. plane figure
17. not necessarily a plane figure
19. not a plane figure
21. not a plane figure
23. a) P is an element of the set of points called a line.
25. Yes, yes.
27. One side of the triangle is a subset of the line.
29. No
31. Angles AQB, BQC; BQC, CQH, CQH, HQA; HQA, AQB.
33. $\overline{AB} \cong \overline{PQ}$
35. $\mathscr{K} = \mathscr{M}$
36. a)   c)

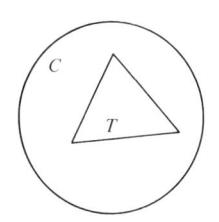

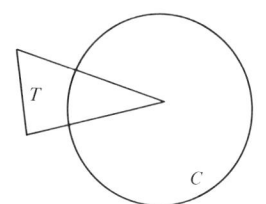

e) 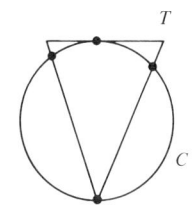   g)

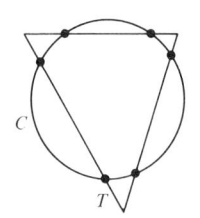

39. ∅
40. a) coplanar   c) non-coplanar

INDEX

INDEX

A bold face page number indicates where the basic properties of the listed concept are first specified, usually the page on which a formal definition can be found.

Absolute value, **200**, 205, **270**
Abstractions, 2, 4, 107
Acute angle, **371** (*see also* Angle)
Addition, 19ff., 228, **230**ff., 247ff., 252ff., 274ff., 306f., 325ff., 329, 300f.
    functions, **189**ff.
Adjunction, rule of, 65
Algebra of Propositions, 103ff.
Algebra of Sets, 156ff.
Algorithms:
    alternative, 329ff.
    arithmetic, 325ff.
Angle, **360**, 367, **369**ff.
    acute, **371**
    adjacent, **370**
    alternate interior, 373f.
    central, **389**
    complementary, **371**
    congruence of, **367**
    supplementary, **371**
Apex, **393** (*see also* Vertex)
Arc, **389**, 395ff.
    major, **389**
    minor, **389**f.
Archimedes, 335
Argument, 6, 25ff., **46**ff., 62ff., 72ff., 160ff.
    conclusion 46, 77
    content versus structure, 27
    contrapositive, 63
    converse, 63

horizontal form, 46
inconsistent, 50f., 78
invalid, 63, 65, 76
inverse, 63
premises, **46**, 76, 82f., 107
proof, 73ff.
quantified, 160ff.
traditional form, 46
valid, 27, **47**ff., 62ff., 73, 77, 82ff., 165
vertical form, 46
Aristotle, 4, 13, 15, 16, 106 (*see also* Logic, Aristotelian)
Associativity, 104, 303
integers, 253
natural numbers, 230f.
propositions, 104
rational numbers, 274
real numbers, 19, 300
sets, 156f.
Augmented premises, 82f.
Axiom, 8, 12, 16f., 91, 106f.
geometry, 337f., 342, 353, 366
integers, 252f.
logic, 29, 91
natural numbers, 230
Peano's, 227
rational numbers, 274
real number, 19f., 300f.
sets, 114
Axiomatic discourse, 47, 96, 101 (*see also* Discourse, deductive)
Axiomatic method, 7ff., 25 (*see also* Pattern of Formal Axiomatics)

Base, **250**, 292, 314, 317ff.
Belongs to (ϵ), 113
Between, 288 (*see also* Betweenness)
Betweenness, 340ff.
Bicondition, **36**, 55
negation, 56
Binary operation, 139
Bisector, **382**
Boole, George 32
Boolean algebra, 158
Bound, **300**f., 304
upper, **300**

least upper, **300**f.
Bracket function, 201 (*see also* Step function)

Cancellation, 234, 238, 240, 257, 262
Carrying, **325**
Cartesian coordinate system, 196
Cartesian products, 179, **180**f. (*see also* Relation)
Casting nines, 332
Center:
  circle, **388**
  sphere, **391**
Central angle, **389**
Chain rule (C.R.), 67f.
Chinese-Japanese mathematics, 310f.
Chord, **388**
Cipherization, **314**, 316
Circle, **388**
  arc, 389
  center, **388**
  chord, **388**
  diameter, **388**
  radius, **388**
Circularity, 11, 13
CLEP, 58
Closure, 303
  division, 278
  integers, 252f.
  natural numbers 230f.
  propositions, 104
  rational numbers, 274
  real numbers 19, 300, 301
  sets, 156f.
  subtraction, 259
Collinear, **345**
Common Divisor, **249**
  greatest, **249**
Common multiple, **250**
  lowest, **250**
Commutativity, 104, 303
  integers, 252f.
  natural numbers, 230f.
  propositions, 60, 104
  rational numbers, 274

real numbers, 19, 300
sets, 156f.
Complement, **135**, 156f. 160
  double, 138, 157
  of an integer, **330**
  relative, **160**
Complementary angle, **371**
Complementation algorithm, 330
Completeness Axiom, 20, 303
Complex numbers **302f.**
Component, 176f.
Composite numbers, **248**
Composition function, **191**
Compounds, logical, 33ff., 53ff.
Conclusion, 25ff., **35**f. 47, 105
Conditional form, 35 (*see also* Implication)
Conditional inference (C.I.), 69, 85ff., 127
Conditional probability, 218 (*see also* Probability)
Condition on a variable, 90 (*see also* Open sentence)
Cone, **393** (*see also* Conical surface)
  circular, **394**
  right circular, **394**
Congruence, 340, 364ff.
  angles, **367**, 370, 377ff.
  circles, **389f.**
  line segments, **366**
  minor arcs, **389**
  triangles **379**ff.
Conical surface, **393** (*see also* Cone)
Conjecture, 6, 25, 194
Conjunction, **34**, 38, 46f., 55, 65, 68, 103ff., 199, 221
  negation, 59
Conjunctive addition (C.A.), 65, 68
Conjunctive simplification (C.S.), 65, 68
Connective, 33
Consistency principle, 12, **17**, 51, 78, 101, 106 (*see also* Discourse)
Constructions, 395ff.
Contradiction, **53**
Contradiction rule, 69
Contradictory propositions, **53**
Contraposition rule (R.C.), 64, 68
Contrapositive, **55**f., 58
Contrapositive argument, 63ff.

Converse, **55f.**
  partial, **97**
Converse argument, 63, 65
Convex set. **357f.**, 360, 362f., 376
Counting, 225ff., 230, 271, 299, 303
Correspondence, one-to-one, **123**, 225, 366
Cross product, **180** (*see also* Cartesian product)
Cube, **393**
Cuneiforms, 307f. (*see also* Sumerian)
Curve, **388**
  closed **388**, 390, 393f.
  simple, **388**, 390, 393f.
Cylinder, **393**
  circular, **394**
  right circular, **394**
Cylindrical surface, **393**

Decagon, **377**
Decimal, 291ff., 316f.
  fraction, 292, **293**ff.
  infinite, 295
  nonrepeating, 297
  repeating, 295ff.
  terminating, 295
Deductive argument 25ff., 46ff. (*see also* Argument and Discourse)
Deductive reasoning, **6f.**, 105 (*see also* Discourse)
Definitions, properties and use, 11, **12**, 16, 106
De Morgan's laws, 59, 60, 105, 157
  conjunction, 59
  disjunction, 60
  intersection, 157
  union, 157
Denseness, **288**, 298f.
Detachment rule (R.D.), **54**, 62, 65, 68
Diameter, **388f.**
Difference, **236f.**
Digit, 314, 316ff., 325ff.
Direction, 340ff.
Discourse, 8, 11ff., 16f.
  axiomatic, 47, 96, 101
  consistent, 17, 101, 106
  deductive 8, 16, 25, 105
Disjoint sets, **121**, 124f., **141**

Disjunction, exclusive, **38**
Disjunction, inclusive, **34**, 38, 60, 103ff, 199, 216f.
   negation, 60
Disjunctive addition (D.A.), **66**, 68
Disjunctive simplification (D.S.), **66f.**, 68
Distance, linear, **271**
Distributivity, 104, 303
   integers, 253
   natural numbers, 230f.
   propositions, 104
   rational numbers, 274
   real numbers, 20, 300
   sets, 157
Division, **239f.**, 262, 278f., 283ff., 328
   functions, 190f.
Dodecagon, **377**
Domain, **178**, 183, 189ff., 204
Double complement, 138, 157
Double negation law (DN), 32, 59, 105

Edge, **358**, **392**
End point, **343**, **344**
Egyptian mathematics, 1f., 305ff., 335
Einstein, Albert, 6
Elements The, (Euclid's), 336f., 339
Empty set, 114ff., 125, 128f., 131f., 135ff., 141, 151ff.
   symbols, 115, 116
Equality, 19
   elements of sets, 19, **133f.**
   ordered pairs, 176f.
   sets, **122ff.**, 127ff.
Equal additions algorithm, 329f.
Equivalence relation, 184ff., 390 (*see also* Relation)
Equivalent sets, 123 (*see also* Matched)
Euclid, 34, 336ff.
Eudoxus, 4, 335
Euler, Leonhard, 119
Event, **210ff.**
   independent, **221**
Exclusive disjunction, 38 (*see also* Disjunction)
Existence, 17, 150
   rational numbers, 275
   real numbers, 19, 301

Existential quantifier, **91**ff., 160ff. (*see also* Quantifier)
  negation, 93f.
Explicit definition, 12ff.
Exponent, 248, **250**f.
  fractional, **281**
  negative, **279**
  zero, **279**
Exterior angles, 373
Exterior of an angle, **361**f.
Exterior of a triangle, **362**

Face of a half plane, 360
Face of a polyhedron, **392**
Factor, **240**f., 248, 249f.
False, 7ff., 17, 25ff., **29**ff., 47ff., 77, 90ff., 96ff., 106, 115, 163
Field, 20
Figure, **340** (*see also* Locus)
Fraction, **279**, 281, 283ff., 307
  common, 292
  decimal, 292, **293**ff.
  exponent, **281**
Function, **187**ff.
  addition, **189**ff.
  composition, **191**
  division, **190**f.
  multiplication, **190**f.
  subtraction, **190**f.
Fundamental theorem:
  arithmetic, 249
  fractions, 284
Fundamental Rule of Inference, **54** (*see also* Rule of Detachment)

Galileo, 5
Generator, **393**f.
Geometry, 335ff.
  Euclidean, 4, 337ff.
  non-Euclidean, 3
  spherical, 338
  table-top, 340ff.
Graph, **196**ff.
Grating, **331** (*see also* Lattice)
Greater than, 243, **259**ff.
Greatest common divisor, **249**

Greatest integer function, 201 (*see also* Step)
Green mathematics, 2f., 105, 335ff.
Griswold's Gremlin, 56
Grundlagen der Analysis, 227, 229
Grundlagen der Geometrie, 336

Half-line, **357**
Half-plane, **358**
Half-space, **360**
Hexagon, **377**
Hilbert, David, 336, 339
Hippocrates of Chios, 4, 335
Hypotenuse, **378**, 381
Hypothesis, **35**, 46, 85, 97, 127 (*see also* Implication)
Hypothetical syllogism, 67

Idempotent property, 105, 157
Identity element: (*see also* Neutral element)
   integers, 254
   natural numbers, 230f.
   propositions, 104
   rational numbers, 276
   real numbers, 22
   sets, 156f.
Implication, **35**f., 46f., 56, 73
   conclusion, **35**
   contrapositive, 55, 58
   converse, 55, 56
   hypothesis, **35**
   inverse, 55, 56
   positive, 55f.,
   reversible, 55
Implicit definition, 11
Inclusive disjunction, **34**, 38, 53, 60, 66, 68, 103ff., 199 (*see also* Disjunction)
   negation, 60
Inconsistent argument, 50f., 78 (*see also* Argument)
Independent events, **221**
Indic mathematics, 1, 316
Indirect inference rule (I.I.) 68, 83f.
Indo-Arabic numeration, 316f.
Induction axiom, 228, 231, 250, 253, 272
Inductive property, 299, 303 (*see also* Counting)
Inductive reasoning **5**f., 99

Inequalities, 199, 243ff. (*see also* Order relation)
  absolute, 243
Inference, fundamental rule, 54 (*see also* Rule of Detachment)
Inference rules, 28, 54, 62ff., 73ff., 83ff., 96 (*see also* Valid argument forms)
  table, 68f.
Infinite repeating decimal, **295ff.**
Infinite set, **344**
Integers, 252-273, 279, 303
  additive solvability, 253ff.
  axiom set, 252f.
  even, 99f., 103, 289
  odd, 99f., 103
  set $\mathscr{I}$, 117
Intercepts, 203ff.
  x-intercept, **203**
  y-intercept, **204**
Interior:
  angle, **361**f.
  polygon, **376**f.
  triangle, **362**
Interior angles, 373
  alternate, 373
Intersection, 345ff., 353ff.
  convex sets, 358
  planes, 355
  sets, **141**ff., 150ff.
Invalidity, 47, 76ff. (*see also* Argument, valid)
Inverse:
  additive, 23, **255**, 261, 263ff.
  conjunction, 104
  disjunction, 104
  intersection, 156f.
  multiplicative, 23, **278**, 286
  union, 156f.
Inverse argument, 63
Inverse of an implication, **55**f.
Inverse relation, 195 (*see also* Reverse relation)
Involution, 228, **250**f., 268, 279, 289
Ionic Greek numeration, 312ff.
Irrational numbers, **297**f., 300f.
Iteration, **308**f., 312ff.

Jalousie, **331** (*see also* Lattice)

Kite, **384**, 386

Landau, Edmund, 227, 229
Lattice, **331**
LEDI, 59, 105
Leg of right triangle, 378
Leibnitz, G. W., 32
Length, **367**
Less than, 199f., **244**ff., **259**ff. (*see also* Order relation)
Line, 3, 18, 337, 344ff., 348, 351ff.
Line of symmetry, 198, 203
Line segment, **343**
   congruence, **366**
Locus, **196**, 199f., 202ff. (*see also* Figure)
Logic, **16**f., 26, 73, 101
   Aristotelian, 17, 29f., 101, 106
Logical discourse, 8 (*see also* Deductive discourse)
Logical equivalence, **30**, 32, 55f., 83
Lowest common multiple, **250**

Matched sets, **123**f., 225
Mayan numerals, 311f.
Measure, **365**ff., **378**ff.
   angles, 369ff.
   line segments, 366ff.
Metric relationships, **364**
Model, mathematical, 18ff., 338
Modus (ponendo) ponens, **54**, 62 (*see also* Rule of Detachment)
Modus (tollendo) tollens, 64 (*see also* Rule of Contraposition)
Modus tollendo ponens, 67 (*see also* Rule of Disjunctive Simplification)
Multiples, **240**
   common, **250**
   least common, **250**
Multiplication, 19ff., 228, **230**ff., 252ff., 274ff., 300f., 327f., 331f.
   functions, 190f.
Multiplicative inverse, **278** (*see also* Inverses)

Natural numbers, 230ff., 303
   axiom set, 230
   counting, 230
   set $\mathcal{N}$, 117, 230
Negation, logical, **31**f., 59
   basic form, 31

bicondition, 56
conjunction, 59
disjunction, 60
implication (NIMP), 60
quantified propositions, 92, **93**
simplified forms, 31, 56, 59, 60, 92
Negation, Double (D.N.), 32, 59
Negative:
number, **260**
of a number, **256**
Neutral element: (*see also* Identity element)
integers, 253f.
natural numbers, 230f.
propositions, 104
rational numbers, 276
real numbers, 22
sets, 156f.
n-gon, **377**
NIMP, 60 (*see also* Negation)
Nines, casting, 332f.
excess, 332
Nonmetric relationships, 364
Null set, 114ff. (*see also* Empty set)
symbols for, 115, 116
Number line, 269ff., 298
Numeral, **305**
Numeration base, 314, 316f.
decimal, 316f.
nondecimal, 317ff.
Numeration system, **305**ff.
Chinese-Japanese, 310f.
Egyptian, 306f.
Indo-Arabic, 316f.
Ionic Greek (later), 312f.
Mayan, 311f.
Sumerian, 307f.

Obtuse angle, **371**
One-to-one correspondence, **123**, 225, 366
Open sentence, **90**ff., 114ff.
Order relation, 20f., 199, 243ff., 259ff., 267f., 287f., 303 (*see also* Less than and Between)
transitive, 244

Ordered counting, 226
Ordered field, 20
  complete, 20
Ordered pairs, 175, **176**ff., 186f.

Parallel lines, **354**, 384, 397
Parallelepiped, **392**
Parallelogram, **384**ff., 392
Partial converse, 97
Partition, **356** (*see also* Separation)
Pasch, Moritz, 11, 336
Pattern of Formal Axiomatics, 16ff., 106, 336f.
Peano, Giuseppe, 32, 227f., 336
Peasant multiplication, 331ff.
Peirce, C.S., 30
Pentagon, 377
Perpendicular, 371, 382, 386, 390, 394, 396f.
Pieri, M., 336
Place holder, 90 (*see also* Variable)
Plane, **351**ff.
Plato, 335
Point, 9, 18, 339ff.
Polygon, **376**ff.
  convex, **376**
  decagon, **377**
  dodecagon, **377**
  hexagon, **377**
  n-gon, **377**
  pentagon, **377**
  regular, **377**
  quadrilateral, **377**
  triangle, **362**
Polyhedron, **391**ff.
Positive number, **260**f., 277, 279
Postulate, 8, 337 (*see also* Axiom)
Power:
  second, 248
  n^{th}, **250**
Premise, 25, **46**ff., 105 (*see also* Discourse and Consistency)
  augmented, 82ff.
Primary statements, 8f. (*see also* Axiom, Postulate)
Primitive terms, **11**f., 15, 22, 29, 106, 113, 229, 252, 274, 300, 339, 342
Prime number, **248**

Principal square root, 205, 279f., 297f., 301ff. (*see also* Root)
Prism, **392**
Probability, 209ff.
  conditional, **218ff.**, 223
  empirical, 220, 223, **229ff.**
  theoretical, 213ff., 223
Proof, 3, 4, **73ff.**, 98, 100f., 106, 130, 158f.
Proper subset, **171**
Propositional form, 90ff. (*see also* Open sentence)
Propositions, algebra, 103ff.
Proposition, **29ff.**
  components, **33ff.**
  compound, 33ff.
  logical equivalence, 30
  negation, 31
  quantified, 91ff.
  simple, 33
Pyramid, **392**
  Egyptian, 1f., 335, 394
Pythagoras, 4, 335

Quadratic solvability, 302f.
Quadrilaterals, **377**, 384ff.
Quantified arguments, 160ff.
Quantified propositions, 89ff. (*see also* Proposition)
  negation, **93**
Quantifier, 90ff.
  existential, 90f.
  universal, 90f.
Quotient, 238ff., 278ff.

Radicals, 205, 279f.
Radius:
  circle, **388f.**
  sphere, **391**
Range, **178**, 204
Ratio, 278f., 292
Rational numbers, 273ff., 296f., 303
  axiom set, 274
  multiplicative solvability, 275ff.
  ratio, 278f.
  repeating decimals, 296
  set $\mathcal{Q}$, 117

Ray, **344**ff., 360ff.
Real numbers, **300**ff., 303
  axioms, 19f., 300f.
  bounds, 300
  quadratic solvability, 302
  set $\mathscr{R}$, 300
Reasoning-forms, 4ff.
Reasoning, deductive, **6**
Rectangle, **384**ff.
Reflexivity, 19, 129, 133, 156, **183**ff. (*see also* Relation)
Region, 204ff., **391**f.
  polygonal, 391f.
Relation, 175, **177**
  Cartesian product, **180**f.
  domain, **178**
  equivalence, **184**f.
  function, **187**ff.
  graphs, 196ff.
  range, **178**
  reflexive, **183**ff.
  reverse, **193**ff.
  symmetric, **183**ff.
  transitive, **184**f.
Relative complement of sets, **160**
Replacement set, **116** (*see also* Scope)
Reverse relation, **193**ff.
Reversible statements, 12, **14**f., 55, 107
Rhombus, **385**f.
Right angle, **370**ff., 375, 378, 381, 384, 386, 394
Root, 279ff.
  n^{th}, **279**
  principal, **279**f.
Roster notation, **115**ff., 124, 145, 170, 172, 177ff., 182, 184
Russell, Bertrand, 32

Sample points, **210**f.
Sample space, **210**ff.
Scope, **116**ff., 177f., 181ff., 196
Semicircle, **389**
Separation, **356** (*see also* Partition)
Sets:
  algebra, 156ff.
  complement, **136**ff.

De Morgan's laws, 153, 157
disjoint, 121, **141**
double complement, 138, 157
element, 113
empty, 115
equality, **122**
equivalent, **123**
finite, 123
infinites, 344
intersection, **141**f.
matched, **123**f.
member, 113
notation, 115f.
   roster, 115
   set builder, 116
null, 115
overlapping, 121
relative complement, **160**
replacement, **116**
scope, **116**ff.
solution, **115**f.
subset, **121**ff.
   proper, **171**
symmetric difference, **160**
truth set, **115**
union, **140**f.
universal, 114

Sides:
   angle, 361
   line, **358**ff.
   polygon, 376
   triangle, 362

Similarity, **340**, 376, 390
Skew, **354**f.
Solution set, **115**ff., 124
Solvability, 19f., 303
   additive, 303
      integers 253ff.
      rational numbers, 273ff.
      real numbers, 19, 21, 300
   multiplicative, 303
      rational numbers, 275ff.
      real numbers, 20, 300

quadratic, 302f.
Space, **354**f., **390**ff.
Specialized terms, 10ff.
   defined, 12ff., 16, 106
   primitive, 11, 16, 106
Sphere, **391**
Square, **385**f.
Square of a number, **248**, 268
Step function, **201**f. (*see also* Bracket, greatest integer)
Statement, 6, 25, 29
Straight line, 341, 343, **344**ff., 352ff., 370ff., 389ff., 395
Structure of arguments, 27, 46ff.
Structure, mathematical, 18ff.
Substitution axiom, 30, 32, 39, 55, 75
Subtraction, **236**ff., 241, 242, 258f., 264, 328, 329ff.
   function, 190f.
Successor, 10, 11, 226ff., 271
Sumerian mathematics, 1f., 307ff., 335
Superset, **122**
Supplementary angle, **371**
Syllogism, hypothetical, 67
Symmetric difference, **160**
Symmetric relations, 19, 129, 133, 156, **183**ff. (*see also* Relation)
Symmetry, 197f., **202**f.

Tangent, **389**
Tautology, **53**f., 65, 69, 73, 98, 108, 128, 129, 131
Terminating decimals, **295**f.
Thales of Miletus, 4, 335
Theorem, 4, **6**, 16, 21, 98, 106f.
Theory, 8, 106
Transitivity, 19f., 130, 133, 156, **184**ff., 244
Transversal, 373f.
Trapezium, **384**
Trapezoid, **384**
Triangle, 97f., **362**f., 376ff.
  acute, **378**
  equilateral, **378**, 382
  isosceles, **378**, 382
  obtuse, **378**
  right, **378**, 381
  scalene, **378**
Trichotomy property:

    integers, 259
    natural numbers, 230f., 242, 245
    rational numbers, 277, 287f., 290
    real numbers, 20
True, 6ff., 13f., 17, 25ff., **29ff.**, 47ff., 73, 76, 90ff., 96ff., 106, 114f., 161ff.
Truth set, **115**ff., 160 (*see also* Solution set)
Truth standard, 7ff., 25ff., 30, 106f.
Truth table, 31ff., 48ff.
Truth value, **29**ff., 39ff., 47ff., 96, 107
    assignment, 76ff.

Unary operation, 139
Union, **140**ff., 156ff.
Uniqueness, 303
    integers, 252f.
    natural numbers, 230f.
    propositions, 104
    rational numbers, 274
    real numbers, 19, 300
    sets, 156f.
Universal quantifier, **90**ff.
Universally quantified proposition, 91ff.
    negation 92, **93**
Universe of discourse, 114f., 119ff., 128, 135ff., 151ff., 160ff.

Valid argument, **47**ff., 54, 62ff., 72ff., 101, 106
Valid argument forms, 62ff.
Valid reasoning, 27, 101
Value of a function, 188ff.
Variable, **89**ff., 114, 116
Venn, John, 119
Vertical angles, **373**
Vertex: (*see also* Apex)
    angle, **361**
    conical surface, **393**
    polygon, **376**
    polyhedron, **392**
    triangle, **362**

Whole numbers, 117, 146
    set $\mathscr{W}$, 117

Zero, 21f., **254**ff., 278, 295, 306, 311f., 316
    exponent, 279f.